Critical Systems Analysis and Design

As systems analysis and design is becoming increasingly concerned with the organization as a whole, systems analysts need to concern themselves with organization design as well as systems design.

This book takes a unique look at systems analysis and design by using an approach that provides learners with a critical personal framework. This enables the reader to develop a personal method for critically considering and developing a knowledge and practice of systems analysis and design by contrasting the real world with the systems world, thus differentiating it from existing systems analysis books.

Each chapter of this book begins by highlighting what can be learned by completion of the chapter and ends with a critical skills development section that contains activities, tasks and discussion questions. Chapters include:

- systems analysis and design in concept and action;
- structured data modelling;
- making systems analysis and design inclusive.

Although the discussion and examples in this text are drawn primarily from business information systems, the lessons apply to both government and healthcare information systems and to systems development in general.

Critical Systems Analysis and Design makes a complex area of study accessible and relevant and as such is an indispensable textbook for both advanced students and professionals concerned with the innovation of information systems.

Nandish V. Patel is Deputy Director of Studies on The Brunel MBA programme at Brunel University, Uxbridge.

Critical Systems Analysis and Design

A personal framework approach

Nandish V. Patel

Routledge
Taylor & Francis Group
LONDON AND NEW YORK

First published 2005
by Routledge
2 Park Square, Milton Park, Abingdon, Oxon OX14 4RN

Simultaneously published in the USA and Canada
by Routledge
270 Madison Ave, New York, NY 10016

Routledge is an imprint of the Taylor & Francis Group

Typeset in Baskerville by
Florence Production Ltd, Stoodleigh, Devon
Printed and bound in Great Britain by
TJ International Ltd, Padstow, Cornwall

British Library Cataloguing in Publication Data
A catalogue record for this book is available from the British Library

Library of Congress Cataloging in Publication Data
Patel, Nandish V., 1959–
Critical systems analysis and design: a personal framework approach/
Nandish V. Patel.
p. cm.
Includes bibliographical references and index.
1. Management information systems. 2. Knowledge management.
3. Systems analysis. 4. System design. 5. Problem solving.
6. Critical thinking. I. Title.
HD30.213.P383 2005
658.4'038'011–dc22 2004011363

ISBN 0–415–33215–X (hbk)
ISBN 0–415–33216–8 (pbk)

My darling daughter Risha
(which means knowledgeable)
for a brighter future

Contents

Illustrations

Figures

Tables

Preface

The purpose of this book is to redress the gap of criticality in systems analysis and design. The development of criticality in learners is the rationale that underpins programme specifications at universities, and increasingly it is the preferred attribute of professional employees in companies. Criticality is important in systems analysis because its scope has vastly increased since the 1960s. Then, it was concerned with the design of Information Systems (IS) for local or departmental functions. It now covers the whole organization in system projects involving business process redesign, eCommerce, eBusiness, and knowledge management systems. Systems analysts are now not just systems designers, they are part of a team involved in organization design too.

IS is an important subject in academic curricula in undergraduate and postgraduate programmes at universities. Learners are exposed to how technical analysis is conducted, how to use formal notations to develop systems models and designs, and how developed systems models and designs are implemented using Information Technology (IT). The emphasis is on providing comprehensive knowledge and the development of necessary problem-solving skills for learners to become practitioners. This emphasis has resulted in an absence of criticality in the subject, which for university degree programmes in business schools and computing departments is probably a significant weakness. Learners and practitioners need to reflect critically on the problem of IS development encompassing organization design, in particular systems analysis and design. Such criticality is necessary in postgraduate programmes and in workplaces where critical thinking is often the impetus for innovation.

Teachers are concerned about the divergence between textbook knowledge of IS and systems analysis and design and its practice in organizations. The formal approaches that constitute knowledge seem at odds with the domains in which they are applied, namely organizations and people. Organizations, and the people who work in them, make it difficult for practitioners to deploy formal IS methodologies or systems analysis and design techniques and tools as intended by their authors and inventors. Often practitioners resign themselves to the fact and stop using formal methodologies, techniques and tools. This is regrettable because formalism and ontological knowledge of systems can make real contributions to practice. Teachers are aware that what they teach their learners in the classroom seems to lack relevance to actual practice in software development companies and in business organizations that develop and use IS. Yet the problem remains. Teaching knowledge is fundamental to understand the vitally revolutionary phenomenon that IS and IT constitutes in the twenty-first-century organization.

To bridge the relevance gap between knowledge and practice it is possible to present knowledge to learners in a stimulating and interesting manner. A Personal Critical Framework (**PCF**) approach will incite learners, and practitioners, to think critically about how they internalize the taught knowledge and how they would put it into practice. The author's ten years of reflective teaching practice and innovations in pedagogy has resulted in an exciting way of teaching that appeals to learners. Taking a Holistic Approach to Learning and Teaching Interaction (Patel, 2003) can make the subject material interesting and relevant to learners' personal and professional development, and most importantly to their notions of the self and their appreciation of the role of knowledge in personal and working lives. The Holistic Approach has been shared with colleagues and nationally at conferences on learning and teaching. Learners and colleagues have applauded it, and learners are satisfied with their levels of performance and knowledge insights achieved.

In the Holistic Approach, learning and teaching interaction is defined as the social context in which a learner, who is seeking to improve the self, is taught specific discipline knowledge by a knowledgeable teacher whose aim is to disseminate discipline knowledge and improve the quality of life of the learner by developing him or her into a critical learner. The Holistic Approach seeks to develop critical, confident, and independent learners capable of action in their professions, and in society generally. The task of the holistic teacher is to enable critical learners to direct themselves and to devise ways of enabling learners to do that.

The purpose of the Holistic Approach is to develop critical thinking skills in learners and practitioners. Cognitive critical thinking skills are difficult to develop in a practical subject like systems analysis and design. The Holistic Approach makes use of PCF to enable learners to *relate* taught material to the actions they perform or would perform. The Holistic Approach, PCF and a **Critical Framework** for studying and analysing the taught material are combined to provide a set of cognitive tools for learners and teachers that engender critical thinking.

The pedagogy of this textbook is termed *critical learning* in which the learner is regarded as a *critical learner*. The traits of a critical learner are criticality, confidence, and independence. The critical learner is an individual who has to act eventually in real situations. Such purposeful *action* requires critical faculties that enable individuals to make decisions about the action, based on knowledge that they learn to assess for its validity in practice. Confidence is required to act. The holistic teacher's task is to provide such confidence by developing independent learners. This independent trait of the critical learner reflects that the individual is responsible for their action. A critical learner can only be independent if they are confident, and confidence only develops if learners are taught to be reflexive. This textbook introduces the development of a PCF as a device to structure knowledge and purposeful action that is based on critical *thought*. The development of a PCF and its critical evaluation can lead to confidence and independence in action.

As Barnett argues:

> The challenges facing the university in developing critique-in-action are, in part, concerned with the ontological questions of what it means to be a self in the world, and about the relationship of the self to the world of work.
>
> (Barnett, 1997: 129)

Barnett's comments have a resonance with the notion of *praxis*. Praxis is the art of acting in given conditions in order to change them. It is precisely this kind of situation that systems analysts and developers of IS encounter. Action in existing situations with given conditions requires critical thinking. The Holistic Approach is intended to develop critical learners who will be capable of such critical praxis. It has been conceptualized as a result of the success of the author's teaching of IS with this approach and appreciative learners' favourable comments.

This textbook is aimed at postgraduate learners undertaking a Master of Science or Master of Business Administration degree. Its purpose is to develop critical learners capable of making judgements on how to use the existing knowledge on IS development in practical work. It can also be used at level three of undergraduate programmes. The traits of criticality, confidence, and independence of a critical learner enable an individual to act in real situations in relation to the discipline knowledge taught. The basis of the teaching approach is praxis – our body of knowledge arises from the very problems that we need to solve to make progress. Knowledge that arises from practice is in turn used to improve personal effectiveness. The discussion and examples in the text are from business IS, though the lessons apply to both government and healthcare IS and to systems development in general.

I am grateful to all my learners who have been exposed to my pedagogical approach to learning critically about IS and practice. They have been, and continue to be, learners on postgraduate programmes in IS and Systems Project Management. Many masters and MBA candidates have taken my modules on IS development. They encouraged me to continue using the Holistic Approach because of the benefits they have derived as practitioners. In particular, graduate trainees on a day-release company-based programme in Information Management have encouraged me to continue developing the approach. All my learners have contributed to my learning on how to teach an essentially problem-solving subject from a critical perspective, and I am pleased and gratified that I have been able to contribute to even highly experienced practitioners too:

Nandish,

May I just say that I enjoyed your lectures a great deal, having worked in IT for 30 years, I understood and appreciated what you were saying and also learned a lot, it was a lecture that I always looked forward to attending!

Best regards

Acknowledgements

I am grateful to Jacqueline Curthoys, Francesca Poynter and Rachel Crookes at Routledge. I would like to thank them especially for their kind consideration.

Abbreviations

ASD	Agile Software Development
BPR	Business Process Re-engineering
CASE	Computer-Aided Software Engineering
CMM	Capability Maturity Model for software development
CoCoMo	Constructive Cost Model
COTS	Commercial-Off-The-Shelf Software
DBMS	Database Management System
ETHICS	Ethical and Technical Implementation of Computer Systems
IS	Information System
IT	Information Technology
JAD	Joint Application Development
JSD	Jackson Systems Development
JSP	Jackson Structured Programming
NCC	National Computing Centre
NIMSAD	Normative Information Model-Based Systems analysis and Design
OSS	Open Source Software
PCF	Personal Critical Framework
RAD	Rapid Application Development
SDLC	Systems Development Life Cycle
SPICE	Software Process Improvement and Capability Determinations
SSADM	Structured Systems Analysis and Design Methodology
SSM	Soft Systems Methodology
STRADIS	Structured Analysis, Design and Implementation of Information System
UML	Unified Modelling Language
XP	eXtreme Programming

Introduction

How to use the book

The aim of this book is to admit learners and practitioners into systems analysis and design. Admission into a subject is called *vishay prevasha* in the Indian Sanskrit language. It means to provide learners themselves with critical faculties for acquiring and developing knowledge in a discipline. The book is not a technically complete text on how to undertake systems analysis and design. Rather, only material sufficient to form the basis for developing criticality is introduced.

Criticality is explored in this textbook through: transformatory critique, refashioning of traditions, reflexive criticality and critical skills, after Barnett (1997). It is to be distinguished from the notion of simply being taught knowledge. The central theme of the textbook is the development of a PCF, either for learners or practitioners. Learners and practitioners can think critically about knowledge and personal action and develop personal constructs based on critical thought. The PCF is to be developed on the basis of critical understanding of knowledge and practice. It should form a set of related cognitive constructs derived through reflective study or reflective practical experience, or both. The value of a PCF is to anticipate or determine future action and, based on personal experience, it is to be continually revised. It serves to enable learners to develop individually methods for critical reflection on knowledge and practice.

The PCF and the Critical Knowledge and Practice Framework provide the lenses for critically studying systems analysis and design. They provide the critical focus. These frameworks serve to encourage learners and practitioners to reflect critically on the notion of systems, methods, techniques and tools and engender reflective and critical discussion aimed at the development of a PCF.

To develop a PCF each chapter serves three purposes. It presents sufficient material to begin thinking on a specific topic. The material is then related to the Critical Knowledge and Practice Framework designed to enable reflection. The PCF is then developed through questions, exercises, and activities on the basis of the reflective study of the material presented and other readings. This section in each chapter is on how the Critical Knowledge and Practice Framework can be used to develop the conceptual and praxis elements of a PCF. Systems analysis and design concepts and analysts' practice based on conceptual and technical knowledge need to be subjected to criticality. The section enables critical evaluation and development of professionalism. Analysts can assess its value in individual and professional terms. In particular, they can consider the impact of knowledge on practice.

Student-centred learning

This textbook can be used in conjunction with the student-centred approach to learning and teaching. As with criticality, student-centred learning requires learners to take responsibility for learning, but it also requires them to be involved in its delivery. It reinforces the transfer of ownership of knowledge from the teacher to learners. It is particularly relevant to postgraduate learning and teaching, where it can draw on learners' experiential knowledge.

The Systems Development Life Cycle (SDLC) is introduced. Learners should be encouraged to do simple, uncomplicated exercises to familiarize themselves with the techniques, tools and methods. Readings of relevant research papers and cases then lead learners to recognize and question premises and assumptions made by the various approaches, SDLC, methodologies, techniques and tools. Questioning leads to the development of a critical perspective on the various problems in structured and object-oriented analyses. Learners can then be encouraged to think how improvements in knowledge and practice can be made in the light of the critical reflection.

Teachers' own seminar exercises, cases, research papers, and seminar discussions can be used to involve learners in the delivery. The teacher can introduce topics via a short interactive lecture and then facilitate seminar discussion. The discussion can be based on cases and research papers. Learners should be encouraged to lead discussions in small groups, which then report to the plenary. The plenary discussion itself will result in a synthesis of learners' contribution to the delivery.

After reading the Delphi Report's definition of critical thinking, self-examination and self-correction, learners should attempt the PCF development section in each chapter. Its purpose is to encourage critical reflection. It contains questions encouraging critical reflection, individual and group discussion exercises, and activities that lead to the development, and subsequent enhancement, of personal constructs. The PCF is thus progressively developed in each chapter.

As there are no right or wrong answers for a particular systems analysis problem no solutions are provided. The teacher can determine when to make use of the student-centred material and provide guidelines to learners on acceptable resolutions. References to research publications in the textbook are not reproduced but a full reference is provided in the further reading sections.

The teacher is encouraged to think of similar activities to those presented in the PCF development sections. Given the interpretive nature of many of the questions and activities in the PCF development sections no model answers are provided. Teachers should use their discretion to judge what constitutes an appropriate response by learners. The author is available to discuss the development of teachers' initiatives and learners' responses via the textbook's associated website.

Supporting web resources

Additional material on the holistic approach to teaching and learning is available at www.routledge.com/textbooks/0415332168. Teachers' activities in developing PCF are:

- Introduce concepts of systems analysis and design and present methods, techniques and tools.
- Use the Critical Framework to develop critical thinking. Relate real and practical

problems in applying systems analysis and design in the real world, and seek pragmatic resolution of the problems.

- Encourage learners to develop a PCF consisting of personal constructs. Personal constructs may be derived from practice or taught material, or from both sources. A PCF should include a list of problems and issues in systems analysis and design, compiled by individuals, groups or the teacher.
- Formulate questions designed to make learners think about clients, systemic problems, social factors, organizational problems, communications problems, knowledge gaps, and other problems and issues. Relate these to learners' PCF by using exercises, discussions and activities in the PCF development sections, thus progressively enhancing PCF.
- Develop their own activities for learners that encourage them to think of how to resolve practical problems in systems analysis and design, ranging from assumptions about organizations, people, IT and IS, covering conceptual deficiencies, difficulties with implementing techniques, and using tools in real contexts.
- Develop activities that encourage learners to relate their resolutions of practical problems to PCF development and to question the effectiveness of their resolutions in terms of the PCF.

Selective chapter readings

A chapter map is given in Figure 0.1. Excepting Part I, the remainder of the book is organized to reflect the phases of the SDLC. This does not justify the SDLC but recognizes the clear conceptual distinctions prevalent in research and practice.

Chapters in Part I are essential reading because they provide the foundation for developing skills in critical thinking and the development of a PCF. Once the notions of a PCF and the Critical Framework are understood, other chapters may be read as required. The selective reading of other chapters can then lead to the development of certain elements of a PCF. Teachers are encouraged to read the supplementary account of the Holistic Approach, available at www.routledge.com/textbooks/0415332168 because it explains the pedagogy underpinning the approach to learning and teaching adopted in this book.

Chapter 1 explains the PAC cycle. Teachers and learners should read it. It details the components of a PCF and discusses how it should be developed. It relates a PCF to action and discusses the development of critical selves through action.

Chapter 2 explains the Critical Framework. Teachers and learners should read it. It details the components of the Critical Framework and explains how to apply them to study critically topics in systems analysis and design. The Critical Framework is a set of constructs to enable criticality.

Part II covers conceptual foundations, project management and systems analysts. A reading of the chapters will develop the intellectual and conceptual elements of a PCF. This coverage is required because criticality assumes conceptual understanding and argument.

Chapter 3 introduces fundamental concepts and discusses their effect on analysts' actions. The SDLC, structured and object-oriented systems analysis and design, and methodologies are covered. A reading of this chapter will lead to the development of fundamental and essential personal constructs in analysis and design for a PCF.

Chapter 4 examines systems project management. It covers how an IS is selected for development, formulating a business case for its development, and project management issues and techniques. Project management as planned action is elaborated. A reading of this chapter will provide the basis for developing reflexive critical thinking and critical skills, and enhancement of the elements in a PCF.

Chapter 5 discusses the role of systems analysts. The qualities and skills required, their tasks, and the techniques and tools available to them are covered. The working relationship between systems analysts and stakeholders and developers is discussed. A reading of this chapter will enable systems analysts to think critically of 'the self', and contribute to considering all the elements in a PCF.

Part III focuses on systems analysis. A reading of the chapters will enable a critical consideration of the instrument elements of PCF. These chapters are included because the development of reflexive criticality and critical skills assumes familiarity with the instruments of practice.

Chapter 6 explains the importance of understanding and developing knowledge of what a new IS is required to do, or system requirements. Structured and object-oriented techniques and alternatives are presented and critically discussed. Chapter 7 introduces structured data modelling techniques. It focuses on logical data modelling, entities, relationships, normalization and documentation. Chapter 8 introduces structured process modelling, covering data flow diagrams and business logic modelling techniques. It also details SSADM technical products. Chapter 9 introduces object-oriented analysis, covering object modelling and UML diagrams.

Part IV focuses on systems design. A reading of the chapters in this part will enable a critical consideration of the design instrumental elements of a PCF. Similar to the previous part, the chapters in this part contribute to reflexive criticality and critical skills.

Chapter 10 examines user interface, input and output design. Chapter 11 considers system and data design, and covers change management because of its importance in introducing IS in an organization. A reading of these chapters will enable an appreciation of the system level design issues and contribute to the personal constructs and instrument elements of PCF.

Part V relates the social context of systems analysis and design. It develops a critical perspective through critical reflection of the previous chapters and paradigmatic analysis. A reading of the chapters in this part will enable critical consideration of the intellectual and conceptual elements of PCF. The development of transformatory critique and the possibility of refashioning traditions assume knowledge of what constitutes knowledge and how it is acquired.

Chapter 12 considers social theory, organizational and political factors that are not covered in structured and object-oriented analyses. Chapter 13 is a critical reflection of the previous chapters in terms of human and organizational factors. A reading of these chapters will contribute significantly to the ethics, assumptions of reality, personal constructs and instrument elements of PCF.

Chapter 14 synthesizes knowledge and action to deepen the knowledge element and enable reflection on praxis in PCF. It considers criticality in terms of refashioning traditions. It provides a paradigmatic analysis of alternative and emerging knowledge on systems analysis and design. A reading of this chapter will contribute to acquiring knowledge of assumptions of reality, personal constructs and instrument elements of PCF.

Chapter 15 takes a future-oriented perspective. A reading of this chapter will provide knowledge of how emerging approaches are making paradigmatic shifts to make systems analysis and design inclusive of social and organizational factors. It gives brief accounts of new

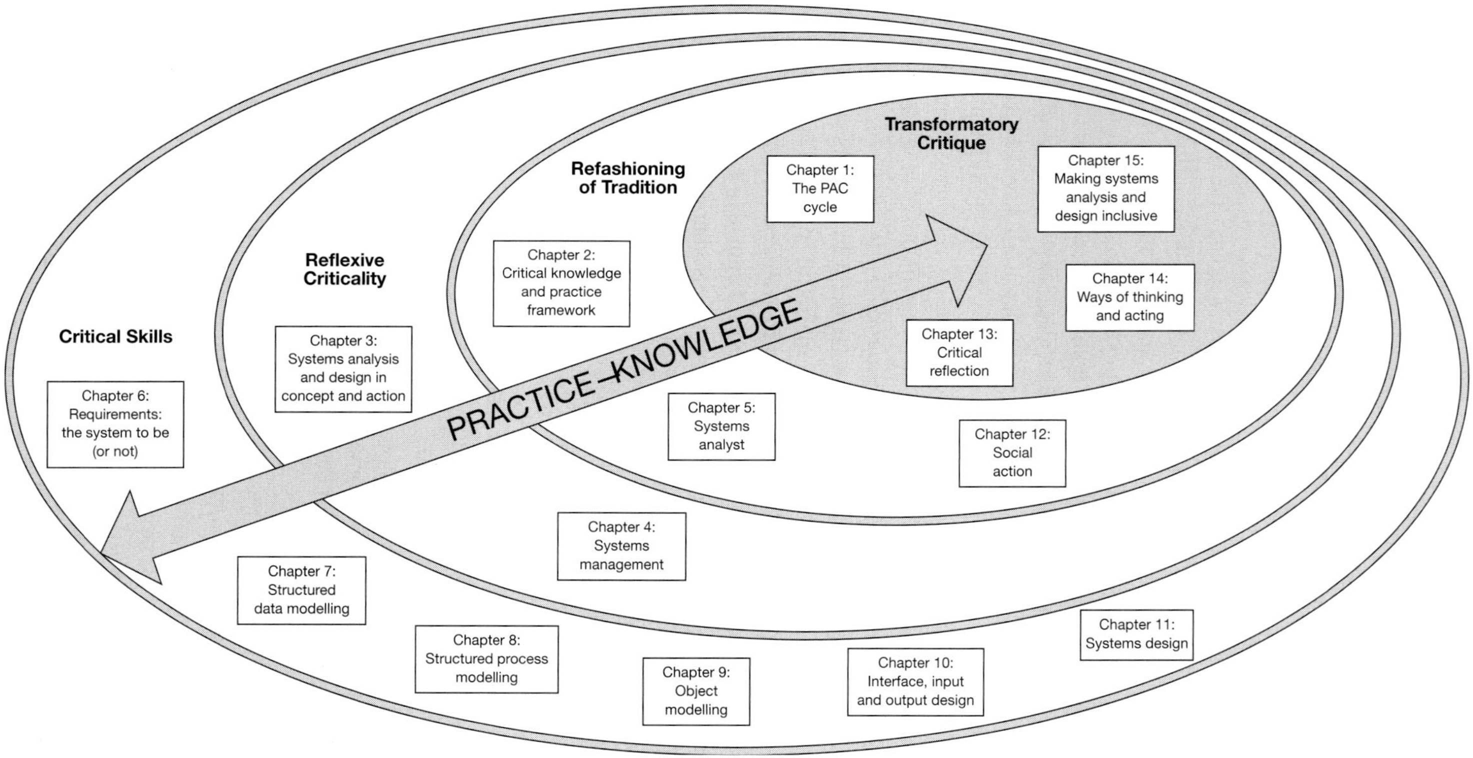

Figure 0.1 Chapter map in the context of criticality

and emerging approaches. Since a PCF is designed to facilitate how learners and practitioners anticipate action, this chapter is important for anticipating future developments in knowledge and practice of systems analysis and design.

Each chapter has a section for the development of a PCF. It contains activities, questions, and tasks that will help to objectify and define personal constructs related to the three major themes and six sub-themes of the Critical Framework. They will help start thinking on a PCF. It can be used to evaluate personal constructs related to knowledge and practice for inclusion into a PCF. Where the questions or activities require reference to an organization, it may be the university where you study or the company where you work. Some activities are alternatives and so do not need to be repeated, though using alternative methods to objectify a PCF can lead to more clarity. For all the activities in the PCF development in each chapter refer to the illustrative PCF in section 1.7 and use the Delphi Report cognitive skills given in Table 1.4 in section 1.8.1.

The Critical Framework figures in each chapter contain much critical information. It is not all discussed or commented upon but selectively addressed. Teachers, learners, and practitioners should study each figure and investigate and critically discuss the points relevant to their interests. The pragmatic resolution component is left with question marks in some chapters to prompt learners to think about their own solutions to the problems depicted.

The chapters are mapped in Figure 0.1 in the context of the critical themes explored. Each concentric circle names the type of criticality covered in the relevant chapters placed in the relevant concentric circle. In accordance with the purpose of this textbook, the chapters are evenly distributed around the themes of transformatory critique and refashioning of traditions for critiques of knowledge, and reflexivity and critical skills for critiques of praxis.

The placing of the chapters within the critical themes is not exact. For instance, all the chapters contain a PCF development section that spans all the critical themes. Chapter 5 on systems analysts should actually be rooted in the core transformatory critique and stretch out through refashioning of traditions and reflexivity to the development of critical skills in the discipline knowledge.

The 'practice–knowledge' bi-directional arrow indicates that praxis is informed by knowledge and that knowledge is informed by praxis. Deep transformatory critique, the inner concentric circle, can have radical effects on practice, the outer concentric circle. Deep critical skills, the outer concentric circle, can have radical effects on knowledge, the inner concentric circle. In the outward direction, transformatory critique and refashioning of traditions in knowledge inform practice, and in the inward direction critical skills and reflexivity in practice inform knowledge.

Learning outcomes

The outcomes of working through the chapters should be critically aware learners and reflexive practitioners. They will be able to:

- Appreciate transformatory critique, refashioning of traditions, reflexive criticality and critical skills as kinds of criticality of knowledge and practice.
- Develop cognitive critical skills in interpretation, analysis, evaluation, inference, explanation and self-regulation.
- Deploy the Critical Framework to identify, question, and critically think of premises and assumptions in systems ontology – structured and object-oriented analyses and design, and other approaches.

- Apply cognitive critical skills to develop, amend and revise continually a PCF.
- Deploy the Critical Framework to consider critically the application of methods, techniques, and tools in real situations and develop pragmatic resolution for practical application problems.

These learning outcomes will lead to the development of a PCF underpinned by critical thinking and criticality of knowledge and practice. They specify the prerequisite cognitive skills required to develop critical thinking and seek to develop deeper and different kinds of criticality.

Part I

Foundations for critical learning and teaching

Part I provides the foundational development for a PCF perspective. It is the pedagogical basis for developing criticality. The contents of a PCF determine how it is used to anticipate action, and how critical thinking is applied to studied and experiential knowledge. This is called the PAC cycle and it is elaborated in Chapters 1 and 2. The PAC cycle enables the structuring of critical thought.

Criticality is only possible if a suitable framework for analysis and evaluation is devised. Criticality requires a critical lens through which systems analysis and design can be examined. Chapter 2 sets out such a Critical Framework used throughout the textbook. In it systems analysis and design is interpreted as human action that can be improved through critical thought. The framework suggests cognitive processes for acquiring, understanding, and assimilating knowledge and its application. It is the requisite lens for developing critical thought and also the basis for insights into how to develop and sustain a PCF for personal effectiveness.

The value of a PCF is its criticality. It is the critical consideration of systems analysis and design and subsequent inclusion in PCF that provides substance. Chapter 2 also explains what a PCF is, how to develop one, and why a PCF is necessary for improving understanding and practice. The question of how to understand something, and assimilate it into a PCF, is addressed in terms of personal cognition and personal constructs.

Chapter 1

The PAC cycle

1.1 Learning outcomes

To engage in the PAC cycle, after completing this chapter you should be able to:

- Develop, revise and amend a PCF based on criticality.
- Apply critical cognitive skills to a PCF.
- Relate a PCF to personal action and effectiveness.
- Interpret formal and practical knowledge in terms of transformatory critique, refashioning of traditions, reflexive practice and critical skills.

1.2 Introduction

Critical systems ontology is the interpretation, analysis, evaluation, inference, explanation, questioning and critical study of systems. It is the notion that current ontological knowledge of systems can be improved, and that it can be 'other than it is' to be practically effective. It is enabled through a **P**ersonal Critical Framework, **A**ction, and **C**riticality or the PAC cycle. The PAC cycle is proposed as a method for systems analysts to develop criticality and critical thought. It is termed a 'cycle' because criticality is a continuous activity. Practitioners need to develop the habit of reflecting on what they do, how they do it, and crucially, why they do it in order to improve practice. These are the preconditions for developing criticality, and require cognitive skills in interpretation, analysis, evaluation, inference, explanation and self-regulation. Learners too need to develop these cognitive skills. They can benefit from learning to reflect on knowledge and develop deeper understanding of systems analysis, systems design, its techniques, **tools** and methods.

The PAC cycle is illustrated in Figure 1.1. Readers are encouraged to develop a PCF marked as (1) in the figure, enactment of the PCF (2), and reflect critically on the effect of the enactment on achieving desired aims (3). A PCF is a qualitative approach to knowledge formation in which **personal constructs** are unique to an individual. The cycle is completed when criticality leads to the revision or amendment of personal constructs and the PCF.

An initial PCF can be formed either through critical study or reflective practice. It is composed of personal constructs on systems analysis and design and the relations between them that

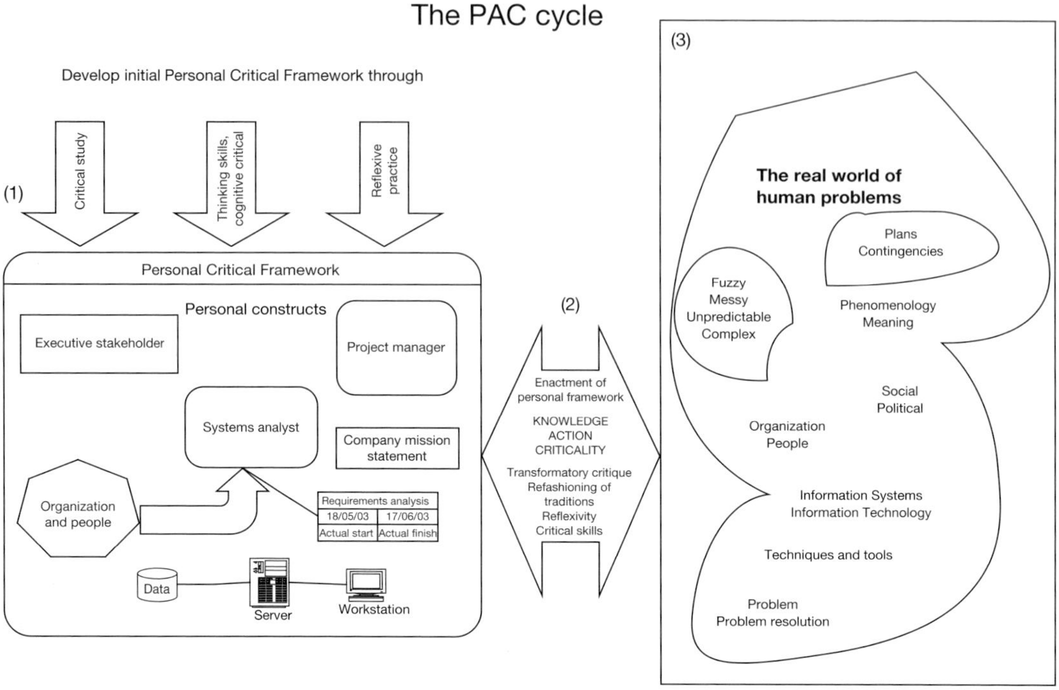

Figure 1.1 The PAC cycle

a systems analyst deems relevant to *anticipate reality*. It is this focus on anticipating reality that makes a PCF relevant for analysts because the development of an IS is such an anticipation of reality. Systems analysis and design is practical and the analysts' actions shape actual IS. The PCF is then enacted in actual IS development situations. The analyst critically considers the effects of the actions in the actual situation to enhance or revise the PCF. The process of forming personal constructs, using them to anticipate events, and revising them is continuous.

The PAC cycle is a cognitive device to assess and improve personal effectiveness and success. Personal effectiveness is the basis for developing professionalism. Assessing personal effectiveness itself requires an objectified process. Objectification is the basis for improving knowledge and understanding practice. Objectified experiential and learnt knowledge is an important part of the PAC cycle. It is necessary for the development of personal constructs and a PCF, reflection on ones actions to develop knowledge further, and the application of criticality to a PCF.

1.3 Personal Critical Framework

A PCF is a significant element of knowledge and professional development. It brings training and education in systems analysis and design together on the basis of the self and the self's need for knowledge to anticipate experience. A framework is: 'typically a mixture of pre-suppositions of correctness, of what is valuable, and of validity. The framework is not purely cognitive; it is not even mainly cognitive. It is invested with values, emotion, commitment, and professional and social identity' (Barnett, 1997).

A PCF is an objectified tool for critical reflection on the self, knowledge, and practice, and it is open to revision, amendment and update. While training in systems analysis and design leads to specific and prescribed behaviour, education results in an awareness of the processes that lead to acquiring knowledge, and cognitive and practical skills. Education engenders personal reflection on these processes.

The practice-orientation of systems analysis and design requires learners themselves to discover ownership of knowledge. Generating ownership is the major aim of a PCF. As individuals have selves, the self is central in the development of a PCF. It is vital for making learning and knowledge relevant and meaningful. Knowledge is gathered and interpreted by the self. Learning that is divorced from the self usually lacks relevance when action is required.

A PCF enables criticality in systems analysis and design. An analyst develops a PCF to improve understanding of knowledge and practice. Its purpose is to improve the self through knowledge and individual action. Knowledge that is devoid of the self creates a vacuity. The self provides ownership and responsibility. In the absence of the self, learnt or experiential knowledge lacks meaning. Such insipid knowledge then bares no relation to actual practice.

1.3.1 Objectifying a Personal Critical Framework

Objectification results in reflection on how knowledge is acquired and used by the self in practice. The process of objectifying knowledge to develop a PCF requires making personal constructs explicit on what constitutes knowledge, how it is acquired, and how it is accepted as valid. Cognitive skills are needed to objectify a PCF and for critically evaluating objectified knowledge. Interpretation, analysis, evaluation, inference, explanation and self-regulation propounded by the Delphi Report (Facione, 1990) serve as requisite cognitive skills to develop a PCF and engage in the PAC cycle.

Objectifying personal knowledge is initially difficult, especially for practitioners immersed in practice. Difficulties in objectifying a PCF arise because of unfamiliarity with reflection. The process of objectifying personal constructs and the relationships between them includes action, reflection, writing, diagramming and discussion. An individual can begin to identify personal constructs through these activities. Objectification may be individual and then discussed with a mentor or trusted colleague, or it may be done in a group. Objectified personal constructs and their relations can then form the foundation for a PCF.

Action Practice is the deployment of knowledge to achieve specific goals. It results in experiential knowledge. Analysts gather and use relevant knowledge for practice. For example, an analyst acquires knowledge of techniques to use to determine system requirements for a new IS. Chosen techniques will be enacted to analyse system requirements.

Reflection Reflection is the process of critical thinking on practice to improve professionalism, and the self. Analysts evaluate and critically assess practice to determine its effectiveness. Practice is scrutinized to understand what was done, how it was done, and whether it achieved predetermined objectives. For example, an analyst reflecting on the effectiveness of interviewing **clients** for functional requirements may assess how the interview was conducted and whether it was suitable for establishing functional requirements.

Writing Making a record or writing is a method for externalizing knowledge and experience. It can be recorded as notes or critical evaluations of practice. For instance, an analyst reflecting on a project to develop a decision support system can record personal activities in the project. The objectified writing can serve as a record for critical analysis of practice and its effectiveness.

Diagramming Drawing diagrams is a method for making a graphical representation of knowledge, concepts, techniques and activities. A diagram provides an overall perspective on knowledge and action. The objectified diagram can be used to further reflect on practice and how to make it more effective.

Discussion The reflection, written records and diagrams can be discussed with trusted colleagues or teachers. They constitute objectified material for further critical understanding of practice through other peoples' perspectives. Other peoples' perspectives help to critically assess, question and evaluate practice and knowledge.

1.3.2 Knowledge and practice elements

A PCF has knowledge and practice elements. Knowledge and practice is conceptually demarcated in the PCF but in practice they can be accumulated separately or jointly. A learner with no practical experience will initially accumulate only knowledge. A practitioner can accumulate knowledge through practical experience and contribute to the development of practice simultaneously.

The knowledge element is required to understand theory and relate it to practice. Theory is an account or explanation of observed phenomena. A theoretical explanation improves understanding, provides explanatory knowledge and informs practice. It consists of ideas that explain the nature of something, its causes that made it possible and how it functions.

It is difficult to find theory in systems analysis and design, but there are paradigms of thinking and acting that seek to explain systems analysis and design, and how it should be performed. Such knowledge is accumulated through formal learning or from practical experience. Such paradigms can contribute to personal explanations of why and how things work. On a personal level a paradigm consists of objectified ideas, concepts, techniques and methods that form personal understanding of systems analysis and design. Theoretical and paradigmatic understanding can be used to enhance and improve personal constructs or redefine relations between elements in a PCF.

Praxis is the practical side of a discipline or profession and it provides practical knowledge. Practice can be conducted in the absence of clear knowledge or understanding of the reasons for acting in a particular manner. It can benefit from clear explanations for acting in a particular manner. Reflexive praxis or reflexive practice can lead to practical knowledge based on critical thinking. Analysts who reflect critically on practice and knowledge can revise and amend personal constructs accordingly. Reflexive practice can be combined with theoretical or paradigmatic explanations to develop deeper knowledge and understanding.

Professionalism, though, needs to combine the knowledge and practice elements in a PCF. Objectifying knowledge and practice personal constructs can improve understanding of practice. Practice can be developed with clear explanations for acting in a particular manner. Most practitioners are content doing their work and do not attempt to explain or understand what underpins practice. A practitioner may find that object modelling does not work well but be unable to explain why it does not work in practice. Reflexive practice can help to explain the method's shortcomings by understanding **systems ontology**. An analyst may be trained to elicit system requirements using structured techniques and may unquestioningly deploy them. The analyst's understanding of the effectiveness of the techniques would improve with knowledge of assumptions made about actual situations, explanation for conducting requirements in the first place, and why structured techniques should be used.

Learners especially need to develop the practice element of a PCF. They are best placed to benefit from the knowledge element and use it to consider how they might act in actual situations. They can be encouraged to develop practice personal constructs through scenarios or cases to determine how they would act. A learner's initial PCF can be used as a guide to praxis when they begin practice, which should be revised subsequently on the basis of reflexive practice.

1.4 Personal effectiveness through a Personal Critical Framework

A PCF is useful for determining the success or effectiveness of practice. The achievement of an objective is a practical measure of success. A PCF can be used to improve personal effectiveness and to determine, through critical reflection, which personal constructs and relations lead to effective practice. It enables practitioners to know how to do things better and learners to understand the reasons for particular action. It can be used to:

- decide how to act in actual situations;
- determine knowledge and critical areas for its improvement;
- explore strengths, develop them further, and exploit them;
- identify weaknesses and take action to improve them.

An objectified PCF can improve the effectiveness of practice. It can be used to determine the efficacy of one's action. It can be the basis for critically reflecting on and understanding one's action, resulting in some personal constructs being revised, others replaced, or new ones introduced.

1.4.1 Reinforcing and stopping action

A PCF can be used to explain why certain actions should be repeated and why others should be stopped. Action that leads to success can obviously be repeated. An explanation of why it is successful can be deduced from a PCF. The explanation will enhance knowledge of practice and deepen understanding. For example, when a method or technique works well in practice it is beneficial to know and understand why and how it works. Developing an understanding of how it works in theory can provide deeper knowledge of its value in practice.

Practice that results in lack of desired levels of success or even failure needs to be stopped. The problems with personal constructs can be identified and amended or entire personal constructs replaced. Simply stopping the action will not improve knowledge or practice. It is vital for professional development to know why certain actions should not be repeated. Stopping action without understanding the reasons for failure may lead to future similar failures.

Explaining practice is important for developing a PCF that contains relevant personal constructs. Actual practice consists of implicit or explicit rules, procedures and assumptions of which a practitioner may not be cognizant. Explaining successful practice will objectify them and develop knowledge that will contribute to better understanding of practice.

Practice of systems analysis and design is an important source for developing criticality. Reflecting on learnt knowledge and practice is important, but reflecting critically adds value. The relation between a PCF and practice is processual and interpretive rather than static and prescriptive. An analyst constructs a PCF processually, as they engage in actual situations. Taking action on the basis of a PCF will either lead to success or not. Successful action should reinforce a PCF and unsuccessful action should lead to revisions.

1.4.2 Understanding human action

Analysts' action underpin PCF development. Objectification of personal constructs related to human action in a PCF, and its use to reflect on action, can improve effectiveness. The utility of knowledge for particular action determines the level of success achievable. Effective action is determined by knowledge of how to act. To understand human action and its effectiveness, researchers and practitioners probe questions concerning the combination of knowledge and practice that leads to effective action and how effectiveness can be improved.

There are two significant strands of research that develop knowledge of human action: **planned action** and **situated action**. Planned action is based on human rationality and its capability for achieving desired objectives. It is closely linked to developments in scientific knowledge and the scientific method. A plan epitomizes human action as rational action (Simon, 1957). A plan is an instrument for achieving desired aims and it explicates activities, rules and procedures to achieve objectives. It assumes that it is possible to control events in actual situations.

Planned action is crystallized as 'methodology' in IS development. An IS development methodology is a detailed plan of how to develop IS. It prescribes systems analysis and design activities for analysts, project managers, and software programmers. Systems project management too epitomizes the planned management of IS development. Systems analysis and design methods and techniques are similarly based on planned action. A plan prescribes actions but is incapable of dealing with anomalies not accounted for in the plan in the real situation. Individuals and groups need to respond to the situations they encounter not accounted for in plans. It is proposed here that they do so as situated action in context.

Situated action is human intention and action informed by context and utilizes embodied skills (Suchman, 1994). It is characterized as the setting of objectives and seeking to achieve them. The process of achieving objectives and the actions taken, depend on and are determined by the context and situational factors.

Situated action has had no influence on systems analysis and design, but analysts need to be aware of it when objectifying personal constructs. Situated action raises questions concerning structured and object-oriented systems ontology. If the ontological assumptions underpinning situated action are valid, then action that is informed by context and utilizes embodied skills, logically cannot be extracted from its context, as required in structured and object-oriented systems ontology. Systems ontology needs to account for such action. Analysts need to be aware of situated action when developing personal constructs, and when resolving systems modelling problems pragmatically. Recent developments in Agile Software Development (**ASD**) seem to be taking account of situated action.

1.5 Components of Personal Critical Frameworks

A PCF should include the knowledge and practice elements detailed above, and the five essential components explained in this section. They are required to ensure that a PCF reflects the knowledge and practice elements. The five components are: ethics, assumptions of reality, personal constructs, instruments, and relations between components. Other components specific to individuals or **organizations** may also be included. For example, an analyst working for a charitable organization may want to include morality as a component, or one working for the government may want to include public service as a component.

1.5.1 Ethics

Ethics is the values component of a PCF. Ethical practice is concerned with making judgements concerning right and wrong behaviour. The purpose of professionalism is to achieve formal objectives, but professionalism that is void of ethical practice often leads to undesirable, illegal, and even immoral practice. Professional bodies in computing and IS provide an ethical code for their members.

IS change the way people work and affect peoples' working lives. Analysts' systems design decisions bind people into working in certain ways, for instance they determine whether people have rewarding working relations. Such design decisions are particularly sensitive in organizations that deal with human healthcare. Researchers have considered the effect of analysts' design decisions on people. The Ethical and Technical Implementation of Computer Systems (**ETHICS**) methodology was developed to encourage ethical behaviour. The British Computing Society have specialist interest groups on ethics. Its Ethics Expert Panel: 'is responsible for providing advice and guidance on ethical issues associated with the development, operation and use of computerised Information Systems (IS).'

The Panel is linked to the International Federation of Information Processing (IFIP) interest group on ethics. IFIP ensures that clients are assured of the professional standard of work of IS professionals. It seeks to provide a code of ethics that: 'acknowledges the professional responsibilities of practitioners to society at large, members of the public, employers, contracting parties and fellow practitioners'.

Analysts need to ethicize personal action and objectify a personal code of ethical behaviour. Analysts' systems design decisions will affect the aims of the organization and the people working in it. Potentially compromised design decisions need to be ethically resolved. They have conflicting demands on them that need to be resolved ethically. These include:

- The tension between thinking in technical terms to comply with the constraints of IT and the needs of the organization and people.
- Potential conflict of interests of paymasters with personal ethics.
- Client contact that may compromise personal ethics and professionalism.
- Behaving as qualified experts who have knowledge of designing IS.

1.5.2 Assumptions of reality

Analysts' assumptions of actual situations they work in constitute the ontological component of a PCF. **Ontology** is knowledge of the nature of a thing, for example organization, people, IS and IT. Such knowledge is based on assumptions, which will be reflected in analysts' personal constructs of these things, and underpin personal practice. As a PCF consists of personal constructs of reality, practitioners and learners need to objectify personal assumptions of reality – which include ideas, notions, concepts and expected behaviours – and relate them to personal constructs.

Analysts make assumptions about organization and people and combine them with actual knowledge to enable practice. As it is not possible for analysts to know the nature of something in its totality, any action will be underpinned by assumptions and the expected effect of the action. Objectifying and understanding these assumptions and expectations results in better knowledge and effective practice. Ironically, assumptions that underpin practice form the practical basis of a PCF.

Objectivism

There are two distinct and mutually exclusive fundamental assumptions an analyst can make. One, that a thing is a fact independent of the analyst's experience and that it exists separately of him- or herself. This is called **objectivism**. Such an analyst would be called an *objectivist*. Objectivists make the assumption that reality can be known and that methods for studying reality result in objective knowledge, or knowledge that is independent of human bias. The process of acquiring objective knowledge is the **scientific method** used by natural scientists. Objectivists in social science are called *positivists*.

Objectivist analysts assume that system 'facts' can be established and that factual knowledge can be used to design systems and 'control' events. They assume that systems analysis and design can result in an objective and factual understanding of current IS and the objective design of new IS. Structured and object-oriented systems ontology assumes this kind of objectivist view of people, organization, IS and IT.

Subjectivism

The other assumption is that a thing is inseparable from the analyst's experience. Such an analyst would be called a *subjectivist*. Subjectivists in social science are called *interpretivists*. Interpretivists assume that knowledge is socially constructed. They make the assumption that people socially construct reality, and that methods for studying reality should lead to knowledge that includes an account of peoples' experience of reality.

An analyst's decision on one of these assumptions about reality has implications for the kinds of personal constructs of knowledge and practice developed. An analyst who assumes there are 'facts out there' has to develop practice that is detached and objective. The analyst will be detached from the organization and people wanting a system, and make design decisions objectively. Such an analyst cannot become involved in the internal politics of the organization or be biased towards particular **stakeholders**. In contrast, an analyst who assumes reality to be subjective has to behave as an involved member of an organization. The analyst will influence the politics of a new IS development, and possibly be biased towards particular stakeholders, or even pursue personal goals.

Researchers are divided into those who believe IS to be objective and those who believe it to be subjective. Objectivists dominate in systems analysis and design, in particular practitioners. Much of the critique on IS development practice stems from the premise of objectivity. For example, Multiview is a framework rather than a methodology, but it acknowledges subjectivism by including Soft Systems Methodology (**SSM**) analysis techniques.

1.5.3 Personal constructs

The individual basis for a PCF is Personal Construct Theory, in which a construct is a way of perceiving, construing, or interpreting events (Kelly, 2000a). Individuals seek to anticipate experience or events, and significantly have the 'freedom to choose' what meaning they attach to their experiences. To do so, they develop a coherent set of personal constructs that are used to interpret and explain events. The theory is based on the principle of 'constructive alternativism', the view that reality does not directly reveal itself to individuals and that each individual construes reality as personal inventions. So, individual analysts will construct images of organization, people, IS, and IT that differ from other analysts' constructions. A personal construct system is a set of personal constructs and the relations between them that provides the *unity* in the experience of individuals.

Personal constructs are necessary because reality cannot be directly known. Individuals experience reality and then develop constructions to help them *anticipate* it. This view is not the same as the subjectivists' view of reality, as personal constructs are open to measurements whereas interpretive knowledge is not. Personal constructs can be elicited and relations between them assessed using the **repertory grid** technique. Personal constructs are:

- Created as interpretations of personal experience and then used as a model of reality.
- Used to understand personal and other peoples' action (understand or make sense of reality).
- Used to repeat, anticipate or predict future personal experience.
- A personal method of acquiring knowledge of reality (a personal epistemology).
- Used to assess the effectiveness of personal action (based on trial and error).

Examples of personal construct types are: social, institutional, knowledge and belief constructs. Knowledge and belief constructs are especially important, because analysts' own understanding of what they regard as knowledge and how it is applied to practical problems are central. By surfacing personal constructs analysts' can begin to *see* themselves *in* the process of knowledge formation, questioning, learning, and critical thinking.

Personal constructs in turn determine action. Personal constructs are subject to revision and personal construct systems change over time. So analysts' experience can be construed differently. Examples of personal constructs are 'structure' from structured systems ontology and 'class' or 'object' from object-orientation. Another is the notion that system requirements can be elicited from clients or that IS can be modelled. A recent example is 'agile' from agile programming. Analysts trained in structured methods create personal constructs that support the 'structured' perspective. They may learn object-oriented analysis in time and so will need to revise their personal constructs to support the 'object' perspective.

Personal constructs underpin PCF because they enable an analyst to elicit a personal construction system or identity (self-characterization). Analysts should objectify personal constructs or a personal construct system to develop a PCF. Personal constructs create professional and personal identity. Objectified personal constructs help analysts to understand knowledge and how it underpins personal action. Analysts experience formal education, training and practice, which is the source for construing personal constructs on systems analysis and design, data, information, knowledge, organization, people, IS and IT. Experience from these sources leads to personal constructs designed to anticipate further experience – practice.

The objectification of personal constructs will enable practitioners, learners, and the teacher, to relate them to systems analysis and design. It leads to deeper knowledge and understanding of human action and reality. Importantly, the objectification process is a way of enabling learners to take ownership of learning and development of a PCF based on knowledge through experience (including practical knowledge) and knowledge acquired through the intellect. The result is the development of a personal **critical systems ontology** through increased complexity and definition of construction systems.

1.5.4 Instruments

Analysts need to objectify the knowledge underpinning the instruments of choice to accomplish objectives in practice. Instruments determine whether the practice is successful or not. The methods, techniques and tools of systems analysis and design form the technical or instrumental component of a PCF. Authors differentiate the terms 'method', 'technique' and 'tool' variously. They are collectively termed 'instruments' here. An instrument is an agency in the achievement of an objective. It embodies the individual's, community of practice, and society's knowledge regarding particular problems and how those problems can be resolved. Structured and object-oriented analyses consist of such instruments to address the problem of developing IS.

Analysts need to explain why certain instruments are preferred and how the chosen instruments address the problem. Assumptions underpinning instruments need to be uncovered and justified in terms of the problem. Experienced analysts will draw on practical knowledge to explain choices. Learners or novices will base explanations on learnt knowledge. Such explanations will enable informed evaluation of personal effectiveness in practice and clarify assumptions made about particular instruments to assess the relevance of those assumptions.

Objectification will further knowledge of how to improve the instruments, their use and selection. Instrument design and choices are determined fundamentally by assumptions analysts make. If an analyst assumes that organization and people exist as independent reality then structured systems analysis and design methods, techniques, and tools will be chosen. If it is assumed that reality is socially constructed, and that the analyst is part of it, then other suitable techniques will be required.

1.5.5 Relations between components

A PCF is only effective when relations between its components are defined. Analysts need to objectify the relations between PCF components to define how personal constructs are applied in practice or how they are used to anticipate experience – deliver required IS. Relations between components integrate them into a practical PCF for developing knowledge and practice. Relationships between components, and how components are affected by other components, need to be explained and justified to ensure the effectiveness of a PCF and subsequent practice based on it.

Objectification may be difficult because the actual relations may not be apparent. Often praxis consists of assumed relations that are hidden and enacted automatically in actual situations. Practitioners interested in delivering results do not reflect on what they do and how they behave in actual situations. Objectifying the validity of particular practice, for example consulting people whose jobs may be threatened by a new IS, is difficult for practitioners because such acts are secondary to the actual task of developing an IS.

The integrative basis of a PCF is clear definitions of relations between components. Relational definitions are essential to improve understanding and the further development of personal constructs in a PCF. They enable further critical thought on whether the relations are appropriate in terms of ethics or assumptions of reality. For example, an analyst needs to define how personal ethical code relates to organization and people, to personal constructs, or to instruments. The analyst may consider it unethical to impose on clients a way of working required in a new IS. The system design would then need to be discussed with clients to determine their preferences.

1.6 Logical and reasonable personal frameworks

The validity of a PCF depends on its content. It needs to be appropriate for making sense of systems analysis and design and practising it. Appropriate content can be assured by checking personal constructs for logicality and reasonableness. Personal constructs need to be reasonable and corroborated by evidence. An analyst cannot expect to determine what a new IS is required to do through observation alone for example. It is unreasonable to expect observation alone to provide an unambiguous and complete set of system requirements.

1.6.1 Evidence-based reasoning and action

Reasoning requires thinking rationally and logically about a problem, and involves rational argument and deductive or inductive logic. It alone is not sufficient to develop a practical PCF, evidence is necessary. **Evidence-based reasoning** results in a valid and practical PCF. Evidence is the experiences that analysts acquire from learnt knowledge and practice. Learners or novices should draw on learnt knowledge as authority or evidence. Such evidence can be objectified in personal constructs and critically appraised for both.

Evidence-based reasoning leads to evidence-based practice. Evidence-based reasoning is possible whether analysts decide on objectivist or subjectivist systems ontology. Both rely on the experiences of the individual to develop knowledge of reality. The reason for including certain instruments or particular ethics should be traceable to evidence in practice or a body of knowledge for validity. Personal constructs derive their validity from the experiences of individuals, and how these experiences are construed to anticipate further experiences. Developing a PCF on the basis of evidence-based reasoning ensures that further practice is itself based on evidence.

1.6.2 Think and act or act and think?

Action to resolve organizational problems can be interpreted in two ways. One, think rationally first about a problem to decide how to act and then follow it through into action, coined here as 'think and act'. A plan is a good example of this kind of action. It draws on rational thinking to detail prescribed action. The **SDLC** and structured systems analysis and design are examples. An IS **methodology** is a particular example of such a plan. The use of systems analysis and design instruments also requires first thinking through what to do and then implementing them.

The other interpretation is to engage with the problem first and then think during its resolution, and afterwards, about how it was solved, coined here as 'act and think'. The action and thinking of the action can be simultaneous. Many inventors state that they solve problems by engaging with them and think through the difficulties as they arise. Analysts too recognize that IS problems are resolved iteratively. ASD is an emerging pertinent example of the act and think strategy.

The think and act strategy is dominant in IS development. In actual practice system project plans are developed or a methodology espoused, but implementing them proves problematical in actual situations. The result is practitioners continually adjusting plans to reflect actuality. The act and think strategy has the advantage of making sense of the real situation in actual time and basing decisions on how to act on such an understanding. It is less likely to be used in practice, especially in a business context, because it does not afford knowledge of what needs to be done and how it should be done.

Analysts need to reflect on how they resolve problems. The choice of the think and act or act and think strategy for resolving problems should be based on evidence. It will determine personal constructs associated with problem resolution and determine what other personal constructs are used and how they are related in a PCF. Models for resolving difficult, complex, real-life problems exist to help individuals to move from the individual level to the organizational level. Action Science for example is the study of such complex problem-solving in organizations.

1.7 Example Personal Critical Frameworks

Analysts should endeavour to begin building a PCF. Three are illustrated in this section. The illustrations serve to show how components in a PCF might be composed and related. They should not form the basis for composing your PCF. They are not meant to be *actual* exemplars. In actuality, a practitioner's PCF will consist of experiential and studied knowledge and will probably include a richer and wider variety of experiential personal constructs. A learner or novice analyst's PCF will contain formally learnt personal constructs, whether through training or educational programmes.

The form of a PCF can be text, tables, diagrams or a combination. The author's PCF for teaching practice is a combination of a diagram and related text to describe and explain the process of learning and teaching (Patel, 2003). Its five main personal constructs are: learning and teaching instruments, discipline knowledge, personal and professional development, the self, and knowledge. The learning and teaching construct concurs with professional teaching bodies. The constructs were objectified as part of reflective and critical practice spanning ten years. The relations between these personal constructs are explained in text form. The author makes use of this PCF to inform practice and to improve it on the basis of evidence from learners of its effectiveness. It is continuously checked in practice with learners to assess its efficacy, and new ways of teaching devised to develop each of the elements in learners.

Three ideal-type PCF are illustrated in tabular form for brevity. The illustrations are in 'pure' form, meaning that actual PCF will have many 'grey' areas where practice and knowledge cannot be neatly compartmentalized as objectivist or subjectivist. One ideal-type is based on an objectivist view of reality, the second is based on a subjectivist view, and the third is simply a personal framework, it is based on an imagined analyst who is unaware of objectivism or subjectivism. The PCF is presented in the first person to make the process of **objectification** easier.

1.7.1 Objectivist Personal Critical Framework

For an analyst who regards reality as objective entities, a PCF may be objectified as Table 1.1. It consists of the five components of a PCF and a textual description of each component.

The description column in Table 1.1 contains the relation sub-element in each component to describe how components relate to each other. The overall relation component is still required to provide a rationale for the whole PCF. The descriptions are minimal and the relations need further elaboration. The PCF reflects practical and theoretical knowledge of systems analysis and design based on objective systems ontology.

1.7.2 Subjectivist Personal Critical Framework

For an analyst who regards reality as subjective entities, a PCF may be objectified as Table 1.2. It consists of the five components and a textual description of each component. The analyst will

Table 1.1 An objectivist Personal Critical Framework

PCF component	*Description*
Ethics	As I am detached from the organization and people for whom I develop IS, I am not concerned with judgements about what is appropriate or inappropriate action. I leave ethical decisions to managers who tell me what to do. *Relations*: As I am detached, my personal ethics does not affect the way I interact with people. I am a professional. Ethics is based on how I personally see reality.
Assumptions of reality	I can give an accurate account of organizations, people, IS amd IT and record this knowledge as facts. Reality can be decomposed to determine elemental facts, which I can use to decide what action I should take. I am detached from the organization and people, which exist separately from me. Knowledge is objective. *Relations*: Instruments can be designed to extract objective system requirements. I am a detached professional. An IS is objective.
Personal constructs	I can identify and objectify these personal constructs: professionalism; problem-solving; Information System; Information Technology; organization; workers; organizational work; organizational politics; rationalism; perfect knowledge. I am detached from the problems I solve and I am a rational person, capable of solving organizational and IS problems rationally. I am not affected by organizational politics. System requirements can be known and elicited, people in the organization should be able to tell me what they require the system to do. I am a professional person with expertise and I know what systems can do better than workers. *Relations*: My personal constructs are based on how I see reality. I can apply instruments to organizations to achieve goals. I can use them to extract system requirements and design systems.
Instruments	I can use my objective knowledge of reality to design and use instruments to solve organizational, informational and knowledge problems. Structured systems analysis techniques can be used to elicit, analyse and model systems. *Relations*: The design and use of my instruments is based on my detached and objective perception of reality.
Relations between components	My view of reality enables me to remain detached. My ethics is detached from the problem domain – organization and people. My view of organizations is that they can be 'engineered' using instruments designed on the basis of an objective knowledge of organizations and people. I am capable of objectively deploying instruments. My personal constructs reflect objective reality.

regard organization and people as meanings and IS and IT as interpretations that are socially constructed, and the analyst personally as involved in that social construction.

Table 1.2 contains brief descriptions. The PCF reflects practical and theoretical knowledge of an interpreted reality. The PCF reflects practical and theoretical knowledge of systems analysis and design based on subjective systems ontology.

1.7.3 A Personal Critical Framework

For an analyst with no awareness of objectivism or subjectivism, a PCF may be objectified as Table 1.3. It contains the five components of a

Table 1.2 A subjectivist Personal Critical Framework

PCF component	*Description*
Ethics	As I am part of the organization, I should behave ethically. I am attached to the organization and am concerned about what is good and bad behaviour. I want to design IS in agreement with people. I am sensitive to design decisions that lead to job loss or reducing skills. *Relations*: My ethical code is related to my personal constructs about organization and people, and other personal constructs.
Assumptions of reality	I alone am unable to account for reality as it is socially constructed. Reality is holistic and cannot be compartmentalized or decomposed. I am part of the organization and one of the people in it. I help to create the meanings underpinning the organization. I can help people to identify their information and knowledge problems and facilitate them to determine what they need from an IS. *Relations*: Instruments can be designed to elicit agreed requirements. My personal constructs are affected by my experiences.
Personal constructs	I can identify and objectify these personal constructs: professionalism; problem-solving; Information System; Information Technology; organization; workers; organizational work; organizational politics; personal involvement and attachment; meaning; interpretation; imperfect knowledge; limitations; social change; collaboration. I have limited rationality and knowledge, as do other people in the organization. People cannot know exactly what they require a system to do. Requirements are socially constructed and I, as a professional, act as a facilitator. *Relations*: My personal constructs are based on my view of reality. I can interact with people to understand what they want through instruments.
Instruments	I can design methods, techniques and tools to help people design appropriate IS. The instruments should be accessible to people and they should be involved in using them to design IS. *Relations*: The instruments I use are based on my view of reality. I apply them in agreement with people.
Relations between components	I am part of reality or organization and people. My ethics shape my personal constructs and determine the use of instruments.

PCF and a textual description of each component. The analyst's view of reality will probably be based on personal experience and knowledge gleaned from it.

The PCF reflects mostly practical knowledge. There is a paucity of theoretical or conceptual knowledge. The 'family' personal construct appears in Table 1.3. Given the holistic approach of PCFs this is normal.

1.7.4 Your personal framework

The reader should develop progressively a PCF by completing the PCF development section in each chapter. Your PCF should be continuously revised as you learn and begin to objectify personal constructs. The sources for your personal constructs will be knowledge or experience from training, formal education and practice.

Table 1.3 A praxis Personal Critical Framework

PCF component	*Description*
Ethics	I simply do my job. The organization is too big for me to do any harm to it. I will stay out of harm's way. I can give value to the business by doing good work. *Relations*: My personal constructs inform my ethics and I can only do the work that my instruments allow.
Assumptions of reality	I believe I can create IS to help people do their work and business to be effective. As I am unable to understand the whole organization, I only focus on a particular problem. I work in a team but the team seems disjointed, and I am only responsible for what I do. I believe the idea of teams does not work for me. *Relations*: The instruments I use enable me to do my work. My personal experience helps me to understand what I can do.
Personal constructs	I can identify and objectify these personal constructs: professionalism; value-added; problem-solving; Information System; Information Technology; organization; workers; organizational work; organizational politics; personal involvement and attachment; knowledge; limitations; family. *Relations*: I can use my technical skills to give value to the business. I use instruments as intended.
Instruments	I use instruments to help me solve problems with which I am concerned. They suffice, they could be improved but I don't know how. *Relations*: The instruments are sufficient for me to do my work. They need to reflect the constraints of the work I do.
Relations between components	I am one person in the organization. I cannot know everything that happens. Others provide the instruments I use. My personal constructs are light, I am only concerned with doing my job.

A PCF *is* to be revised. It should grow in knowledge and understanding. You should revise and amend your personal constructs by critically evaluating your knowledge. As you apply your PCF or reflect on it you will develop better understanding of how your knowledge and practice are related, and amend your PCF accordingly.

You should use your PCF to reflect deeply on the efficacy of personal action and its effectiveness. As an example, an analyst who has established a set of requirements from clients for a new IS may find that the interviewing techniques used failed to capture some functional requirements that only arise in the context of clients' work. The analyst should question the efficacy of the interviewing technique for contextual requirements capture, and think of alternative ways of capturing such requirements. You should base your reflection of knowledge and practice on **criticality**.

1.8 Criticality

Criticality is concerned with making judgements on the basis of critical thought. In pragmatic terms, criticality requires questioning and assessing or interpreting a current situation and reflecting on how it could be better. It results in doing something differently. It is not sufficient for practitioners to simply reflect on what they do, how they do it, and on the results of their actions. Reflective thinking supposes criticality. Reflective and critical analysts make judgements on what constitutes appropriate knowledge and practice to achieve objectives. Such reflection requires criticality to make a difference to practice and the accumulation of relevant knowledge.

Systems analysts should reflect deeply on knowledge and practice and analytically evaluate the efficacy of concepts and instruments. Such evaluation should lead to improvements in knowledge and practice. Criticality can be developed from formal learning or through reflexive practice.

A PCF should be based on *being* critical. Being critical of personal knowledge and actions should provide relevance to objectified components of a PCF. The objectified components should be subjected to further **critical thinking**.

1.8.1 Critical thinking

Critical thinking is essentially a cognitive act that benefits individuals and organizations. Understanding and developing critical thinking requires personal effort aided by formal education. The Delphi Report provides a list of cognitive skills and sub-skills needed to develop critical thinking (Facione, 1990). PCF development requires these cognitive skills elaborated in Table 1.4. They will enable analysts to reflect critically on knowledge and practice.

Critical cognitive skills are required to initially compose, and enable continuous revision of, a PCF. To compose a PCF initial reflective thinking is required to determine relevant assumptions about reality, ethical issues,

Table 1.4 Critical thinking skills

Cognitive skill	*Description*
Interpretation	To comprehend and express the meaning or significance of a wide variety of experiences, situations, data, events, judgements, conventions, beliefs, rules, procedures or criteria.
Analysis	To identify the intended and actual inferential relationships among statements, questions, concepts, descriptions or other forms of representation intended to express belief, judgements, experiences, reasons, information or opinions.
Evaluation	To assess the credibility of statements or other representations which are accounts or descriptions of a person's perception, experience, situation, judgement, belief or opinion; and to assess the logical strength of the actual or intend inferential relationships among statements, descriptions, questions or other forms of representations.
Inference	To identify and secure elements needed to draw reasonable conclusions; to form conjectures and hypotheses; to consider relevant information and to educe the consequences flowing from data, statements, principles, evidence, judgements, beliefs, opinions, concepts, descriptions, questions or other forms of representation.
Explanation	To state the results of one's reasoning; to justify that reasoning in terms of the evidential, conceptual, methodological, criteriological and contextual considerations upon which one's results were based; and to present one's reasoning in the form of cogent arguments.
Self-regulation	Self-consciously to monitor one's cognitive activities, the elements used in those activities, and the results educed, particularly by applying skills in analysis and evaluation to one's own inferential judgements with a view toward questioning, confirming, validating or correcting either one's reasoning or one's results.

instruments for practice, personal constructs and relationships between constructs. Once a PCF is composed, it needs to be continuously revised to reflect advances in knowledge in systems analysis, development of new techniques, and experiences from practice. Such revisions require critical thinking to justify new personal constructs or to justify retention or amendment of existing ones.

The Delphi Report defines a critical thinker:

> To the experts, a good critical thinker, the paradigm case, is habitually disposed to engage in, and to encourage others to engage in, critical judgement. She is able to make such judgements in a wide range of contexts and for a wide variety of purposes. Although perhaps not always uppermost in mind, the rational justification for cultivating those affective dispositions which characterise the paradigm critical thinker are soundly grounded in CT's personal and civic value. CT is known to contribute to the fair-minded analysis and resolution of questions. CT is a powerful tool in the search for knowledge. CT can help people overcome the blind, sophistic, or irrational defence of intellectually defective or biased opinions. CT promotes rational autonomy, intellectual freedom and the objective, reasoned and evidence based investigation of a very wide range of personal and social issues and concerns.
>
> (p. 13)

Self-regulation is crucial for developing and continuously revising a PCF. The Delphi Report states two associated sub-skills: self-examination and self-correction (Facione, 1990). Self-examination entails activities like reflecting on 'one's own reasoning' or 'motivations, values, attitudes and interests', being 'objective', judge 'deficiencies in one's knowledge', being 'unbiased and fair-minded', and 'respectful of the truth'.

Barnett (1997) defines criticality as: 'a human disposition of engagement where it is recognised that the object of attention could be other than it is.' There are three forms of criticality in relation to its three *domains* of expression: '*critical reason*, *critical self-reflection*, and *critical action*' (p. 179 italics in the original). He provides ways of being critical in the form of a schema for critical being and identifies four levels of criticality:

Transformatory critique At the knowledge dimension, in transformatory critique knowledge itself is reframed. For example, practitioners in the field collectively developed and accepted knowledge of structured systems analysis that departed from traditional systems analysis. Recently, there has been collective reconstruction of knowledge as object-oriented systems analysis and design, as deficiencies in knowledge of the structured approach are identified.

Refashioning of traditions In refashioning of traditions critical thought is applied to 'malleable traditions of thought'. For example, practitioners and professional bodies in the field developed mutual understanding of the roles of developers as central in IS development. Gradually the role of 'users' as participants in IS development was recognized. Transformatory critique and refashioning of traditions may be considered as higher levels of criticality requiring intellectual insights aimed at redefining accepted knowledge.

Reflexivity Reflexivity is critical thinking based on one's experience and understanding of situations. It is reflective practice. For example, experience of structured systems modelling may reveal weaknesses in separating data and process models. The analyst may adapt practice by using object-oriented systems modelling to create integrated or encapsulated data and process models. Here criticality results in adaptations to and flexibility in practice.

Critical skills Critical skill is the fourth form where 'discipline-specific critical skills' are developed. For example, analysts need to develop systemic problem-solving or modelling skills. The development of discipline-specific critical skills is termed 'means-end instrumentalism'. Reflexivity and critical skills are regarded as personal-oriented criticality. This is termed 'critical reason'.

The four levels of criticality are applied to three dimensions: knowledge, the self and the world. The need for criticality arises from doubting the effectiveness of certain systems analysis and design concepts, instruments, or from reflecting on ineffective practice. It may arise from making an assessment of personal practice that leads to critical observations of applied knowledge or professionalism. There are various ways of being critical of praxis. Question the efficacy of certain instruments or doubt the usefulness of a particular approach to systems analysis. Trying different approaches and instruments leads to critical observations about their effectiveness. While these forms of criticality are important for the development of a PCF, Barnett's three forms of criticality provide the foundations for its development. Critical reason, critical action, and critical self-reflection are necessary to develop a PCF.

1.9 A critical learning journey

The analogy of a journey is used in learning to indicate what knowledge the learner will acquire. Learning can be likened to a journey that has a starting point and an end point. The learner begins with a starting state of mind and through the study of the discipline gains and appreciates knowledge that will create a different state of mind. In this textbook the learner's starting state of mind is the desire to acquire knowledge of systems analysis and design that leads to personal effectiveness. The end point is the development of criticality and critical thinking cognitive skills and their application to knowledge and practice to develop a PCF for personal effectiveness.

The journey's fundamental milestone will be transformatory critique in the self-dimension that leads to the 'reconstruction of self'. The task of the actual reconstruction of the self is left to learners and practitioners as they develop a PCF to objectify personal knowledge and to reconstruct it on the basis of criticality. Analysts can reconstruct the self by understanding the role of the self in learning, objectifying personal knowledge and reflecting on practice.

The journey's practical prominent milestone will be transformatory critique in the world-dimension that leads to 'critique in action' or 'collective reconstruction'. Chapter 2 is a framework for critically appraising systemic knowledge and practice. Analysts can draw on it to develop the components of a PCF. The collective reconstruction of knowledge is exemplified in Part V in terms of critique of systems analysis and design. This critique will enable analysts to understand deeply assumptions of reality and personal constructs. An alternative term paradigm shift for transformatory critique in Chapter 14 is illustrated in terms of IS development knowledge. A cursory example of transformatory critique here is the creation of knowledge in 'agile' IS development and associated agile systems analysis techniques.

During the journey, analysts will tread traditions to draw on refashioning of traditions in the self-dimension to develop the self within traditions, as covered in Parts II, III and IV. This coverage will include reflexivity in the self-dimension that leads to 'reflection on one's own projects'. It will also include critical skills in the self-dimension that leads to 'self-monitoring to given standards', termed 'critical self-reflection'. This coverage will enable analysts to develop their assumptions of reality, personal constructs and instrumental elements of a PCF.

The learning journey will expose analysts to the problem of developing IS, in particular to systems analysis and design. It will show how structured and object-oriented methods are used to analyse and design IS, and how they presume and make assumptions about what constitutes knowledge of the **problem domain** – the workplace, the organization, people – where IT is to be applied.

The end point of the journey is developed criticality. The critical perspective on structured and object-oriented systems ontology will focus on the assumptions they make about organization, people, and the work they do, and on what analysts and designers can know about them. The adequacy and efficacy of the instruments used to analyse and design IS will be questioned. This critical perspective will reveal that some assumptions of the problem domain and the efficacy of instruments used are debatable. It will reveal the limits of knowledge that analysts can establish about the design domain and raise awareness of what is still problematical in IS.

At the end of the journey your PCF should contain understanding and appreciation that knowledge and practice is progressive, and that your PCF should be similarly progressive. A final PCF is a dead PCF. Your PCF should be richer for understanding that some researchers and analysts regard the problem domain as 'objective facts' that can be established and used to develop IS. Others regard it as a complex mix of subjective, social and organizational issues that need to be understood and incorporated in systems analysis and design. It should recognize that the methods, techniques and tools in structured and object-oriented analyses have limits, and that there are political, organizational and cultural obstacles for analysts to overcome if they are to deliver relevant systems models and designs.

1.10 Personal Critical Framework development

1.10.1 Repertory grid technique

Personal constructs should be elicited using the repertory grid, a technique to make learning more relevant to individual needs. The technique is used to elicit and analyse knowledge, and can be used for self-help as in a PCF development. Its use should challenge beliefs and self-images of the self that are not tested but held to be 'true'. True learning happens when the self is somehow affected.

The repertory grid technique consists of compiling a table with a left-hand column and a right-hand column that contain polar opposite concepts, and up to 21 other columns containing the concepts, ideas or objects related to systems analysis and systems design. It is not necessary to have 21 columns. Use a word processor table or spreadsheet for marking out the grid.

The personal constructs on the horizontal bar are suggestions only. You may add more or change them to suit your ideas on systems analysis and design. Compile your own elements and polar opposites to make it personally relevant. The right and left bars show the two ends of a pole. Make out cards for each of the concepts, ideas and objects that make up the columns – personal constructs. Table 1.5 illustrates the technique and an elaborate grid is shown in section 5.9.1, Table 5.4.

Each concept, idea and object that makes up the 21 columns is also written on a separate card. Working with a peer (the tester), complete the grid to reveal your personal constructs. Groups of three cards are shown by the tester to the individual whose personal constructs are to be elicited, and the tester asks the same question each time: 'How are two of these similar and the third different?' For example, for the rational–emergent polar dimension in the first

Table 1.5 Repertory grid technique

Pole 1	*Systems analyst*	*Organization*	*Workers*	*Project team*	*Interviewing*	*Pole 2*
Rational		✓		✗	✓	Emergent
Efficient		✓	✗			Problematic
Objective		✓		✗		Subjective
Diplomatic	✗		✓			Authoritative
Ethical			✗		✗	Unethical

row, the three cards shown are organization, project team and interviewing. Mark a stroke for each card dealt and for the odd one out (the one identified as different) mark with a cross-stroke.

Your answers will be your personal 'constructs' for systems analysis, or a dimension along which you have divided up knowledge and experience. Once the grid is complete it is to be analysed using one of many ranking methods to reveal which are the core personal constructs that you use to construct and anticipate events in the real world. The personal constructs will indicate how you have divided up your knowledge and experiences of systems analysis and design.

You can then evaluate the usefulness and validity of these objectified personal constructs for your PCF. You can decide whether the personal constructs you identified form the foundation for professionalism, or which systems ontology you prefer, or help you decide which instruments to use.

Types of repertory grid are available. The one used in this book is visual. It can be analysed using a visual technique. Table 1.5 is reproduced in Table 1.6 to demonstrate how to analyse the content of a completed repertory grid. Table 1.6 shows the elements (in columns) and constructs (in rows) and the completed tick/cross by an assumed analyst.

To analyse the completed grid the method of visual focusing is used. In a separate new table of three columns, the pattern in the first element 'systems analysis' from the completed grid is copied into the first column of the new table. This column is then compared in turn with each other element. For example, the 'Systems analyst' element is compared with the 'users' element first, then with the 'Organization' element, and so on, in turn. Each time there is an agreement, defined as either two ticks matching or two crosses matching, '1' is recorded in the third total column. The column-by-column comparison results in various strips as shown in Tables 1.7 and 1.8. The maximum agreement score is 5, which is the number of constructs in the grid and the minimum is zero.

When all the possible pairs of elements have been so analysed they are then drawn as a matrix. (All the pairs are not reproduced here.) It is not necessary to use the original element names, they can be coded as E1, E2, etc. for ease. The pattern that emerges is the significant understanding of personal constructs. The matrix with the agreement scores for all the pairs might look like Table 1.9.

This analysis reveals elements and their associated meaning for a person. To discover the significant elements, Table 1.9 is then initially analysed along the rows to find the matching

Table 1.6 An assumed analyst's repertory grid

Pole 1	*Systems analyst*	*Organization*	*Users*	*Project team*	*Interviewing*	*Pole 2*
Rational	✓	✓	✓	✗	✓	Emergent
Efficient	✗	✗	✗	✓	✓	Problematic
Objective	✓	✓	✓	✗	✓	Subjective
Diplomatic	✗	✗	✓	✓	✓	Authoritative
Ethical	✓	✓	✗	✓	✗	Unethical

Table 1.7 Visual focusing (A)

Systems analyst	*Users*	*Total*
✓	✓	= 1
✗	✗	= 1
✓	✓	= 1
✗	✗	= 1
✓	✓	= 1
Total		5

Table 1.8 Visual focusing (B)

Systems analyst	*Organization*	*Total*
✓	✓	= 1
✗	✗	= 1
✓	✓	= 1
✗	✓	= 0
✓	✗	= 0
Total		3

Table 1.9 Agreement scores

	Systems analyst	*Organization*	*Users*	*Project team*	*Interviewing*
Systems analyst	✗	3	2	5	3
Organization		✗	3	2	4
Users			✗	5	3
Project team				✗	2
Interviewing					✗

higher numbers. The project team and users elements are in agreement, with each scoring 5. Where elements are in agreement they share significant meaning for the supposed analyst. It means that 5 out of 5 times the supposed analyst agrees on these two elements as equally valid or important. The analyst sees no difference.

The significance of these numbers becomes apparent when they are considered in the context of an actual or supposed problem. Suppose that the analyst is engaged in gathering system requirements information in an organization. He or she does not differentiate between the project team and users, because

5 out of 5 times they are given the same rating by the analyst. There is also an agreement score of 3 out of 6 between users and interviewing.

The analyst can use the repertory grid results to evaluate the impact of these meaning on practice. The analyst can assess whether the equivalence afforded to users and project team is practical. The analyst's project team and users elements is significant because in practice the analyst will behave equally towards users and the project team, which may cause difficulties when system project priorities clash with users' priorities. Such reflection will result in additional personal constructs, which can be added to the repertory grid and scored.

To decipher the significance of elements they are rearranged with similar elements replaced together in a different matrix, as shown in Table 1.10. This grid shows the similarity between elements and constructs in terms of how near they are in physical space – placed adjacent to each other.

The users and project team are placed together (first two columns) because they are in total agreement. The systems analyst and organization elements are nearly in agreement so they are placed together. The interviewing and instruments elements are similarly placed together because they are nearly in total agreement. The significance of the precise re-sorting depends on the analyst's concern. If only similarities are sought rather then exact matches the re-sorting does not have to be exact. If exact matches between particular elements are sought then only those need to be placed together exactly.

A similar ordering of constructs can be arranged. For example, the diplomatic–authoritative and ethical–unethical constructs can be compared as in Table 1.11

This shows that there is lack of agreement between diplomacy and ethical behaviour because on three scores they do not match. This result suggests that the supposed systems analyst is influenced in diplomacy by ethics. A matrix for all the personal constructs can be constructed and analysed. Further references for understanding and compiling a personal construct system are given in section 1.10.3. Use the references to seek further guidance on analysing a repertory grid.

Table 1.10 Similar elements

Pole 1	*Users*	*Project team*	*Interviewing*	*Instruments*	*Systems analyst*	*Organization*	*Pole 2*
Rational	✓	✓	✓	✓	✓	✓	Emergent
Objective	✓	✓	✗	✓	✗	✓	Subjective
Efficient	✗	✗	✓	✓	✗	✗	Problematic
Diplomatic	✓	✓	✓	✓	✓	✓	Authoritative
Ethical	✗	✗	✗	✓	✓	✓	Unethical

Table 1.11 Similar–dissimilar personal constructs

Diplomatic	✓	✓	✓	✓	✓	✓	Authoritative
Ethical	✗	✗	✗	✓	✓	✓	Unethical

It is assumed that readers will attempt some kind of personal construct elicitation. If the repertory grid **technique** is not used the personal constructs can be 'conceptual' and a PCF can still be developed using any of the activities A to E below, and similar activities in other chapters. It is the PCF that is more important than a correct method for eliciting personal constructs.

Activity A

For this activity refer to the internet sources for personal construct theory detailed in section 1.10.3 and further reading in section 1.10.4. Copy Table 1.5 onto a spreadsheet and add your own ideas for personal constructs (columns) and polar opposites (rows). Using the repertory grid technique determine your personal constructs. Work with someone to help you deal the cards and fill in the grid. Discuss with your colleague or peer whether the results surprise you or confirm your understanding of systems analysis. Evaluate whether the personal constructs you identified are appropriate for the successful achievement of the work you do?

Activity B

This activity provides a staged approach to starting to think about a PCF. It makes use of writing and diagramming as expression. You may want to complete the following:

- Identify and write down concepts, phrases, words and things you do in practice.
- If you have not yet practised systems analysis, identify and write down concepts, phrases, words and things you would do in practice.
- Draw a diagram placing these items into boxes, rectangles and triangles, etc.
- Draw arrows, lines and other connectors to show how these items are related to each other.
- Analyse the diagram to assess whether it accurately describes what you do or what you would do.

Activity C

Develop a tabular PCF similar to the illustrations in section 1.7 to describe your approach to systems analysis. Use the Delphi Report cognitive skills to think about your ethics, assumptions of reality, personal constructs, and the instruments you use to achieve goals. Describe the relations between these elements.

Activity D

Find a trusted colleague or peer and discuss your ideas on the five components of the PCF. Make notes of your ideas as you discuss them. Elaborate your ideas in textual prose.

Activity E

Either draw a diagram of how you act to complete an assignment or draw a diagram of how you acted in an IS development project. Begin by jotting down words that come to your mind. Organize these words into a logical order to describe your knowledge and practice and then illustrate them in a diagram.

1.10.2 Experiences, evidence and criticality

Activity A

Barnett's schema for criticality consists of transformatory critique, refashioning of traditions, reflexivity and critical skills.

- Describe one example either from personal life experience or practice for each type.
- Describe an example of reflexivity in your practice.
- Make a list of critical skills an analyst would need to develop to improve practice.

Activity B

In groups of two or more, individually write 300 words on your interpretation of systems analysis and design. Share your text with the group and discuss how they interpret your views. Is your interpretation a shared view or different? In either case discuss the importance of interpretation and shared knowledge in understanding and establishing knowledge.

Activity C

For all the activities in the previous section and this section, identify the elements that are based on evidence of systems analysis and design. What kind of problems might arise from non-evidence-based reasoning in practice? How is the validity of critical observation improved with evidence?

1.10.3 Internet sources

Critical thinking

Definitions of critical thinking can be found at: http://www.philosophy.unimelb.edu.au/reason/critical/pages/definitions.html (accessed 24 April 2003).

The Education for Thinking Project examines the development of thinking and learning skills as a goal of education. This site contains a series of short 'meditations' on thinking, knowledge, argument, etc. from one of the world's leading psychologists of critical thinking. http://www.tc.edu/centers/eft/ (accessed 24 April 2003).

To explore software for critical thinking see: http://www.philosophy.unimelb.edu.au/reason/critical/pages/software.html (accessed 24 April 2003).

Ethics

Search www.bcs.org for the ethical code it prescribes for its members. Think about how you would include it into your ethical practice.

For an internationally accepted source for ethics and professionalism surf IFIP's SIG 9.2.2 'Framework on the Ethics of Computing' http://www.ifip.or.at/ (accessed 21 May 2003).

Personal constructs

Surf http://www.brint.com/PCT.htm for an easy to read overview of personal constructs (accessed 21 May 2003).

For a brief overview of how personal constructs are established using the repertory grid technique see: Atherton, J. S. (2002) *Learning and Teaching: Personal Construct Psychology* [Online] UK: Available at http://www.dmu.ac.uk/~jamesa/learning/personal.htm (accessed 21 May 2003).

For a business application of the repertory grid technique see: http://www.enquirewithin.co.nz/BUS_APP/business.htm (accessed 21 May 2003).

1.10.4 Further reading

For an elaboration of the role of criticality in education see: Barnett, R. (1997) *Higher Education: a Critical Business*, Buckingham: Oxford University Press.

For the development of critical thinking in learners see the Delphi Report: Facione, P. A. (1990) *Critical Thinking: a Statement of Expert Consensus for Purposes of Educational Assessment and Instruction*, Millbrae: Santa Clara University, pp. 1–19.

An elaboration of the holistic approach to learning and teaching, upon which this book is based, can be found in: Patel, N. V. (2003) 'A Holistic Approach to Learning and Teaching Interaction: Factors in the Development of Critical Learners', *International Journal of Educational Management*, 17(6): 243–247.

A rather old edition but the original contains a fuller elaboration of the repertory grid technique and how to interpret the hidden meanings within it, is: Thomas, L. F. and Harri-Augstein, E. S. (1985) *Self-Organised Learning, Foundations of a Conversational Science of Psychology*, London: Routledge & Kegan Paul.

Chapter 2

Critical knowledge and practice framework

2.1 Learning outcomes

To engage in the PAC cycle, after completing this chapter you should be able to:

- Interpret the Critical Knowledge and Practice Framework (Critical Framework for short) to critique, evaluate and analyse knowledge of systems analysis and design.
- Apply the Critical Framework to interpret the application of formalism in systems analysis and design to actual IS problems.
- Interpret the major themes and sub-themes in the Critical Framework to evaluate critically knowledge and practice.
- Explain how the Critical Framework relates to the development of the components of a PCF.
- Use the Critical Framework for 'self-regulation' in terms of the Delphi Report's cognitive critical thinking skills.

2.2 Introduction

Structured systems analysis and design is here termed 'structured systems ontology' and object-oriented systems analysis and design is termed 'object systems ontology'. They are two current dominant strands in systems analysis and design. They seem partial and unsatisfactory because organizations and analysts experience a variety of problems applying them in practice. Large organizations experience problems in developing and managing inflexible IT and IS that pose a significant overhead cost to business operations. Problems arise in determining what IS to select for development and how to develop them, determining system functionality, or keeping developed IS relevant to organizational needs. Consequently, analysts need to approach knowledge and practice with a reflexive perspective. They work in an environment where the complexity of IS is increasing but improvements in knowledge and practice is lagging significantly.

Professional analysts can be *trained* to apply instruments to develop IS. Trained analysts' successes will be limited to smaller IS and may not be scalable to larger IS and organizations. While training provides knowledge of how to apply instruments, it is incapable of engendering criticality. The ever-increasing complex problem of developing IS in organizations requires trained and *educated* professionals. Analysts interact with professional accountants, marketers, business executives and other business professionals, so they need to learn to understand business problems such as 'efficiency drives', 'quality improvement' and 'competitiveness'. Simple training is insufficient in this kind of environment.

A Critical Framework is elaborated in this chapter to engender a critical perspective. A critical perspective means to learn to regard (personal) knowledge and practice as progressive rather than established. It is used throughout the chapters to develop criticality. Criticality is necessary to understand systems ontology, instruments, the business context, and determine how to proceed within it. It is an important attribute of a reflective practitioner who has to deal with ever-increasing IS problems.

2.3 Human action and criticality

Computer-based IS are required to manage data, information and knowledge in modern business organizations. The development of IS is a human activity that encompasses organization, people, IS and IT.

The quality and delivery of a developed IS is likely to improve if analysts develop a critical perspective. Systems analysis and systems design are key activities that determine the success of the development. Analysts' actions are central and critical in the development process. Their knowledge and understanding of systems analysis and design and the organizational problem to be solved determine the actions they themselves take.

Such human action is based on knowledge and understanding of the actual situation and knowledge of systems analysis and design. Both depend on ontological knowledge. Such ontological knowledge contains an explanation of the actual situation that determines what action is taken. For example, if human intelligence is understood and explained as genetic inheritance, then there would be an absence of educational policy actions designed to develop intelligence in children. If the genetic inheritance explanation for intelligence is accepted uncritically then it can lead to major disadvantages for people who do not have the 'intelligent' genes. Similarly, if systems analysis and design is accepted as objective it will determine how an analyst acts.

Analysts need to think critically about knowledge and how it informs practice. Unquestioned acceptance of systems ontology will result in rote practice, and lead to developed IS lacking quality and relevance. Critical evaluation of knowledge and its practical validity to achieve desired objectives will improve action and lead to the design of the most appropriate IS.

2.3.1 Systems analysts, knowledge and practice

The Critical Framework is formulated to enable criticality. Analysts need to develop cognitive skills to evaluate a variety of issues and develop critique. Issues range from deficiencies in knowledge, grey areas in practice, and adequacy of instruments. Such issues can be evaluated by developing critical thinking skills.

Critique is developed by uncovering and questioning assumptions about reality made in methodologies and systems ontology. The Critical Framework is a point of reference to develop significant critique. Its purpose is to:

- contribute to the development of PCF;
- enable the development of critique;
- engender reflexivity;
- improve knowledge, practice and understanding of context and instruments;
- critically understand and critically apply instruments.

2.3.2 The critical knowledge and practice framework

The Critical Framework, shown in Figure 2.1, is thematic. Analysts can use the themes to critically scrutinize knowledge, conceptual underpinnings, methods, instruments and practice. The Critical Framework is a set of themes, ideas, concepts, and learning tools arranged in three layers to develop critical thinking. The thick arrows indicate the logical moves from one theme to the next and what is required to make the move.

The first layer of the Critical Framework consists of three major themes and six subthemes through which knowledge and practice can be explored. The major themes are systems ontology, the real world of human problems, and pragmatic resolution, and the six subthemes provide analytic foci for critically exploring the major themes. The major themes depict the object of study (real world of human problems), the nature of the object and how it is studied (systems ontology), and pragmatic actions required to analyse and design IS (pragmatic resolution). The combined major themes

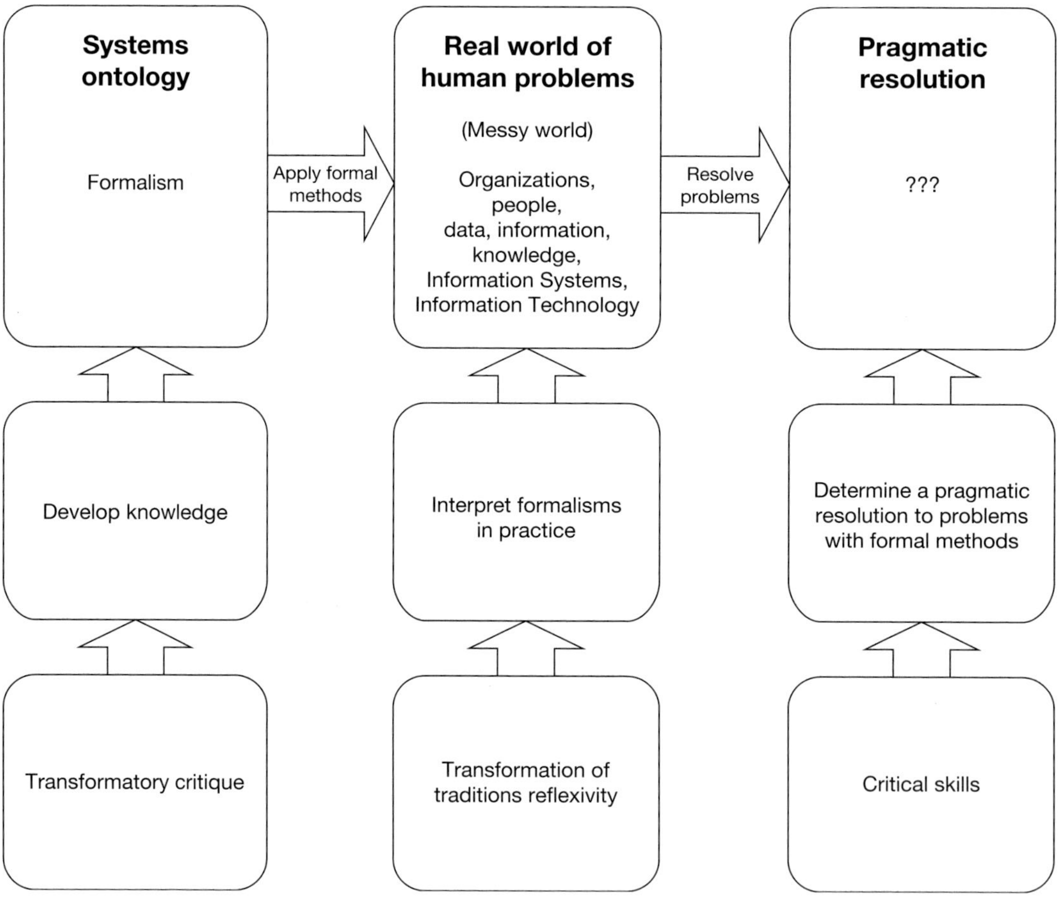

Figure 2.1 Critical knowledge and practice framework

and the sub-themes provide a critical lens for analysts to examine knowledge and practice. They enable analysts to think critically and should be referred to in enacting the PAC cycle to interpret and analytically evaluate knowledge and practice to include in a PCF.

The second layer explains the rational for each major theme. The systems ontology theme is to develop formal knowledge of systems or **formalism**, including theory. Such knowledge may be objective or subjective. This theme is knowledge oriented. The real world of human problems theme is to understand the application of formalism and its effectiveness in organizational context. The pragmatic resolution theme is to resolve contextual problems with the application of formalism to real problems. The human problems and pragmatic resolution themes are practice-oriented.

The third layer correlates the type of criticality that might normally be associated with each of the three major themes. Transformatory critique is associated with the development of knowledge of systems ontology. Reflexivity and refashioning of traditions criticality is associated with practice, the human problems theme. Critical skills are associated with resolving practical problems with applying formalism, the pragmatic resolution theme. The critical thinking skills in Table 1.4 (section 1.8.1) are required in, and common to, all three themes.

The first layer of the Critical Framework is underpinned with six sub-themes. They provide the foci for analysts to interpret and evaluate knowledge, practice, and know-how. The sub-themes are:

- Reality and knowledge, of organization and people, and IS and IT.
- Theoretical, conceptual, and formal knowledge of systems.
- Pragmatism, or practical knowledge of systems analysis and design.
- Planned action and plans.
- Situated action, and situations and context.
- Methods for acquiring knowledge.

Reality and knowledge IS researchers seek to develop knowledge of IT and IS and its use in organization and society. Their aim is to develop valid knowledge that is established through agreed methods of inquiry. Knowledge from research is referred to as 'knowledge claim'. Knowledge is also established through praxis or practical experience.

Analysts need to evaluate this kind of knowledge. As they work in or for organizations, they need to understand them, develop knowledge of people who work in them, and how they want to use IS and IT. Analysts should interpret and evaluate knowledge claims concerning these issues, and uncover assumptions underpinning the knowledge claims to assess reliability and validity.

Knowledge needs to be critically examined to uncover assumptions that may not be valid. Researchers and practitioners make certain assumptions when developing knowledge because the subject of study cannot be known in its totality. These assumptions may weaken the reliability of the knowledge claim or its effectiveness for practical purposes.

Theoretical and conceptual knowledge Theory is concerned with providing explanations of observed phenomena. The explanations can be conjectural and abstract or based on empirical data gathered by research. Sometimes theoretical explanations can be speculative. For example, organization theory is concerned with describing and explaining what organizations are and how they work, it seeks to understand and explain people in organized activity.

There is no substantive theory of systems analysis and design pertinent to IS. There is conceptual knowledge and formalism, but formalism does not provide explanation. So researchers draw on organization theory, systems theory, social theory, communication theory, and information and situation theory. They do so to understand the role of information, knowledge, IS and IT in organizations and its effect on people.

This is a fundamental sub-theme for developing critical thinking. Concepts and instruments in systems analysis and thinking stem from theoretic explanations of organization, people, IS and IT. As practice is affected by such knowledge it is necessary to consider conceptual knowledge and formalism critically.

Pragmatism The actual achievement of an objective is a pragmatic process focusing on the practical rather than the conceptual or formalism. IS development is a pragmatic activity and systems analysis and design is concerned with practical action – the analysis of human problems, development of systems models and their implementation.

The actual application of formalism in context requires critical thinking. This sub-theme is related to actual practice of systems analysis and design. Pragmatic behaviour is contextually rich. Analysts come to know details about work, workflows, processes, business data, information and knowledge, people and communications, and many other elements of organization in context. They need such detailed knowledge because they have to develop systems analysis and design systems models in such a contextually rich setting. Contextual, pragmatic knowledge helps analyst to find ways of resolving problems with formalism, or practical knowledge, when applied in context.

Planned action A plan is a rational, conceptual construct detailing and prescribing human behaviour designed to achieve objectives. Plans epitomize human action as rational action. They contain predetermined activities that specifically prescribe behaviour and describe desired outcomes. Planned action is closely linked to developments in science and the scientific method. It is crystallized as 'methodology' in IS development. This sub-theme covers IS methodologies, methods, techniques and tools.

Planned action is contextually poor because it cannot foresee the actual situation in which the plan will be implemented. The SDLC, structured systems analysis and design (SSADM), and IS methodologies are examples of planned action. They are together here termed the structured systems ontology. Object-orientation and object-oriented analysis is not entirely planned action because it is flexible in how it is carried out. It is termed here as object-oriented systems ontology. This sub-theme is a central source for critically analysing the efficacy of systems analysis and design based on planned action.

Situated action Situated action is human intention that cannot be explicated in terms of rules and procedures. It is informed by context and utilizes 'embodied skills' (Suchman, 1994). Unlike planned action, situated action cannot be detailed or described in advance of action.

Situated action is a comparative critical thinking sub-theme. It can be compared with planned action to understand how analysts could act. Situated action makes use of contextual and situational information and draws on embodied skills. Arguably, systems ontology needs to incorporate situated action.

Knowledge of action as situated action is not discussed much in IS methodologies and systems analysis and design. It can be said to be emerging in methodologies, and in some IS development approaches like ASD and **JAD**.

Methods of knowing This sub-theme is to evaluate analytically methods for acquiring knowledge on the nature of organizations and people, IS and IT, and how to analyse and design IS. Deciding what IS to build, determining system functionality, deciding on user interfaces design, determining data inputs, process and outputs, is all dependent on the methods used to acquire knowledge. Knowledge to develop a new IS can be regarded as 'objective fact' or as 'interpretation'. Analysts should understand both forms of knowledge, and interpret and evaluate a personal position.

This sub-theme is important for the development of criticality in analysts. Analysts are concerned with the systematic attempt to shape the future in a coherent way – 'praxeology'. To shape the future it is necessary to acquire knowledge about the nature of the subject acted upon. The nature of something is called its 'ontology'. How we know something, such as how organizations work or the relation between information and people, is called 'epistemology' – or how to acquire knowledge. Analysts are particularly concerned with pragmatic action and how they act is based on the kinds of epistemology and ontology they accept as valid and reliable.

This sub-theme is difficult to relate specifically to systems analysis and design or IS methodologies because their inventors do not explicitly state how they acquire knowledge or the nature of their subject. SSM, which is advocated for systems analysis, is based on 'systems thinking'. Its underlying method of knowing is interpretation based on 'holism'. To establish 'objective facts' for systems models other methodologies use 'reductionism'.

2.3.3 The ontology of systems

Transformatory critique is applied to systems ontology. Systems ontology is the knowledge theme in the Critical Framework. Systems ontology is knowledge of the nature of systems. System ontological knowledge covers the problem of defining, describing and explaining IS in terms of the capacity and limits of IT, and how humans interact with it. Ontological system knowledge underpins how the IS development problem is framed and resolved, and underpins pragmatic response. This theme can be extended to cover practical knowledge on how an IS is selected for development, what systems analysis and design conceptual knowledge is generated, and what approach and instruments to use.

Systems analysis and design can be explained in terms of systems ontology. The different approaches, methods, techniques and tools are based on various explicit or implicit ontological assumptions. This theme is to make explicit the systems ontology, its assumptions, account for how it is constructed and formalized, enable comparative analysis of systems ontology and facilitate the inclusion of an appropriate systems ontology in a PCF.

Ontology is the philosophical study of the nature of being. It has implications for knowledge and practice of systems analysis and design. The systems ontology theme is concerned with the nature or knowledge of systems. The nature of systems can be regarded as purely technological or a combination of the social and technological. Whichever systems ontology is accepted it contains assumptions about organizations, people, IT, and IS and the interrelations between them.

Philosophy is an important issue in systems ontology. Software programming methods and IS methodologies, including systems analysis and design approaches, have 'philosophical' basis, which may not be obvious to analysts. It needs to be made explicit. Philosophical underpinnings are important because of implications for practice in real situations. For instance, object-oriented programming is based on a philosophy that characterizes organizations and people in terms of 'classes' or 'objects' that have 'attributes' and provide 'services' to other related objects. It influences analysts' perceptions of reality.

Ontological knowledge is acquired with formal research methods and from the reflections of experienced practitioners. Formal methods of inquiry are termed **epistemology**. Some formal methods of inquiry describe and explain systems ontology as objective. These methods lead to practice that makes analysts detached and independent of the human problems they seek to resolve. Other formal methods lead to knowledge that makes analysts inclusive and subjected to the human problems they seek to resolve.

Epistemology in social science that leads to objective knowledge is termed positivism. Positivism acquires knowledge about organizations and society by reducing the problem down into manageable, observable and measurable elements. Reducing the problem into manageable elements is called reductionism. Scientists use it to develop scientific knowledge. Reductionism as a problem-solving strategy is also used in systems analysis and design, particularly in structured systems ontology.

Epistemology in social science that leads to subjective knowledge is termed phenomenology. It acquires knowledge about organizations and people by understanding peoples' interpretations and perceptions of their actions. Understanding the problem in terms of people's experiences and their interpretations is called **interpretivism**. There is marginal recognition of interpretivism as a problem-solving strategy in systems analysis and design, though NIMSAD is normative, and Open Source Software (**OSS**) and ASD may be construed as interpretivism. Interpretivism is an accepted method for research in IS. Positivism and interpretivism are used to acquire IS knowledge and instruments for IS development. These epistemologies have resulted in knowledge of the problems in, and investigation of, IS development.

Structured systems ontology is based on human rational capability. Rationality is the use of reason and logic to think on a problem and its resolution. It has resulted in formalism in systems analysis and design. One form of rationality is planned action. In organizations plans underpin IS strategy, systems analysis, systems design, governance and management mechanisms. Planners and plans assume an organization is a rational and optimizing entity, and that once developed, plans can be implemented to achieve stated objectives. Structured systems ontology assumes objective knowledge and relies on rational planning. Rationalism underpins its instruments, but they pose practical problems in practice and are insufficient for developing complex organizational IS. The SDLC and systems project management similarly rely on rational behaviour.

Systems ontology based on interpretivism views reality as humans' interpretations. It centralizes individuals and the meanings they attach to their actions. Reality is socially constructed in interpretive systems ontology. SSM and situated action are examples of interpretive systems ontology. The use of stories in ASD acknowledges interpretive systems ontology.

Systems ontology is a representation of real situations. It develops knowledge of and structures problems in IS, but the actual problems exist in the real world of human problems.

2.3.4 Real world of human problems

Refashioning of traditions and reflexivity critically is applied to the real world of human problems. The real world of human problems is the first practice theme in the Critical Framework. A human problem is any organized effort to achieve an objective. Business and technical problems combined constitute actual human problems. Business problems range from strategic issues on competitiveness to operational issues on business process efficiencies or cost reduction.

The application of IT to resolve business problems is a separate technical problem. IT is applied on the basis of ontological system knowledge. Systems ontology provides the formalism and instrumental means to apply IT. The actual application of IT underpinned with system ontological knowledge is problematic. Actual organizational business context and technical problems differ from an assumed systems ontology or formalism. For example, in contrast to optimum solutions based on complete information proposed in structured systems ontology, in actual situations organizations seek non-optimum solutions to problems and do so with limited available information. People have intentions and are capable of influencing plans. Organization and organizing do not easily succumb to the application of formalism. Uncertainty in real situations and decision-making affect the application of formalism.

Analysts' apply systems analysis and systems design formalism in practice. They are concerned with determining how to respond to actual problems by applying ontological and technical system knowledge. Interpretation is significant in the application of knowledge because of varying perceptions of individual or groups of analysts. Formalism encounters problems when applied to actual situations, which in turn necessitates understanding organizations ontologically, and leads to revisions in systems ontology or partial deployment of formalism.

Similarly, the resolution of the technical problem necessitates analysts to understand the business and organizational context. Analysts have to interact with people and appreciate issues in business strategy, competition and business process improvement. They have to understand organizational interaction and communication designed to achieve tasks and objectives. People pose particular problems when applying instruments and require understanding:

- How human purpose is established and its dynamism.
- Organization and communication and its dynamism.
- Human interpretation and meaning attached to actions and the results of actions.
- Social aspects of human interaction and communication.
- Political aspects of human interaction and power relationships in organizations.

Real situations are referred to as the 'messy world' in the Critical Framework. The description of real situations as 'untidy' or 'messy' is used to compare actual human problems with how they are characterized, and the solutions proffered, in systems ontology, or formalism. This messy world is not susceptible to the direct translation of rational formalism of systems analysis and design. The application of structured and object-oriented systems analysis and design in the messy world is problematical.

The messy world theme enables critical evaluation of formalism in the context of actual problems in organizations and in relation to people. For example, the use of questionnaires to elicit requirements for a new IS may raise suspicion in people who are required to complete them. The suspicion may lead to refusal to comply or at worst result in sabotage. The development of a PCF requires an understanding of these organizational, people and application problems. Novice analysts need to be aware of them and experienced analysts need to reflect critically on them to develop relevant personal constructs.

2.3.5 The pragmatic resolution

The development of critical skills is realized in pragmatic situations to enable critical thinking on praxis. The pragmatic resolution theme is concerned with developing practical knowledge of applying formalism in real situations. It is the second practice theme. Its purpose is to encourage analysts to overcome practically the problems with applying formalism in organizations and in relation to people, and to develop critical skills.

Adjustments to system ontological knowledge and the application of instruments are often necessary in practice. Changes to instruments or prescribed application processes need to be made to achieve required objectives. Innovative ways have to be found to interact with people to apply the prescribed formalism. Ironically, the changes and innovations made to apply formalism in actual context may be characterized as situated action.

Practical knowledge is a prerequisite of professionalism. Analysts need to make decisions on how to resolve problems with applying formalism practically in actual situations. Pragmatic solutions to application problems are necessary and may result in amending or adjusting formalism and how it is applied. Changes to how instruments are applied can be based on clearer understanding of the problems in the messy world.

Critical understanding of formalism, planned action and situated action sub-themes can result from the pragmatic context. For example, the questionnaire technique may be part of an analyst's plan to elicit requirements. Its actual implementation may create anxiety in people for fear of change. So the analyst should be prepared to think of an alternative that does not cause anxiety to people who's work is being analysed. Another example is an analyst communicating gathered system requirements as E–R or DFD diagrams to 'users', who fail to provide intelligible comments because they do not understand the diagrams. The analyst can resolve the problem pragmatically by developing prototype user interfaces to show to users. 'Seeing' the system will help users to make useful comments.

2.3.6 Comments on the Critical Framework

The Critical Framework is necessarily laden with the author's subjective values phenomenological ontological position. The choice of themes and sub-themes is personal, but they do reflect the discourse in the wider literature on systems analysis and design and IS. The issues they raise are pertinent to knowledge and practice. It is assumed that the major themes and sub-themes in the Critical Framework will be sufficient to develop criticality.

The Critical Framework is to be used to improve *personal* knowledge and practice. It is not proposed (at present) as a model for critical systems ontology in the field. It offers scope and a framework for criticality because it is arguable whether systems analysis and design can be practised objectively. It is plausible to argue that it is phenomenological, as many analysts make choices based on familiarity or successful prior use of formalism or instruments. Such epistemological positions and counter positions pose a validity issue for positivist researchers.

The Critical Framework will enable critical analysis and evaluation of various issues. It enables fundamental assumptions made in systems ontology to be revealed, analysed and evaluated in terms of actual situations and pragmatic application. Systems analysis and design derives its *system* formalism from systems theory, systems thinking, and positivism. The major themes will enable critical consideration of the knowledge value and practicality of notions like 'structured', 'objects', and 'engineering systems'. In particular, the Critical Framework will enable interpretation, analysis, explanation and evaluation of:

- Philosophies and assumptions underpinning structured and object-oriented systems ontology.
- Formalisms like the SDLC and its application.
- Structured and object-oriented systems analysis and design (and other approaches like prototyping and JAD).

It will also enable:

- Personal inferences to be drawn for developing a PCF.
- 'Self-Regulation' because the Critical Framework acts as a device to 'monitor cognitive activities, the elements used in those activities, and the results educed' (Facione, 1990).

The primary purpose of the Critical Framework is to develop critical learners and practitioners. It facilitates criticality in two ways. One, it will enable a critique of ontological system knowledge. It is an aid for making critical observations on knowledge and practice, preparing project managers and analysts, and developing reflective and critical skills in them when selecting and implementing formalism in practice. Two, it will enable critical thinking on IS projects, requirements analysis and system specification, data and process modelling, object modelling, user interface and data design, system interfaces. Both these levels of criticality will inform the development of a PCF.

2.3.7 From the particular to the general

Human action in actual situations is unique and particular. Ideally, each situation requires its own particular response suitable to it only. In small system projects a unique response is feasible, but in larger projects practitioners draw on established generalized knowledge. The systems ontology theme constitutes such generalized knowledge to facilitate criticality. It is possible to be critical of individual situations, but not of all particular situations and experiences. The latter is possible with generalized knowledge.

Though human action is particular, it is informed by, and often predicated on, generalized knowledge. Analysts infer generalized knowledge from particular instances. Similarly, IS researchers draw general inferences from samples of empirical data. Generalized knowledge of human action, social factors, organization, people, systems, IS and IT should be interpreted and analytically evaluated in terms of the Critical Framework. The following three sections is an outline of such generalized knowledge.

2.4 Social and technical factors

Knowledge of organization itself is complex. Knowledge of organizational application of IT for IS to achieve objectives is a higher order of complexity. Organizational application of IT is the context for systems analysis and design. Analysts need to appreciate and understand this complexity.

The combination of organizational social factors and technological technical factors makes systems analysis and design uniquely problematical. The social factors require analysts to consider how humans organize themselves in a collective to determine purpose and strive to achieve it, how people communicate, cooperate, and interpret their actions in organizations, and what meanings people attach to information and knowledge. The technical factors include systemic knowledge, IS methodologies, formal methods, instruments, choice of software and hardware, problem-solving and other skills of analysts, programmers and project managers. A significant element of the technical decisions made concern assessment of the suitability and capability of the available IT to produce required IS.

The dichotomy between social and technical factors is useful when analytically evaluating formalism. Some formalisms account for only social or technical factors, others consider both. For example, Information Engineering and early versions of SSADM are orientated towards developing technically optimal systems models. Others have developed a **socio-technical** perspective. The ETHICS methodology is the seminal socio-technical perspective. It is based on the notion that both social and technical factors need to be understood to develop effective, socially responsible IS.

2.5 Human action

Human action can be categorized and explained as the planned action and situated action ideal types. Both types provide theoretical, conceptual and generalized knowledge of human action. IS methodologies and formalism in systems analysis and design is interpreted here as planned action. In contrast, situated action emphasizes the context and situation, and explains how humans use certain 'embodied' knowledge that arises in situations. It can be used to counter-argue planned action and explain many of the problems that arise in applying formalism in practice.

Planned action and situated action subthemes are intellectual traditions to explain human action. They need to be critically considered in systems analysis and design and can be analytically evaluated in terms of the systems ontology theme. They proffer descriptions and explanations of human action, which need to be analysed and their basic assumptions raised and discussed.

2.5.1 Planned action

In planned action people and organizations are characterized as rational and goal-oriented. Planned action is action that is predetermined with expected outcomes to achieve known objectives and enacted in actual situations to realize the expected outcomes. A plan is a formal instrument of organizational systems strategy that details what IT to apply and how IS should be developed. Formal plans are central in business organizations. Their scope is wide, covering plans of business strategy, IS and IT strategy, and systems project management. Planned action assumes it is possible to plan and control organized human activity, and do so rationally.

Systems analysis and design draws on planned action. IS methodologies are detailed plans of how to define and develop IS. Systems project management epitomize planned management of IS development. Planned action of this kind characterizes organization, people, information and knowledge, as entities that can be analysed, objectified and modelled with precision rationally. Structured and object-oriented instruments attempt precisely such planned, rational systems modelling and design.

The generic structured SDLC is characterized here as planned action. It supports the view that planning IS development will result in successful and relevant IS. It prescribes planned stages for systems developers to follow systematically and is composed of staged activities and detailed activities in each stage. The focus of the activities is the completion of the technical tasks of systems analysis, systems design, testing and implementation. Structured, and to some extent object-oriented, systems ontology assume that planned action can lead to successful IS development, and relevant and required IS.

2.5.2 Situated action

Situated action is not explicit in any formalism in systems analysis and design or methodology. Nevertheless, it is a significant sub-theme within systems ontology and human problem themes that can be the basis of action and explanation. eXtreme Programming (XP) may be interpreted as drawing on situated action to elicit system requirements. It uses 'stories' recounted by people to understand system requirements. Such stories are necessarily situated.

In situated action, organization and people are characterized as goal-oriented, but contextual, and situated. They draw on situational knowledge and embodied skills to achieve goals. When problems arise in pursuing goals, they draw on any relevant knowledge, situated factors, and embodied skills to overcome them.

Systems analysis and design can be interpreted in terms of situated action. The interpretation can explain problems with applying formalism or lead to better formalism sensitive to context. Analysts can draw on embodied skills and situated knowledge to resolve problems with applying formalism to actual situations.

2.6 Systems theory

Systems theory is used to inform human action. The system concept is an intellectual tool for rigorous and thorough problem definition and problem resolution. It enables human problems from actual situations to be abstracted in system terms for reasoning and resolution. Systemic features facilitate such problem framing and resolution. They are:

- Purposeful activity or goals.
- The whole is greater than the sum of the parts.
- Predictability of systems processes and relations between the subsystems.
- The distinction system enables between the system and its environment.
- Related to the latter, is the notion of open and closed system.

A **system** consists of a set of interrelated elements or components that has a purpose and the system is greater than the sum total of its parts. It has a boundary that defines and distinguishes the system from its environment with which it communicates inputs and outputs. Processes in the system transform inputs into outputs. A system can have subsystems with the same characteristics. The system is coherent because it has structure and processes, which are maintained by feedback and control. A system can be open, meaning that its elements interact freely with its environment, or closed if the interaction with the environment is controlled by the set boundary.

An example system is shown in Figure 2.2. The system is set in the environment with which it communicates inputs and outputs. Its purpose is determined by events in the environment. It contains subsystems that communicate with each other, and the subsystems and the communication compose the processes that transform the inputs into the required outputs. Feedback from the environment on the effectiveness of the outputs is used to adjust the inputs and processes to ensure the required outputs are produced. Control, shown as echo waves, pervades the whole system and is used to direct the processes towards the goal.

Systems theory has influenced how the problem of IS development is defined, structured and resolved. As in Figure 2.2, in its most fundamental form it informs the view of IS as a systemic entity with inputs, subsystems, relations, processes, outputs, a system boundary, and a system environment. Earliest applications of computers to business organization mirror these system elements. Structured systems

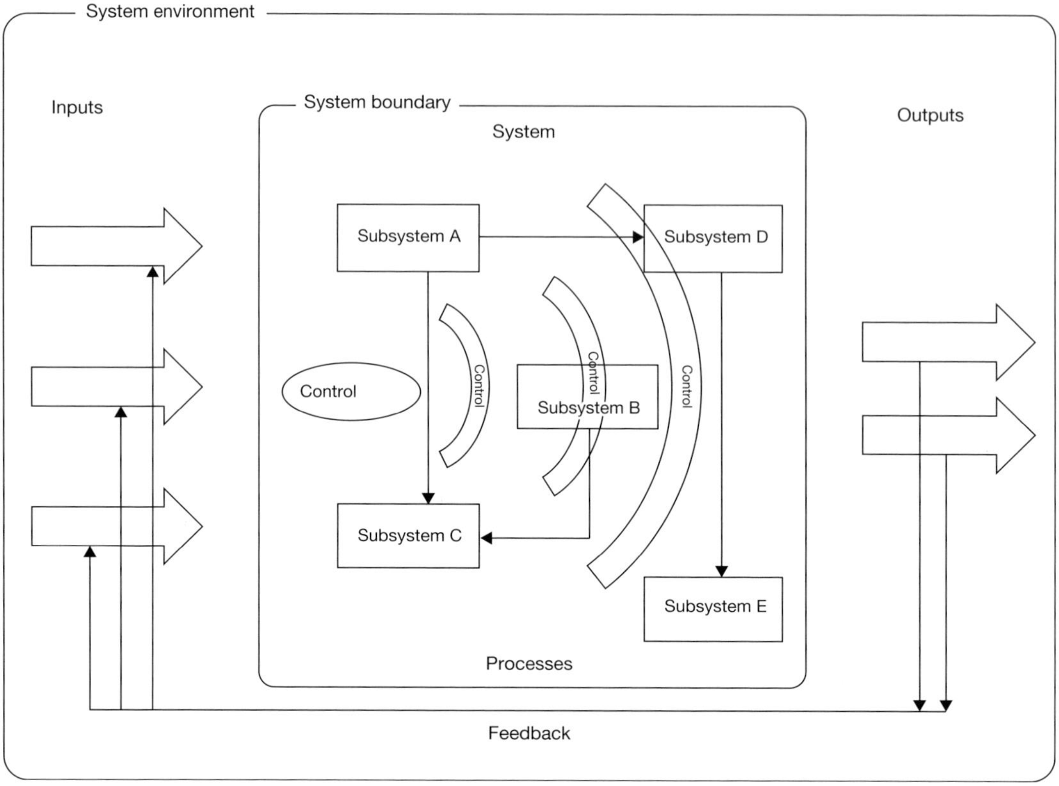

Figure 2.2 The system concept

analysis with its focus on data processing is strongly informed by systems theory.

2.6.1 Systems thinking

Systems thinking draws on systems theory. It is applied to characterize organizational problems in systemic terms. An organizational problem, for instance improving the management efficiency of a NHS Trust, can be interpreted in systemic terms as consisting of purposeful behaviour, with elements that communicate with each other and with the environment to achieve specific purpose. This type of system is termed a human activity system. A human activity system has purposeful control where decision makers make choices about the system.

The systemic attribute of the whole being greater than the sum of the parts is significant. The combination of the elements and subsystems produces an ontologically different entity, the system. This is termed the emergent property of system. This means that the whole cannot be understood simply by explaining its parts. The emergent property of systems is significant for understanding human activity. Systems thinking has several uses:

Method of enquiry Systems thinking is a method for acquiring knowledge. It is a formal interpretive method for understanding, problematizing, and solving human problems. Systems thinking is used to develop knowledge and understanding of organizations and people, and its proponents recommend its use in conceptualizing IS and in IS development.

Conceptualize real world problems Systems thinking can be used to think about real situations and human problems. Complex organization problems can be interpreted in systemic terms that would otherwise be impossible to conceptualize. They can be conceptualized as wholes with interrelated subsystems that have emergent properties, and purposes that are achieved through feedback and control mechanisms.

Analysis technique It can also be used as a method or technique to analyse organizational problems that would otherwise be difficult to do. Systems thinking is used to develop an appreciation of real world problems and to conduct an analysis of subsystems, constraints, and explore solutions.

Management method Systems thinking can be used as a method to inform management practice. It can be used to conceptualize and analyse quite complex organizational problems, and through systemic analysis provide management with clearer understanding of problems and possible solution actions.

Systems thinking is normally used to structure problems in organizations and bring to the surface how people perceive these problems. It can be applied to IS to *structure* thinking. The emergent property of systems is clear in IS, since the combination of the constituent parts of an IS produce an Information *System* that is different in nature from its compositional parts. It is ontologically different. Modelling these components in systemic terms will depict the elements, interrelations, the boundary and the system's environment. The various components that constitute an IS can then be interpreted in systemic terms to *organize* ideas and concepts about organization and people, data, information, IT and IS.

The descriptive and analytical capability of systems thinking is useful for systems analysis and design. It can be used to describe the actual situation and perform an analysis of it in systemic terms. Various systems analysis and design formalism in IS methodologies make use of systems thinking. For example, in Multiview the SSM 'rich picture' technique is used. A rich picture is used to develop an appreciation of the problem as perceived by people in the organization. In Multiview it is proposed as a systems analysis technique to help determine system requirements.

2.6.2 Organization, people and systems thinking

Organizations use IT to process business data to produce information and knowledge. They capture, process, store and manage data for several reasons:

- Legal requirements: compile statutory accounts, reports for shareholders.
- Improve organizational efficiency and effectiveness: reduce costs or improve business processes.
- Determine business strategy: executives develop business strategies to compete with other businesses.
- Provide information for decision-making: increase or decrease production, make an investment in capital machinery.
- **Organizational knowledge**: manage organizational knowledge and build knowledge bases for innovative product design.

Analysts need to appreciate organizations and what IT and IS can contribute. Organizations exist in business, government, healthcare and other human interests like charities. An organization is defined as having goals, a boundary and human activity. It is composed of people collaborating, sharing and communicating to achieve set goals. The goal may be to produce an aircraft or car, or to provide a financial service, or to retail goods.

IT is used to 'transform' organization, in particular business processes. Production processes or management processes can be improved to produce better products or services by applying IT innovatively. Transforming business processes with IT though is a complex task. Systems thinking can be used to describe and analyse organizational uses of IS. It can be used to understand conceptually:

- How people in organizations communicate with each other, what information they share, process or exchange.
- How various departments, sections, groups (elements in systems thinking) communicate information to achieve organizational goals.
- How to demarcate a particular area of the organization (draw a boundary in systems thinking).
- How to determine what external factors affect the organization (determine how the elements interact with the environment in system terms).

The conceptual knowledge can then be used to manage the actual problem. Understanding the nature of organizations is an element of the systems ontology theme and a significant factor in applying systems analysis and design formalism.

Systems theory and systems thinking has fundamental implications for the systems ontology theme. Systems theory regards people as the source of problems in the system and it is concerned with responsibility, or moral and ethical issues. It creates a *systemic* view of organization, people, information, and IS. It separates knowledge about work from the subject or individual and it is concerned with how we gain knowledge, or epistemological issues. This is regarded as valuable in systems thinking because such knowledge can then be stored and processed and used to improve organizational efficiency and effectiveness.

2.7 Focus on criticality in systems analysis and design

Analysts need to develop a critical disposition because of the centrality of their role. They have an important multi-purpose role. The core of their work is the creation of systems models for which they communicate with people for whom the system is to be developed. Analysts communicate the systems models to systems designers and software programmers.

This core work of investigation and creation of systems models constitutes systems analysis and design. Their design decisions affect people, workflow patterns, business processes and information provided to managers. The ideas for these creative systems models come from investigation of organization and people and the resultant understanding of system requirements. The purpose of the Critical Framework is to engender criticality and adduction of personal constructs. The emphasis on the personal approach is to enable knowledge and practice to be developed for practical application.

Criticality in the field of systems analysis and design has lead to improvements in conceptual knowledge and instruments. It has progressed systems analysis from 'art' to 'structured' and from 'object-orientation' to 'eXtreme Programming'. Critical thinking resulted in systematic 'problem-solving' and produced methods to structure an IS problem and solve it. It resulted in the concept of systems analysis and design as a rational process, where perfect knowledge is assumed, and depicted the role of analysts as rational problem-solver.

2.7.1 Systems analysts, problem-solving and creativity

A critical disposition is important because analysts' work is future-oriented. Analysts examine current problem situations to determine a better future organization. A critical disposition is required to:

- think of innovative strategic and operational systems;
- use IT creatively;
- improve approaches, methods and techniques;
- improve problem-solving techniques.

IT is a fundamental prerequisite for modern information and knowledge management in organizations. It has created the modern notions of 'data' and 'information', and it is important for organizational knowledge management. Computer network technology, especially the internet, enables knowledge to be managed electronically and enables eCommerce. So analysts need to develop methods to think of innovative strategic and operational systems. A critical disposition can engender creative uses of IT. Their work requires creatively addressing business and technical problems. At a strategic level companies now expect IT to deliver innovative business processes and eCommerce. IT has enabled radical innovations in strategic IS, and recently in BPR and knowledge management. A strategic IS is difficult to innovate because it should be inimitable, and its precursor is often an operational system that has the potential to be strategically important.

Analysts require two types of **problem-solving** skills: business problem resolution and technical problem-solving. Criticality is important in both types. It is required to define the business problem because it affects the organization. It determines how the available resources will be deployed and whether the resultant IS will benefit the organization. Analysts work with business analysts to define the business problem. Strategic business problems concerning competition require IT to be applied creatively. IT, the internet and communication technologies are combined to enable companies to create new **business models**. A business model is a set of interrelated business ideas on how to produce products or services better than competitors to the satisfaction of customers. To create business models existing ways of working and organizing need to be analysed and radical new designs created. Systems analysts need to base systems models on business models that give an organization a competitive advantage or improve efficiency.

Analysts need to ensure that IS complement innovation policies. Business organizations have innovation policies for products, services, organization of the production of goods and provision of services. Analysts should critically assess whether structured and object-oriented systems ontology engenders innovative thinking. Significantly, they should evaluate whether the systems ontology they assume is relevant in practice.

Technical problem-solving requires systematic or logical thinking. To solve systemic problems analysts first need to define or structure problems as clearly as possible. They need to consider issues internal to the organization and external issues, like competitors and technological know-how. If definition of the problem is inappropriate or completely misses real business needs the resultant IS will be of little use, resulting in obsolete investment. Analysts should evaluate available problem-solving strategies to choose the most appropriate. One strategy is breaking the problem down into manageable components and solving the components separately. Another is holism in which the problem is modelled as a whole. These skills are needed for systems modelling.

Problem-solving has an element of creativity. Analysts need to develop lateral thinking skills. What can be done with IT and how IS problems can be solved should not be constrained by existing knowledge and practice. Creativity in problem definition and its resolution may involve brainstorming and role-play which may result in radically different ideas. Electronic devices and other tools can be used in the creative process. For example, video recordings can be made of system requirements elicitation to facilitate better analysis.

2.8 Relating the Critical Framework to Personal Critical Framework

The Critical Framework is the basis for developing a PCF. The three major themes and six sub-themes in the Critical Framework provide substance for the PAC cycle activities that lead to the development of personal constructs and their continuous revision.

2.8.1 Objectification of personal constructs

The Critical Framework themes enable analysts to objectify relevant personal constructs and to evaluate them critically. The themes enable analysts to explain personal constructs in the context of knowledge and practice. Personal constructs are shaped by systems ontology knowledge and practical application of formalism in actual situations, or human problems theme.

Personal constructs need to be objectified and relations between them explained to improve understanding and personal effectiveness. Implicit personal constructs will not be understood and may lead to personal action that is ineffective. Such action is never explained. The Critical Framework provides the basis for the objectification of relevant personal constructs by exploration of its themes critically in relation to knowledge and practice.

2.8.2 Critical evaluation of personal constructs

Critical thinking needs to be applied to the Critical Framework themes. Objectified personal constructs can be reflected upon and decisions regarding inclusion in a PCF can be made on the basis of the themes. This process can aid analysts to objectify and evaluate personal constructs critically rather than accept them uncritically and implicitly.

The process of objectifying personal constructs and assessing their relevance once objectified requires critical thinking. The Critical Framework is a cognitive device to aid analysts to analytically evaluate personal constructs. It combines cognitive critical thinking skills, the three major themes and the six sub-themes, and Barnett's (1997) three criticality types to provide a framework to develop criticality.

The Critical Framework and PCF complement one another. The Critical Framework is to be used to develop personal constructs on structured and object-oriented analyses and systems ontology generally. It will develop analysts' reflexivity skills to:

- Bring to the surface problem definitions, premises and assumptions made in structured and object-oriented analyses.
- Enable critical reflection that leads to a personal interpretation of structured and object-oriented analyses.
- Enable analysis of structured and object-oriented analyses through the three major themes and six sub-themes.
- Provide a basis to evaluate the validity and reliability of structured and object-oriented

analyses, its instruments, and analysts' related personal constructs.
- Facilitate discussion and provide a context, especially in terms of human problems, in which to compare structured and object-oriented analyses and other approaches.
- Support the development of evidence-based critical thought and reasoned comments and opinions based on experiences and readings.

The Critical Framework enables thematic critical thinking on practical IS development problems. Its critical perspective is required to improve effectiveness of project team members, especially analysts and project managers, and gives them the confidence to evaluate approaches and techniques for particular IS projects.

2.8.3 Requirements: an example objectification

An example of the process of objectifying personal constructs is an analyst who uses questionnaires to elicit system requirements. The analyst diligently prepares a questionnaire by reading widely to gain knowledge on questionnaires, and identifies the right people to probe by examining the organization chart and speaking to managers of departments. The questionnaire is then administered to these people with equal diligence. The resulting system requirements information is disappointing because the analyst is unable to clearly identify the new IS functions.

In this situation the Critical Framework can be used for reflexive questioning and development of personal constructs. The following questions and sub-themes may be explored: (a) What assumptions are made about peoples' knowledge of requirements and the practical application of the questionnaire? This covers the three major themes. (b) Is the questionnaire an appropriate method to capture what people know? This covers the method of inquiry sub-theme. (c) Are there situational factors that need to be investigated too? This covers the situated action sub-theme.

The analyst can use knowledge from the practical experience to probe these themes to revise personal constructs and determine an appropriate pragmatic resolution. The analyst should begin the objectification process by asking questions:

- Why did I choose questionnaires to elicit requirements?
- What evidence is there that questionnaires are successful?
- Can people relate to questionnaires or did it distance people?
- How do questionnaires relate to my other personal constructs about people's knowledge and their ability to communicate it?

2.9 Applying the Critical Framework

An example application of the Critical Framework will illustrate its relevance. It will demonstrate how it can be used to think critically of knowledge and practice, and how it is related to developing a PCF.

The example application is on 'data' and 'information'. They form a significant contribution to knowledge and practice in IS development. They are investigated during systems analysis and systems design. To enable analysts to understand how to objectify personal constructs related to data and information, the Critical Framework needs to be populated with knowledge and practice on data and information. An outline example of data and information in terms of the Critical Framework is shown in Figure 2.3. Performing the following activities populates the Critical Framework:

- Examine the knowledge base of the concept, method, technique, tool or methodology.
- Explore its practical relevance for human problem resolution.
- Determine a pragmatic resolution to practical application problems.
- Underpin the above with criticality.

2.9.1 Considering major themes

The first layer of the Critical Framework in Figure 2.3 illustrates how the systems ontology theme is populated with ontological knowledge of data and information. Following the arrow across, they are an interpretation of organization and communication issues in the real world of human problems. Formalism related to data and information is applied in an actual situation. Following the second arrow across, problems with application of formalism are resolved pragmatically.

The second layer is the rationale, assumptions and knowledge on data and information. An example in the second layer, first column, shows data is transformed into information by processing. Through application to practice or reflective study, the second column of the second layer, analysts can evaluate validity of information as processed data and assess whether it confirms their interpretation, knowledge or experiences. The validity of the assumption and practicality of instruments based on it are then resolved pragmatically, as shown in the third column.

The third layer engenders the three types of criticality. The first column in the third layer is then used to develop a critique of systems ontology by asking relevant questions based on extant knowledge and practical experience. For example, is information objective, or is it interpreted. This is transformatory critique. The second column in the third layer is populated with questions concerning practice to engender reflexive criticality. For example, what is the role of 'meaning' in information, or does an analyst need to be objective, and how can this be realized in the messy world. This is reflexivity. The third column in the third layer is populated with questions concerning the pragmatic resolution of application problems. It engenders critical skills to enable the task to be accomplished. For example, to resolve the problem of interpreted information, develop ways of helping people to objectify what they already know and incorporate into requirements analysis.

2.9.2 Other issues

Analysts can use the Critical Framework to interpret other issues:

- the term 'Information System';
- the idea of 'system';
- information flows in organization.

The term 'Information System' itself is debated academically. Many monologues on what it means have been written. Some focus on the sub-term 'information' and others on 'system'. An Information System may be regarded as an objective or interpreted entity. As structured and object-oriented analyses assumes an objectivist ontology, an IS may be accepted as an objective entity. An analyst's personal construct may assume either ontology, but the choice of the one will negate the other, and the choice can be made by critically evaluating both through the themes of the Critical Framework.

Similarly, the use of the term 'system' generates much academic debate. Some researchers in systems thinking argue that systems need to be central to research and practice in IS development. Other researchers focus on the human and social factors in IS. The social context, namely organization and people, of IS development has resulted in

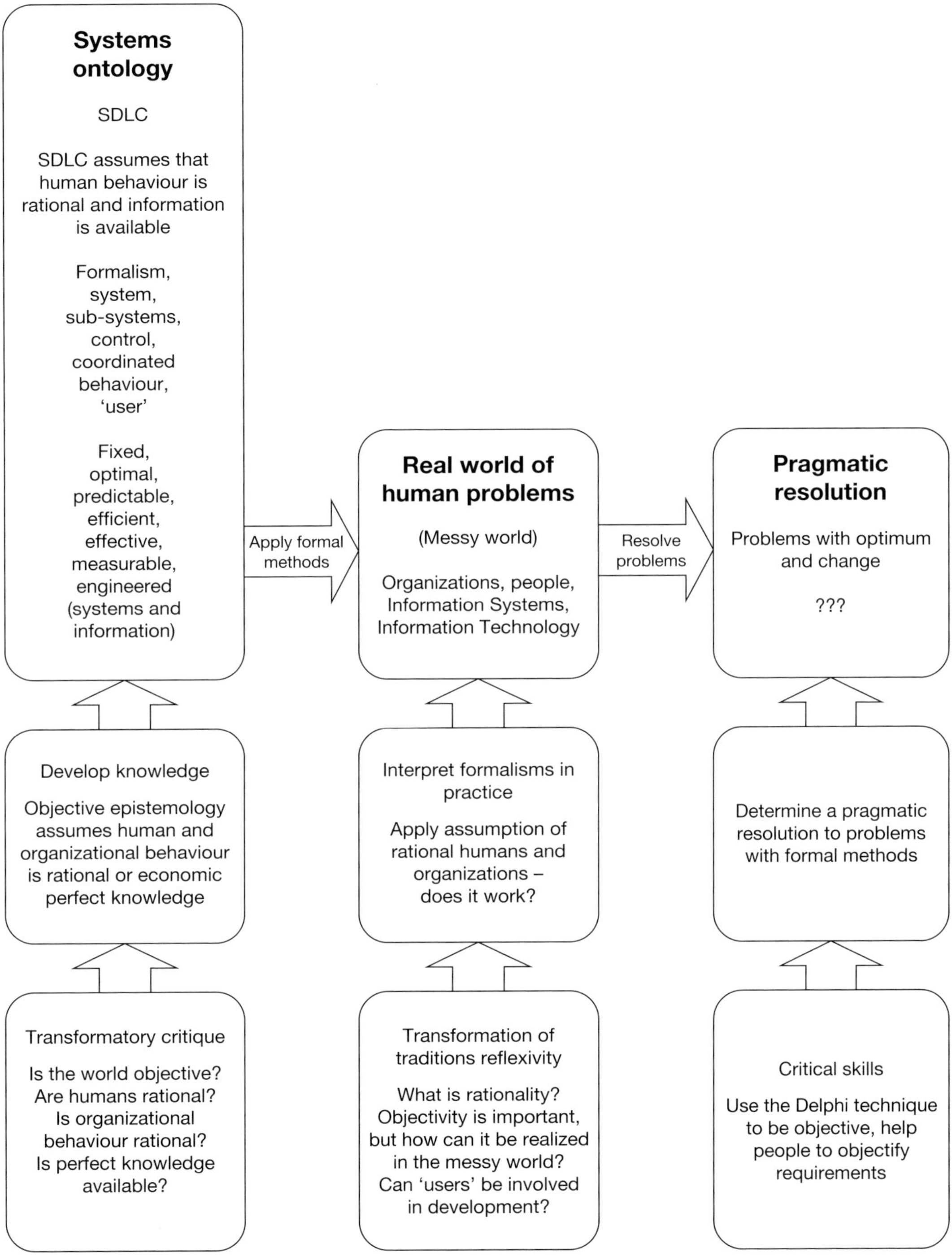

Figure 2.3 The SDLC – an example of a populated critical framework

socio-technical methodology. The human element of IS has resulted in researchers examining information as phenomenological or interpreted.

Structured and object-oriented systems analyses are used commercially to develop IS, though some business organizations do not make use of methodologies. For these organizations, the lack of contextual analysis in structured techniques makes them unsuitable. The context of IS development is important and analysts need to consider appropriate personal constructs to account for it.

2.10 Personal Critical Framework development

2.10.1 Education and training

Questions

1 How often do you reflect on knowledge that you possess or you practice? What is the topic of your reflection? What mechanisms do you use to reflect? How does reflection improve your knowledge or professional practice?

2 Define training. Why is training not suited for reflection? How does it differ from education?

3 How can the Delphi Report's set of cognitive skills be used to think critically about systems analysis and design?

Activity A

To paraphrase the education psychologist Skinner, education is what survives after what has been learnt has been forgotten. Make a list of facts or themes you remember from your first university degree or earlier formal study to ascertain what you remember. You may find you have forgotten the details. If memory is not a good measure of education, then what do you think should be used? Address this question and then revisit the Delphi Report's cognitive skills for critical thinking. Now make a list of cognitive capabilities that you think are the benefits of an education. What kind of critical cognitive capabilities have you identified?

Activity B

- Write down the attributes that you think describe you as 'educated'.
- Make a list of the attributes that someone trained in systems analysis and design might posses.
- Compare (1) with (2). Select your preference and explain why.

2.10.2 Fundamental elements of Information Technology

Question

Discuss how IT can be used to develop IS that: (a) improve operational efficiency, (b) provide information for decision-making and (c) give organizations a strategic advantage.

Activity A

Identify an IS in your organization. What data does it require as inputs to produce information as outputs? Who in the organization uses the information and for what organizational purpose is it used? How do you think the IS helps people who use it to interpret events in the organization? How does the IS contribute to making their work more efficient or productive?

Tasks

- Differentiate between data, information, and knowledge in terms of IT.
- In a group, determine a code of ethical behaviour for systems analysts.

2.10.3 Systems thinking

Questions

1 Evaluate the theoretic view of organization as consisting of a purpose, boundary and activities. Discuss its value for systems analysis.
2 Critically discuss whether the system concept makes unique human problem situations homogenous.

Activity A

Choose an organization you are familiar with, your place of study or work:

- What is its boundary?
- What is its purpose?
- What are its main activities?
- How did you ascertain each of the above? Discuss whether a formal method would improve the process?

Activity B

Suppose you were required to develop an IS for supply chain management for a wholesaler of pharmaceutical products. Make whatever assumptions you think necessary:

- What is the purpose of the IS?
- Define its boundary?
- What inputs would be required from its environment?
- What outputs would the system generate?
- What would be the main activities (processes) required to generate the outputs?

Activity C

Systems theory has influenced approaches to IS development. In its most fundamental form it has led to the view of an IS as a systemic entity with a boundary, environment, inputs, subsystems, relations, processes, control, feedback and outputs.

- Choose an IS you are familiar with, for example a student record system or a payroll system. Describe it in systemic terms as in Figure 2.2.
- Discuss whether a systemic ontology is sufficient. What other factors do analysts need to consider.
- What aspects of the organization and people are not present in the systemic model? How would the success of the IS be affected if these aspects were absent?

2.10.4 Problem solving and reasoning

Questions

1 What particular interfaces between an IS and its application domain are difficult to design and implement?
2 Critically evaluate the interpretation of systems analysis and design as a problem-solving activity in system theoretical terms.
3 Explain what you understand by the term 'problem-solving'. What is the role of logic in it?
4 Explain the differences between data, information and knowledge in your organization. Make a list of data, information and knowledge that your organization requires to conduct its business.
5 Explain what is meant by 'description', 'analysis', and 'criticality'. Which adds more value to problem-solving?
6 What is your understanding of the following terms used in problem-solving:
    - task definition
    - information-seeking strategies
    - location and access
    - use of information

- synthesis
- evaluation.

Activity A

Identify an information management issue in your organization. Define it as an IS problem. Think creatively about how IT can be used to manage the problem.

Activity B

How would you characterize the problem of IS development? Apply systems theory to characterize the problem. Evaluate whether system theoretic characterization adds value to understanding the problem. Does systems theory enable making inferences on the role of IT in organization?

Activity C

Various techniques can be used for problem-solving. Brainstorming and role-play are two examples. Brainstorming is used to do some initial thinking on a problem. It consists of a group of people objectifying whatever comes into their minds, despite its relevance, and someone recording the ideas on a whiteboard. Apply the brainstorming technique to the problem of developing a website for an organization. Analyse the feasibility of the resultant ideas.

Activity D

Compare 'trial and error' problem-solving, also called 'exponential learning' with formal problem-solving (for example top-down decomposition in systems analysis). Why might trail and error be more appropriate in a social context? How can formal problem-solving support trail and error?

2.10.5 Systems ontology

Activity A

Identify a management information need, for example information on students at a university or information on customer purchasing habits in a company.

Characterize the information need in system terms.

- Identify the activities in the system.
- Determine the inputs and outputs.
- Determine the control mechanism.
- Describe feedback mechanisms.
- What is the value added by characterizing the information need problem in system terms?
- What elements of the real situation are missing in the system interpretation? What might be the effect on the information need if the problem is resolved in their absence?

2.10.6 Plans and situated action

Activity A

Identify any task you were required to complete. Describe the task and how you completed it. Include in your description your approach and:

- Any previous knowledge that you brought to bear on the task.
- Gaps in your knowledge to complete the task successfully.
- New knowledge that you had to develop in the situation.
- Discuss whether you would categorize your action to complete the task as planned action or situated action.

Activity B

Based on the result from Activity A, if you only used planned action, make a list of activities that you did to complete the task that might be construed as situated action. If your activities were dominated with planned action or situated action explain why it is so.

2.10.7 Applying the critical framework

Question

Discuss 'criticality' in systems analysis and design in terms of transformatory critique, refashioning of traditions, reflexive criticality and critical skills.

Activity A

Using the three major themes and six-sub themes of the Critical Framework, describe what you understand by the term 'system'. Use the three columns and three-layer layout in Figure 2.3 to plot your knowledge of system. How might you use the system concept in your PCF?

2.10.8 Internet sources

The *Journal of Vocational Education and Training* at http://www.triangle.co.uk/vae/ publishes research on the development of practice and theory in work-related education (accessed 25 March 2004).

For tutors and learners the Information Technology Fundamentals site at http://elearning.asu.edu/itf is a valuable resource covering computer hardware and networking (accessed 25 March 2004).

The Thinking Page at http://www.thinking.net covers systems thinking, creativity and other psychological aspects (accessed 25 March 2004).

Pegasus Communication inc. at http://pegasuscom.com/aboutst.html provides purchasable resources on systems thinking (accessed 25 March 2005).

2.10.9 Further reading

Checkland's work is seminal in systems thinking. See: Checkland, P. (1999) *Systems Thinking, Systems Practice*, Chichester: Wiley.

For the relevance of systems thinking for IS see: Checkland, P. and Hollwell, S. (1998) *Information, Systems, and Information Systems*, Chichester: Wiley.

The seminal work on situated action is by Suchman, L. (1994) *Plans and Situated Actions*, Cambridge: Cambridge University Press.

For recent developments in situated contextual factors in software development see the insightful work by Dourish, P. (2001) *Where the Action Is: Foundations for Embodied Interaction*, Cambridge, MA: Bradford Book, MIT Press.

Part II

IS, projects and application domains

Three vital issues in IS development are considered in Part II on the basis of the PCF and Critical Framework developed in Part I. They are:

- The conceptual basis for systems analysis and design.
- The use of system projects to develop IS.
- The IS application or problem domain.

The first two issues are important because they form the knowledge and basis for practice. The third issue is important because the application domain determines what a new IS will become and affects how the system project is managed. These issues need to be critically considered because they contribute to the success of an IS.

The SDLC, making a business case for a new IS, and the role of analysts will be covered in Part II. The focus of formalism in the SDLC is on the structured expression of information. The SDLC, role of analysts and development of business cases are formalisms in IS development. Understanding the role of formalism in IS development will enhance the development of PCF. Analysts can determine the value of formalism in a PCF through critical evaluation. Validity and effectiveness issues in formalism will be explored.

The conceptual basis of systems analysis and design is covered in Chapter 3. In terms of criticality, it contributes to the development of reflexive criticality. The SDLC, structured and object-orientation concepts, instruments and IS methodologies, are covered in Chapter 4. In terms of the Critical Framework, analysts need to evaluate the effectiveness of current knowledge for practice. They need to interpret and analytically evaluate current knowledge so that it can be incorporated into a PCF.

An IS is usually developed as a project. Project management concepts and techniques are covered in Chapter 4. Systems project management is analysed in terms of planned action. Making a business case for an IS is covered in Chapter 4. In terms of criticality, it contributes to the development of reflexive criticality. Business organizations invest in IT and IS because it can reduce operating costs and improve organizational effectiveness. Significantly, IT can provide strategic advantages over rival companies. Making a business case is a fundamental practical element of IS development.

The important role of analysts is covered in Chapter 5. In terms of criticality it contributes to the development of refashioning of traditions and critical skills. Systems analysts have a central role in IS development. They conduct three vital stages in the SDLC: the feasibility study, systems analysis, and systems design, and contribute to making the business case too.

Chapter 3

Systems analysis and design in concept and action

3.1 Learning outcomes

After completing this chapter you should be able to:

- Explain the SDLC, evaluate its generalizability and describe the role of systems analysis and design in it.
- Interpret and evaluate structured and object-oriented systems analysis concepts and instruments.
- Interpret and analyse problem-solving strategies used in systems analysis and design.
- Describe ontological characteristics of IS methodologies.
- Apply reflexive criticality to the conceptual basis of personal practice in systems analysis and design.

Knowledge and practice arise from practitioners' experiences and investigations by researchers. Practitioners codify their knowledge and disseminate it as practical knowledge. IS researchers conduct empirical research to discover ontological and applied knowledge and disseminate it as generalizable knowledge. Interpretive IS researchers do not seek generalizable knowledge.

IS is a complex socio-technical phenomenon because it encompasses social and technical factors. Whether it can be theoretically explained challenges researchers. They debate whether IS can be called a discipline and the possibility of formal theory is keenly discussed. Presently, there are no formal theories of IS, though some researchers attempt to develop theory. Whether theories of systems analysis and design are possible is also open to question.

In the absence of theories, systems analysis and design has concepts, methods, techniques and tools, methodologies and frameworks that constitute knowledge. They are the result of reflective practice and research to develop knowledge, especially generalizable knowledge. In this chapter the SDLC, structured and object-orientation concepts, methods, techniques and tools, and IS methodologies are introduced and critically discussed. They underpin and constitute knowledge of systems analysis and design, which translates into practice in whole or in parts.

3.2 Introduction

Systems analysis and design is central in IS development, eCommerce systems, eBusiness and knowledge management systems. Its prime aim is to develop systems models to inform the design and implementation of suites of software programs that constitute a system. Systems analysis focuses on determining what a new IS is required to do. Analysts investigate, understand, and develop organizational knowledge about business areas where IT is to be applied and develop systems models. Systems models are used to determine appropriate suites of software programs to write and their functions.

Systems analysis and design knowledge and practice has evolved. Practitioners have introduced new ideas, concepts and instruments to resolve pragmatic problems as they have encountered them. Researchers have studied IS development and the IS field and contributed epistemological and ontological knowledge. Early knowledge and practice may be described as traditional or informal, where the aim of practitioners was to produce an IS but the means were not clearly articulated, objectified or codified.

The prevalent current ideas and concepts are **structured systems analysis** and **structured systems design** and **object-oriented analysis** and design. Structured analysis and design was introduced in the early 1970s because traditional knowledge and practice did not meet business needs, and IS were more expensive than budgeted and took longer to develop then expected. Some reflective practitioners began to think that structure in the process was needed to improve the quality and success of IS. They introduced structured systems analysis and structured systems design methods, techniques and tools. They defined clearly roles for organizational employees, project managers, systems analysts and designers, and software programmers. This added structure to the process, but despite structuring the process, the quality and success of IS remained problematical.

The concept of object-orientation in software programming appeared around 1990. It addressed the problems of quality and production of software programs. The idea logically spread to systems analysis and design. Consequently, object-oriented systems analysis and design emerged as an alternative to structured analysis and design.

3.2.1 Data, information and knowledge

Organizations generate **data**, **information** and *knowledge* from their activities. IT is now a significant means to make them into resources and assets. Its algorithmic processing capacity contributes to the notion of information assets. Figure 3.1 depicts the roles of IT and humans in transforming data into information, and information into knowledge. IT is used to capture, store, process, analyse, structure and sort data to produce information. Organizations can regard information and knowledge as assets because they can be stored and retrieved from databases and knowledge bases.

IT enables the creation of information assets and knowledge assets. An asset in business is normally used to produce a product or service. Data on customers can be processed to yield information assets for marketing products or services. It can be processed to generate knowledge on customers' purchasing habits. Their value is reflected in their status as resources or assets, similar to human or capital resources. Organizations use them to become more efficient and effective, and to create competitive advantage over rivals. Information resource is now as equally important as human or capital resources.

Data and information is distinguishable in IT terms. In IS data is converted into information by software algorithms that process the data for specific purposes, shown by the first horizontal arrow in Figure 3.1. Data is unprocessed information. Data is captured, stored, and processed by digital computers and information is the result of processed data. A software **algorithm** is the sequence in a computer program for processing data. Data processing involves the analysis, structuring and sorting of data to provide information.

In human terms data becomes information when a human interprets data for a specific purpose, shown as the second column in Figure 3.1. For example, railway companies produce train timetables detailing the routes and times for trains. Train timetables are data. When someone wants to travel to a specific place they would use the timetable to plan a journey. A person processes the timetable for information about their journey. After determining their destination they would look-up their starting point and trace it to the destination to provide information about the departure and arrival times of the trains.

Figure 3.1 illustrates the case of improving customer service, shown in the third layer, the business rational for using IT. In the first column, data on customers is collected. It is then used to develop models of customer behaviour, the second column. This is information.

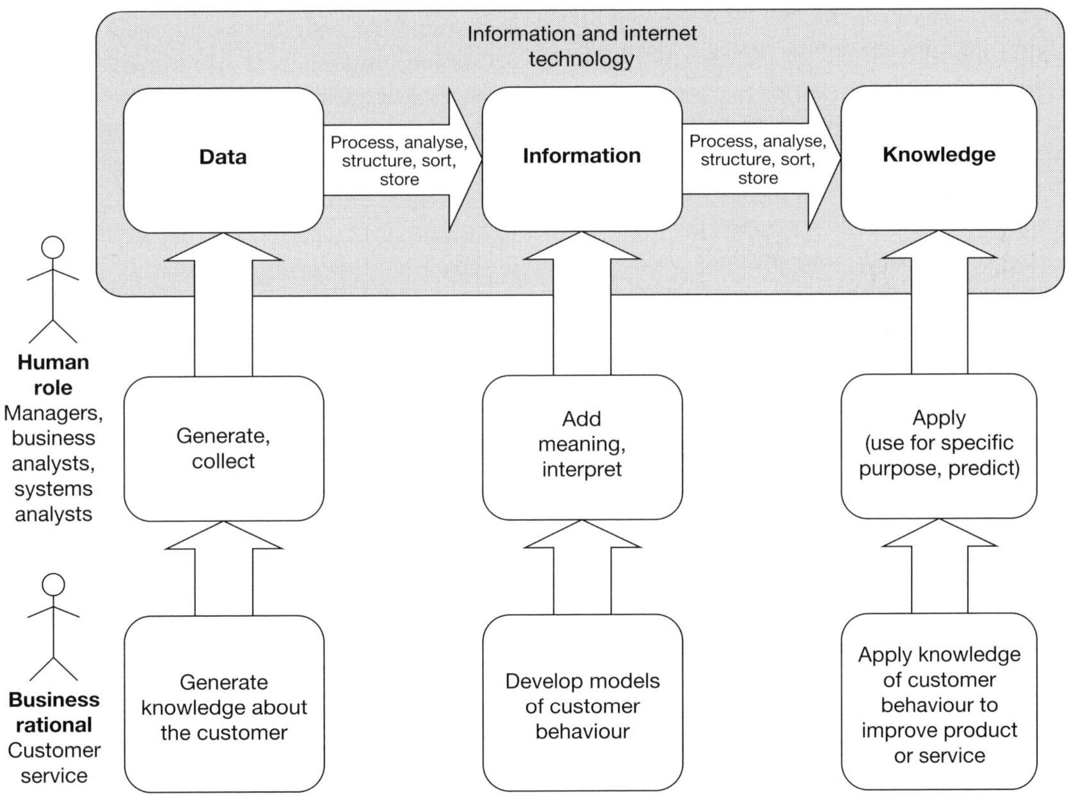

Figure 3.1 Data, information and knowledge in technological and business contexts

The third column shows that knowledge is then applied to improve products or services to provide better customer service. When information is applied in a certain way to achieve a goal it constitutes knowledge. For example, applying information about customer behaviour in terms of customer relationship management is knowledge.

The role of humans, managers, business analysts and systems analysts in organizations is shown in the first column in the second layer in Figure 3.1. They address the significant problem of determining what IS are required. They determine what data to collect to produce information for organizational processes, tasks and decision-making relevant to improve customer service. In the second column humans, business people, interpret the information to add meaning to it. In the third column they apply information to the specific purpose of improving customer service.

Analysts aid managers to determine what data to collect and how it should be processed to provide relevant information. The processing capacity of IT enables organizations to redesign organization and business processes to produce quality products and services, and transform organizations, through computer networks, into networked organizations.

The role of IT is shown in the top layer. It can capture the required data and process it to produce information on customer service improvement. This is shown as a process of converting data into information and, through application, to knowledge. An IS is the combination of all three layers. The role of IT for managing organizational knowledge is similar to managing information. The same functions of collection, storage, processing, analysis, structuring and sorting are applied to knowledge possessed by individuals or groups to produce, store and retrieve organizational knowledge.

3.2.2 Application domains, problem domains and Information Systems

In IS development a specific business area targeted for the application of IT is called an **application domain**. The application domain consists of people, peoples' intention and purpose, the work they do and the constraints of their work, how and who they collaborate and communicate with, the business or task objectives they need to fulfil, products or services they seek to make or provide and the physical resources like documents and files.

The range of application domains or business problems for which IT is used to create IS is wide. Organizations apply IT to specific business problems, business decision-making, business processes or whole organization. It is applied to major business functions ranging from customer billing, payroll, stock control, accounting and financial planning to human resource management and knowledge management. It is vital to business marketing and is used to enable electronic customer relationships management and enterprise resource planning. In eBusiness, IT and the internet are combined to design networked organization.

Analysts investigate the application domain. The information needs of people in the application domain are called system requirements. System requirements are rooted in the application domain and new IS need to be designed to satisfy the requirements. Analysts' task is to acquire knowledge and understanding of the application domain to establish the system requirements, which are passed to software programmers who write the actual programs for the IS.

The application domain seen from the perspective of systems analysts and IS developers is called a problem domain. It is so called because systems developers seek to define it in order to provide a computer-based 'solution'. It is the specific articulation or framing of a problem in the application domain for which a new IS is required. Framing the IS problem is a critical task for analysts and it affects the selection of instruments for IS development. The framed problem provides project managers with information to determine what resources will be required and to develop work breakdown and product breakdown structures.

Various systems ontology characterize the problem domain differently. The adequacy of this characterization is the issue critically explored in the systems ontology theme of the Critical Framework. The very characterization of the application domain as the 'problem domain', for which a 'solution' is needed, is accepted system ontological knowledge by practitioners. It is part of structured and object-oriented systems ontology. Such a characterization in turn affects how an IS is conceptualized and developed.

3.3 Systems development life cycle

A new IS is the product of articulating the business problem, understanding the application domain and framing the IS problem. The generic SDLC is applied to understand application domains and define IS problems and develop IS. The 'life' metaphor is to indicate that a computer system has a beginning, middle and an end. Each of the stages in the SDLC consists of the 'life' of IS development. Once an IS is developed, people will normally want different information, and the process of development would begin again. Hence the term 'cycle'.

The SDLC influences thinking on how to develop IS and consists of phases of IS development, shown in Figure 3.2. It is a systematic method to solve the problem of IS development, consisting of the following phases:

- feasibility study
- systems investigation
- systems analysis
- systems design
- implementation
- post-implementation review, evaluation and maintenance.

The SDLC seeks to address the problem of fulfilling the system requirements of the application domain. The SDLC is shown in Figure 3.2 in the context of an example business problem, namely improving customer service, the top layer. The phases of the SDLC are shown as rectangle boxes below it and demarcated in terms of the business problem.

The feasibility study and systems investigation phases identify and define the business problem and provide alternative IT or non-IT solutions. The systems analysis and systems design phases are concerned with making improvements to the business problem. The implementation and maintenance phase is concerned with developing a new IS. The final review and evaluation stage is concerned with assessing whether a new IS adds value. Activities in the SDLC phases and some deliverables of each phase are shown as rounded rectangles below the SDLC phase.

The SDLC is regarded as generalizable to all organizations. It was designed as a general solution for any organization intending to develop IS. It is supposed to work even when analysts may be unfamiliar with the organization and lack experience of its processes, culture and people. All the SDLC phases are normally required to develop an IS. A system project normally encompasses the SDLC phases. IS development methodologies reflect the SDLC phases in varying degrees. Some methodologies emphasize certain phases more than other phases.

The SDLC contains systems analysis, systems design and implementation as major phases.

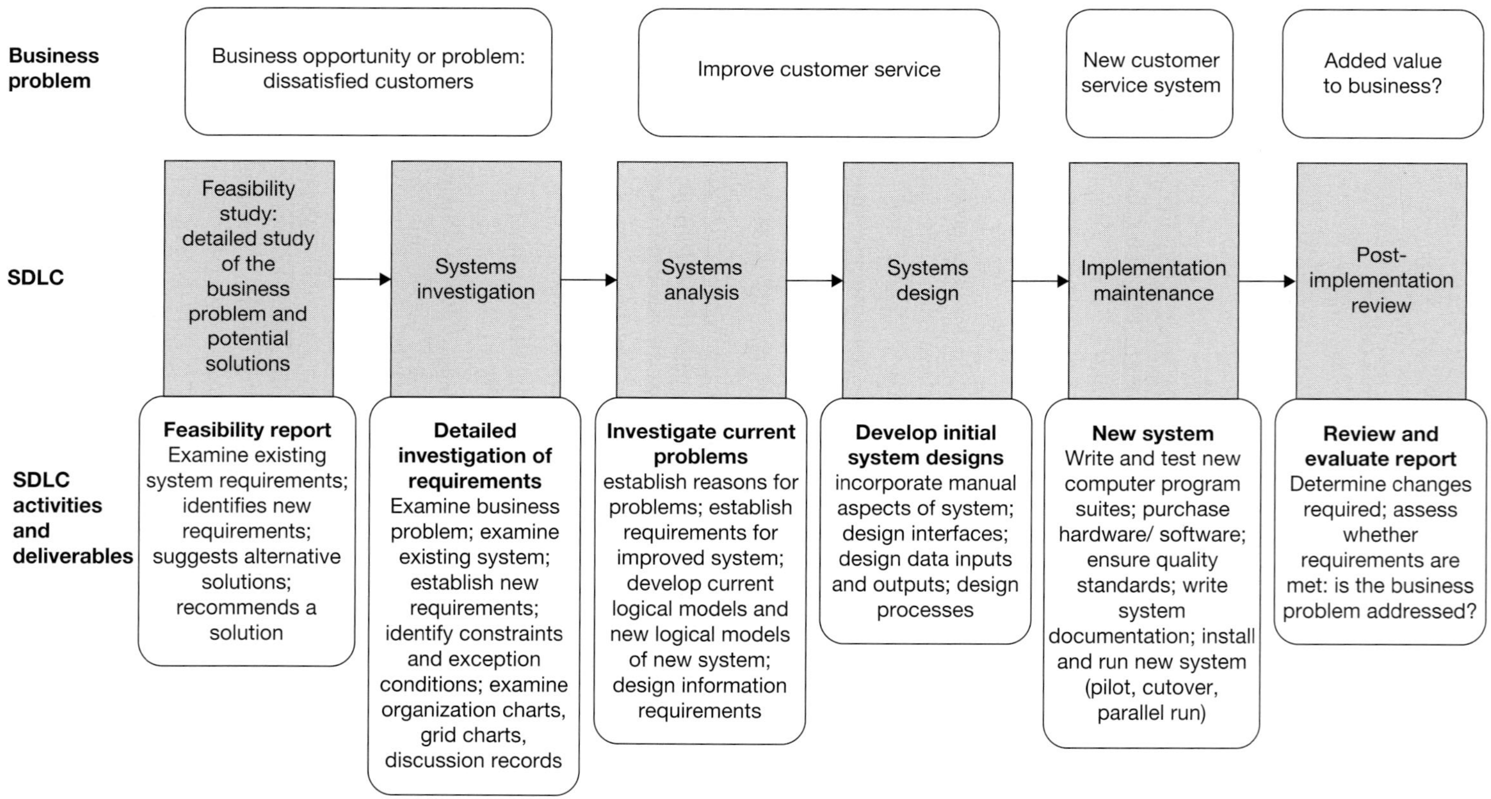

Figure 3.2 Systems development life cycle in the context of business organisations

The activity of systems analysis and design, and consequently the role of analysts, is an important and central feature in it. Although these phases are shown as part of the SDLC, they can be practised using various systems analysis and design methods. Among them are the traditional, structured and object-oriented methods.

3.3.1 Feasibility study

Analysts conduct the feasibility study. The problem domains vary from operational issues to matters concerning business strategy. Operational issues include poor customer service, such as complaints from customers that their orders are not fulfilled, or the costs of producing an item are too high. They may be to do with improving efficiency of business or management processes or ensuring the effectiveness of certain activities. Strategic issues include the re-design of business processes to make greater and better use of IT, the development of knowledge management systems to manage organizational knowledge, or networking collaborative work, all to improve competitiveness.

Analysts assess the problem domain – the work people do, whether manually or with current IS – with respect to the current business problem to produce a feasibility report. The assessment is to define and understand current problems and whether the current situation is acceptable or requires improving. The report contains a statement of the new needs, whether IT-based or manual, of the organization, people and the business.

The feasibility study is to determine whether it is both feasible and beneficial to produce a computer-based IS for a particular business problem. Activities to produce the feasibility study are to determine and define the business problem and assess the contribution that a computer-based system, or other alternative, would make. Analysts interact and communicate with managers and other employees to investigate the problem domain. Existing workflows and processes, associated decisions and the information available is analysed. An initial assessment of system requirements is made.

The problem domain is investigated with the aid of systems analysis instruments. Systems analysts use interviews, document and process analysis and observation. Managers may be interviewed and the documents they use analysed. People whose work will be affected by a new IS may be observed and records of the observation made. Analysts would normally wish to include senior managers, supervisors, clerks and systems developers.

Analysts investigate alternatives available, the costs to resolve the current business problems, and determine the benefits that can be obtained from each of the alternatives. Based on this evaluation, they would recommend a solution to management decision-makers. The next phase would ensue if senior management decides to proceed with the project.

3.3.2 Systems investigation

The purpose of the systems investigation phase is to determine requirements of the current and new IS. It is a systematic and detailed investigation of the problem domain. The analyst would normally use probing and searching techniques to establish the requirements. These include interviews, observation and examination of documents used in the feasibility study phase.

Analysts interview and record the responses of people in the problem domain who perceive the business problem and who will use a new IS. Organizational issues that impinge on the problem domain would be examined too, for example re-designing workflows or processes.

3.3.3 Systems analysis

In the systems analysis phase, analysts conduct a thorough and detailed investigation of the current system and the new IS based on the previous two phases. Analysis is concerned with providing a set of system requirements. This is achieved by developing systems models to:

- understand how business tasks are done and how current work is organized;
- define the data to be captured;
- determine the functions of the new IS to produce required information.

Systems modelling activities focus on understanding the present way work is done to develop new systems models. The systems models are often very different from existing workflow patterns. If there is an existing IS, applied models are identified and an assessment is made of how they are used. Analysts would need to consult people whose work and processes are to be modelled. Managers would also be consulted to help verify the systems models.

Systems models are developed with a modelling notation or language. A modelling notation is a formal set of symbols with precise meanings to represent the current and new IS. Modelling notations help to objectify the problem domain and facilitate communication between analysts themselves, and with clients and software programmers.

3.3.4 Systems design

Analysts design a new IS based on the information gathered in the previous phases. Systems design activities are concerned with the design of user interfaces, inputs, outputs and interfaces with other systems. The designs may be improvements in existing IS or radically new designs that redefine organization workflows and processes. The computer hardware equipment required to implement a new IS is identified and evaluated. Relevant software needs like databases are determined.

System security and integrity issues would be foremost in the minds of designers. The designs would be in the form of diagrammatic models. Systems design models would be developed using a modelling notation or language to represent the inputs, functionality and outputs. The notation would normally be consistent with that used in the systems analysis phase.

The systems models would be normally shown to relevant people affected by a new IS. Their comments would be considered, though major changes to the designed systems models would be resisted because of time and cost considerations.

3.3.5 Implementation

The implementation stage is concerned with activities to implement the systems design models from the analysis and design phases. They are implemented using IT identified in the systems design phase. Analysts are involved in determining the computer hardware and software required to implement the designed system. They work closely with the project manager, software programmers, database administrators and people in the organization. They are a conduit between software programmers and people, and work with software programmers to implement the systems models.

In addition to computer hardware and software considerations in the previous phase, database software and programming languages, where necessary, need to be evaluated, purchased and installed. Software programs need to be written and tested, and suites of programs need to be integrated to check the integrity of the system. Documentation on how to operate the new IS and guide maintenance would have to be written, normally by technical writers.

3.3.6 Post implementation evaluation

Once the IS has been developed and is operational, analysts normally conduct an evaluation of it to assess whether it meets the predetermined business needs. The evaluation is to ensure that the specified requirements of the problem domain are being fulfilled by the new IS.

The length of time it takes to develop an IS is usually problematical. As business organizations need to respond to competitors and markets, it is possible that the evaluation would reveal the need for changes or enhancements to the developed system, which the analyst would have to suggest and for which revised systems design be developed. The project manager would need to ensure appropriate staffing to perform corrective or enhancement maintenance, and that ongoing maintenance is adequately resourced.

3.4 The traditional approach

Practical IS development and systems analysis is concerned with solving problems: how to select an IS project for development, how business information needs can be established, finding ways of designing proposed IS or developing data and process models, and then converting the designed models into implementations.

The traditional approach preceded the SDLC. Systems analysis had no clear conceptual basis or a central concept on which practice was based. It relied on the knowledge and skills of experienced and competent practitioners. The business problem and the IS problem were investigated and studied simultaneously. There was no clear temporal or phased progress between analysis, design and implementation demarcated.

Systems analysis and design consisted of ad hoc use of techniques and entailed the design of a suite of computer programs and data structures. There was no, or limited, involvement of people with the result that requirements gathering was inadequate. The analyst's task was to specify the data structures and detail how they were to be manipulated and stored on magnetic media. Programs were designed and implemented as and when they were specified, and analysis conducted concurrently.

Practitioners would make no distinction between the analysis of the problem domain and the design of the system. Though it does not contain phases similar to the SDLC, the traditional approach normally included requirements analysis, requirements specification, and high-level design. It does not include many of the other phases in the SDLC nor does it produce the deliverables shown in the SDLC Figure 3.2.

3.5 The structured concept

Since the first application of a computer to business in the 1950s, researchers and practitioners have addressed the knowledge and practice elements differently. Early practitioners in the traditional approach conceptualized IS development as writing computer software programs. It was seen as a 'craft', and software programmers were thought to be creative, introverted professionals. The analyst would provide a system design and specification and the software programmer would write the required systems programs.

The resultant software products in the traditional approach lacked quality, failed to meet user requirements, cost more than was expected and were delivered later than required. Reflective practitioners began to question the 'craft' analogy, and proposed a 'systematic' approach. Craft-based programming practices such as the 'GoTo' statements were challenged. 'The software crises' was coined to depict these problems, and researchers and practitioners collaborated to improve the process of developing software. The debate was broadened to include the **engineering metaphor** to improve software production. The major outcome of this debate was the idea of formulating and using systematic and structured methods to develop software culminating in the SDLC.

Structure is the central concept on which practice is based in structured systems analysis and structured design. It complements the SDLC. In the structured approach the term **users** refers to people who will use the developed IS. The structured concept emerged from software programmers' search to improve the quality of software for IS, which led to the notion of structure in software programs. Practitioners who wanted to improve the success of IS development introduced the structured concept in systems analysis and systems design. Its purpose is to organize or 'structure' knowledge of system requirements through functional decomposition, rather than trust the process solely to experienced and knowledgeable practitioners, as in the traditional approach.

3.5.1 The structured concept

The central premise in structured systems analysis and design is that greater structure and formalism based on the scientific method or the engineering metaphor improves the success of IS. The structured knowledge is then used to inform practitioners' action. It aims to produce concise, unambiguous system specification and reduce data redundancy. This is achieved by making the IS development process thorough, formal and precise. Software programmers can then use the system specification to implement required system functionality.

Concise system specification System specifications were produced in the traditional approach. The difference in the structured approach is that the system specification is based on a formal **notation language**. A notation language enables precise and concise definition of the current and new IS in terms of notation that is common and communicable. It is used to make systems models and draw up designs.

Unambiguous specification The traditional approach would often result in unclear or ambiguous system specification. Ambiguous system specification meant that software programmers often misinterpreted system requirements and implemented unwanted system functionality or missed required functionality. A prime cause of ambiguity in system specification is lack of communication between analysts and users. Formal notation languages are intended to facilitate the communication and so enable the production of unambiguous system specification.

Nonredundant The traditional approach would often lead to systems design and implementation that had much redundant data that produced inefficient use of computers. Inefficient use of computer processing and data storage made it costly to use computers. Formalism was introduced to reduce data redundancy by objectifying data, data stores, dataflows and processes, usually in diagrammatic systems models. Analysis of the objectified data reveals data duplications that can then be removed.

Thorough As the traditional approach relied on the experience and knowledge of IS practitioners, they made decisions about when to seek further information on the business problem and how to design and write the code. This practice led to arbitrary decisions about the problem domain and system functionality. Through formal notation languages and their systematic usage, the structured approach aims to reduce arbitrary systems design decisions and make systems analysis and design a thorough process.

The benefits claimed of structured analysis are that it produces a nonredundant, unambiguous and complete specification, and a specification that is maintainable. Its aim is to produce a formal document normally called the system requirement document. This document consists of clients' requirements, alternative solutions available and costs, and the analyst's recommendation for the most appropriate solution.

Formalism underpins structured systems analysis and design. Formalism in structured systems analysis makes analysts' tasks formal, prescribed and official. It is based on notation languages to develop systems models of a new IS. Structured analysis makes use of diagrammatic notations, and structured English, to provide a common communication medium between analysts and users, and between users and software programmers and database administrators. To develop systems models analysts have to consult people whose work is to be modelled, so formalism 'forces' communication between analysts and people.

As structured systems analysis and design closely reflects the SDLC, it seeks to determine what the current organizational processes or systems do and specify what a new IS should do, and who will benefit from it.

Logical and physical systems models of the current system and a new IS are drawn in structured systems analysis and design. A logical systems model is a description of the actual situation – problem domain – in terms of data and processes. Logical models are separated from the physical computer implementation because they contain no references to the means (physical things) for achieving the result. A physical systems model is a description of the means for achieving the required result – computer implemented data and processes in a system.

The problem domain in structured analysis is characterized as business transactions data, dataflows, relations between data, processes and business logic. These form the basic units of analysis. Analysts develop current systems and new system logical and physical models of these units. Logical and physical models are produced of:

- data in entity-relationship diagrams;
- process flows in data flow diagrams;
- logic processes in process descriptions.

The purpose of structured systems design is to create a blueprint for a new IS that satisfies the requirements recorded in the system requirements document, resulting from the systems analysis phase. Structured systems design will result in systematic documentation of inputs, outputs, interfaces and systems processes. The systems design models are used as a communication tool between analysts and software programmers.

Roles are created for people in structured analysis and design. It demarcates roles for people and describes the responsibilities for each role as part of structuring the process of IS development. The formalism in structured analysis encourages people in the problem domain to be recognized during systems analysis. The important roles are:

- project manager
- systems analyst
- user.

The project manager is responsible for planning the development of the IS, allocating human resources to the planned tasks and for monitoring and controlling the system project. The project manager allocates appropriate resources, including analysts, to enable users and analysts to work together to determine system requirements and specification. The project manager is the main communication channel with sponsors of the project and senior business managers.

The analyst's task is to investigate the problem domain to establish system requirements. They work closely with users to understand what current IS do and what a new IS is required to do. Both current IS and a new IS are documented. Analysts develop a complete specification of a new IS by keeping a catalogue of data, its flow, and the transformations that happen to data, as data and process models. These analysis systems models are used as a communication tool between the analyst and users to check correct requirements are recorded. Analysts make use of instruments based on structured formalism to establish a system specification, which is then converted to modular structure charts during systems design. These structure charts are then transformed into structured programs in the implementation phase.

Users are involved during systems analysis to enable analysts to gather knowledge on the problem domain. They are encouraged to take an active interest in changing the current system, and it is assumed they are open to change. They are also assumed to be accessible and available to analysts for interviews and discussions to ascertain system requirements and discuss developed systems models. Users are involved because it is assumed that they know, and can communicate, system requirements and details of the business problem to be analysed. It is further assumed that they can help to define clear and objective goals for a new IS.

3.6 The object-oriented concept

Structure certainly improved IS development, but critical practitioners recognized that it could be better. Object-orientation in systems analysis and design emerged from efforts to improve software programming in the early 1980s. Its initial focus was on the design of software programming languages to resolve coding or programming inefficiencies. So object-oriented systems analysis and design has links with object-oriented programming languages like Smalltalk.

Object-orientation is a set of concepts and ideas on how to develop complex software systems and IS, it is collectively called object technology. Software objects are a representation of things, actual or conceptual, in real situations – human problems. The things or objects may be conceptual or actual – business invoices or product orders, files, records, processes, suppliers, products or customers.

Object-orientation has spread to all the SDLC stages. Object technology is significant for IS development and it is now a prominent approach to systems analysis and design. 'Object' is the central concept on which practice is based and encompasses the systems analysis, systems design and systems implementation phases of the SDLC. Practitioners have developed object-oriented systems analysis and design methodologies and instruments. These include Object-Oriented Analysis (OOA), Object Modelling Technology (OMT), Objectory, and Booch and Coad's object-oriented methodology. Object-oriented systems analysis and design is practised using instruments developed by many practitioners. Among them are dynamic and static object-oriented models, state-transition diagrams, case scenarios and Unified Modelling Language (**UML**).

3.6.1 Object-oriented systems analysis

An object-oriented system is a set of related and interacting objects organized into classes. Object-oriented systems analysis involves the development of integrated systems models that encompass functional, informational and behavioural views of a new IS.

In object-oriented systems analysis and design the problem domain is characterized as consisting of objects, relations, patterns, responsibilities and scenarios. The data, relationships and processes in the problem domain are integrated into *one* class model. Analysts' tasks are to:

- find class-and-object
- identify structures
- identify subjects
- define attributes
- define services.

Analysts' investigate the problem domain to identify relevant objects, patterns, responsibilities and scenarios descriptive of a required new IS. Analysts identify and name relevant objects and their responsibilities. An object's responsibility is its attributes, relationships with other objects and the services it provides to other objects. An object can be business transactions, data or things that are relevant to completing business objectives. It is an item for which data needs to be processed in a new IS. Analysts consult with people, or users, to define data objects, their attributes, and the way in which the objects need to be interrogated or changed. Objects do not have to be created anew. Skilled analysts are capable of software reuse, making use of existing objects from a library of objects.

Encapsulation is an important concept in object systems ontology. Encapsulation is gathering of data and operations in an object. An object contains both the data and instructions on what needs to be done to process the data, or its operations. Analysts conduct an object structure analysis to analyse the structure of the object types and an object behaviour analysis of the object types to define operations for objects.

As in structured analysis, project managers, users and analysts are involved in object-oriented analysis. Formalism is central in object-oriented analysis too. Object systems models are created with object notation languages and these models are used to communicate with users and systems designers. As one integrated systems class model is created, which contains analysis and design models, it makes communication between analysts, users and software programmers simpler. The same class model is progressively built through analysis, design, coding, testing and implementation.

3.7 Methods, techniques and tools

Methods, techniques, and tools – or instruments – are embodiments of system knowledge. Knowledge of organizations, people, application domains, IT, IS and design is represented in the instruments used to develop IS. Instruments available to investigate problem domains and to create systems models depend on knowledge and understanding of what constitutes IS. Automation, decision support and BPR exemplify such knowledge, which has progressively determined how IT is applied to organizations. Recent applications are eCommerce, eBusiness and the internet.

Practical system knowledge is acquired by developing conceptual and theoretical knowledge. In the absence of substantive theory in systems analysis and design, conceptual explanations determine practice. Concepts for phenomena like organizations and IS determine how managers or analysts act. They also determine the kinds of instruments devised to enable practice. Concepts like hierarchical or flat structure in organization theory and automation or decision support in IS have influenced the design of organization and IS with appropriate instruments.

An example of the relation between conceptual knowledge and its practical embodiment in systems analysis and design is systems theory (section 2.6). The ideas of interrelated elements, communication and boundaries in systems theory have influenced systems modelling techniques in structured and object-oriented systems analyses.

3.7.1 Problem solving strategies

Behaviourist theories use previous experiences as the unit of analysis to explain how people frame and solve problems, such theories are classified as reproductive. It proposes that people draw on previous experience to solve a problem. The weakness in this explanation is that previous experience can be a hindrance to finding a solution in certain situations and people could 'fixate' around such experience.

The Gestalt theory of problem solving arose because of weaknesses in behaviourist theories. Gestalt theorists argue that people solve problems through insight and structuring. The problem is first structured or framed and insight is important in the process. Such theories are classified as productive.

Another problem-solving strategy is the General Problem Solver Model. It is based on the argument that there is a 'space' of problem solving, in which a problem has various states. The problem-solver uses heuristics to frame the problem and explore a solution.

IS problems can be divided into the two categories of business problems and systems problems. The various problem-solving strategies are used to address both types of problems. Systems analysts help organizations and people to identify, describe, understand and define business problems related to the application of IT. They work with business analysts whose job is to identify, investigate and resolve business problems. An example business problem is dissatisfied customers or poor quality products. Often the initial business problem is experienced or identified by people in the business area and analysts work with business analysts to improve the situation.

IS problems are concerned with how IT can be applied to business problems. A business organization can benefit from supporting or integrated IS. For example, IT can be applied to the problem of improving customer service. Databases and relevant systems models can be designed to capture and process data on customer behaviour, and information on customer behaviour can be provided to business decision makers.

Problem-solving is systems analysts' major task. Problem-solving is rooted in systems theory and science. In science and systems theory a boundary is drawn around the problem and separate elements identified to solve the problem. Separating out elements makes it easier to solve individual elements, which are then combined to solve the problem. This process is called reductionism, a process of isolating a problem and breaking it down into simpler elements that can each be analysed separately.

IS development is a problem-solving activity. At a high level, the SDLC is an IS problem-solving method and methodologies are detailed planned approaches to solving the problem of developing an IS. The SDLC phases can be compared with reductionism because the problem is broken down into separate phases and each phase is progressively resolved.

Different systems ontology advocates particular problem-solving strategies. Structured systems ontology emphasizes the functions a new IS needs to perform, and consequently focuses problem-solving on identifying functions from the problem domain. Object-oriented systems ontology focuses on identifying classes, objects and patterns. The Information Engineering methodology focuses on the information required by business people, and so the focus is on the information or data in the problem domain.

The details of IS problems are addressed in the phases of the SDCL. They concern deciding what systems models to design and how to design them. Formalism is used to solve system level problems. A formal problem-solving strategy may be distinguished from an ad hoc approach that may lack a structure and vary each time a similar problem arises. Most formalism consists of diagramming techniques that express and structure knowledge about the problem domain as graphical systems models.

The formal problem-solving strategy used in structured systems ontology is functional decomposition. It is based on scientific reductionism. Functional decomposition requires breaking a problem down into smaller manageable problems or components and solving each component of the problem separately. The separate solutions are then combined to provide a complete solution for the whole problem. Functional decomposition results in treating business transaction data and business processes separately to develop data and process systems models.

The formal problem-solving strategy used in object systems ontology is classification theory. It regards the problem domain as consisting of human experience that can be classified as objects and relations identified between the set of objects or classes. The objects have attributes and provide services to other objects, both of which need to be identified. This is a holistic view of the problem domain that treats business transaction data and business processes as integrated and leads to the development of an integrated systems class model. In object-oriented analysis classes, their definitions and attributes are relatively stable over time. A class once identified usually remains constant in systems class model.

3.7.2 Notation languages

Problem-solving strategies rely on notation languages to express, structure and manage the process of solving problems. Notations are formalism and are used in systems analysis and design to develop systems models. A notation language is a set of defined symbols to describe and represent an IS problem. Some notations cover both business and IS problems. Problem-solving in systems analysis depends on the power of notation languages to represent business problems.

The development of an IS involves organizations, people and IT. These elements cannot be described precisely in a natural language like English. A notation language is used to represent the problem domain as systems models. Its prime purpose is to describe and represent the problem domain and to enable *manipulation* of the notation symbols to determine an appropriate resolution of the problem. Analysts can manipulate the symbols until a desired solution or required systems models are found.

The utility of a notation language is determined by how well it can express (describe) the problem domain and how accurately it can represent it. Its power is in its notation symbols and in how they can be manipulated to resolve a problem. The symbols should be sufficient enough to capture or represent elements of real situations that interest the analyst and simple to use. The manipulability of symbols also should be sufficient to arrange the symbols to achieve a desired solution.

Notation symbols and their manipulability determine the kinds of IS problems that can be addressed. Improvements in notation languages enable more complex IS problems to be solved. An example early notation is a flowchart, used to design computer programs and extended to systems analysis. Flowcharts are limited to solving logically sequential problems. In IS reliant on databases the entity relationship notation is used to develop database models.

Data flow diagrams and process models in structured analysis separate out the elements of a problem according to functional decomposition, whereas a problem domain is treated holistically in object-oriented notations. In object-oriented analysis notation languages focus on an integrated or holistic problem-solving strategy. The data and process in the problem domain are viewed as a whole, and the problem is resolved holistically by developing an integrated systems class model. The UML is an example of an object-oriented notation.

3.7.3 Techniques and tools

Instruments make it possible to develop systems models. Techniques are used to develop systems models and tools are used to support the process. Prior to computer-based tools becoming available systems models were developed manually with pencil, paper and notation templates. Both structured and object-oriented systems analyses provide techniques to develop systems models and computer-based tools to support the process.

Structured techniques and tools

The practitioner-originators of structured analysis aimed to provide a systematic and structured method for developing IS. They developed diagramming techniques for IS modelling and tools to support the development of stipulated systems models. Figure 3.3 shows the deployment of techniques within the SDLC. Many of the structured techniques are incorporated into the SDLC. It provides structured systems modelling techniques across its major phases.

Cost benefit analysis used during feasibility study addresses business problems that assess the benefits of investing in IT and development of IS, though it originates in financial accounting practices. The entity relationship modelling technique is significant during the requirements analysis phase. It is used to identify things in the problem domain for which data needs to be captured, stored and processed. Consequently, related techniques like entity/event modelling and entity life histories are used to check for completeness and correctness of entities.

SSADM modules are feasibility study, requirements analysis, requirements specification, logical system specification and physical design. Each module contains a number of stages. The components of SSADM are: structures, techniques and documentation. Structure is concerned with systematic and staged activities, and the inputs and outputs required to perform them. Techniques is concerned with how the staged activities are to be performed. Documentation is concerned with presenting the products of the activities. SSADM makes use of three techniques, logical data modelling, data flow modelling and entity/event modelling. The systems models produced provide three interdependent views of the problem domain. The different systems models are checked to ensure consistency, accuracy and completeness.

There are other structured techniques, listed in Table 3.1, that can be used to execute SSADM phases or used as part of other IS methodologies.

Table 3.1 Structured systems analysis and design techniques

Structured techniques
Data dictionaries
Decision tables/trees
Structured English
User/user role modelling
Function definition
Relational data analysis
Logical database process design
Update and enquiry update process models

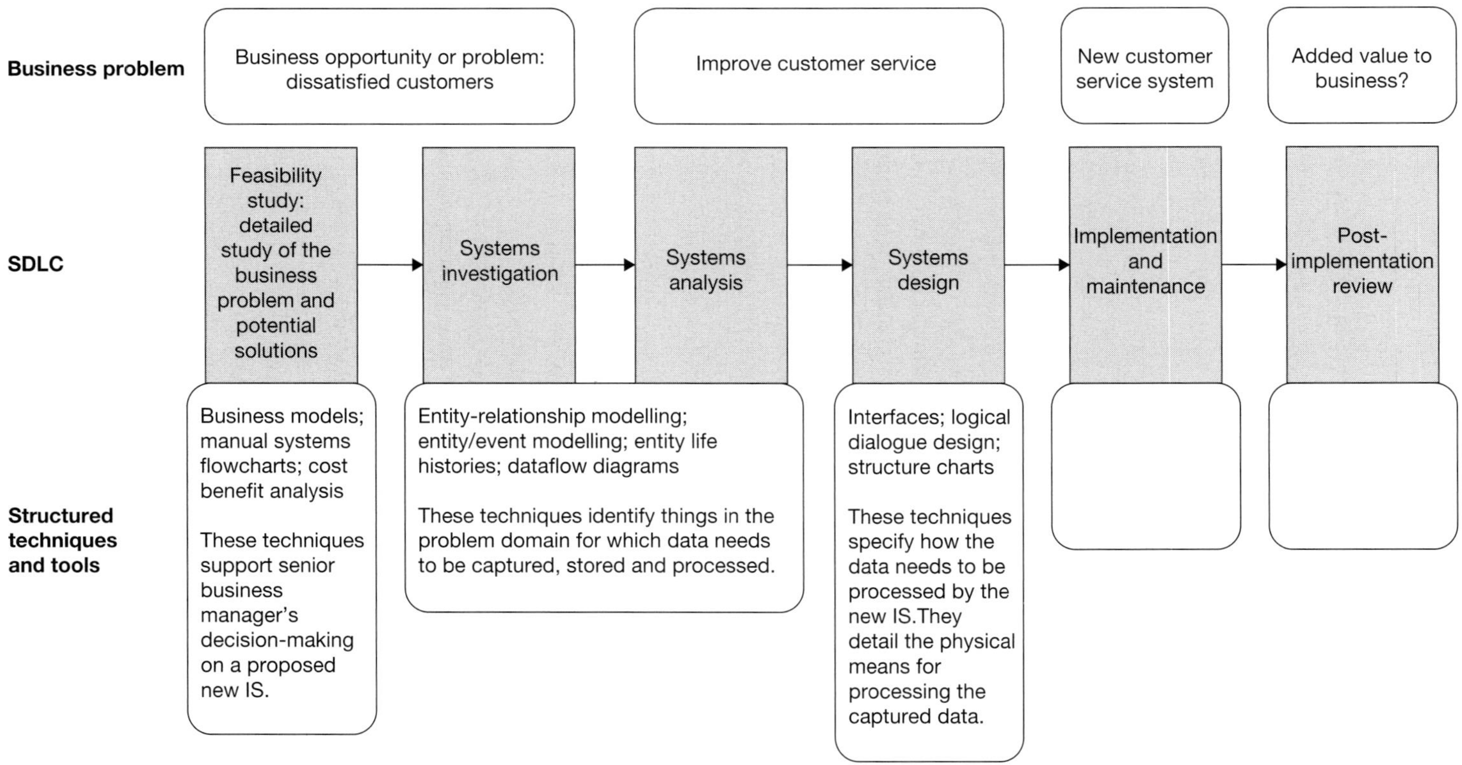

Figure 3.3 Structured IS modelling techniques within SDLC phases

Structured tools

Computer-Aided Software Engineering (**CASE**) tools are used to support the development of software. They were designed either to support specific steps in software programming or to act as generic support tools that could be chosen as required. They support analysts to build correct and integrated systems models. Correctness here refers to complying with the technical stipulations of structured notations for systems modelling. They are mostly used for requirements specification.

Horizontal integration CASE tools are used to integrate the use of tools within the same SDLC phase, for example within systems analysis or systems design. Such integration enables models to share data and to be validated for accuracy. Horizontally integrated CASE tools are available for entity relationship modelling and dataflow diagramming. Database design and business process design tools are examples of vertical integration CASE tools. Vertical integration is the use of CASE tools to integrate adjacent SDLC phases, for example analysis and design.

A data dictionary is a fundamental concept in structured analysis. It records all textual and numeric information about a system that cannot be captured in system diagrams. Some CASE tools integrate data dictionaries with diagrams to make checks for consistency easier. The quality of CASE tools to support SSADM has been questioned, but there are diagram editors and consistency checkers to improve quality of systems models.

Object-oriented techniques and tools

Object-oriented analysis and design provides techniques or strategies for finding objects in the problem domain. Object-oriented techniques are provided for many of the SDLC phases, as shown in Figure 3.4. The use case technique is used to develop a high level model of the system and its functional uses. The object class diagram is developed for both the existing system and new IS.

Two techniques that are easy to understand intuitively make use of nouns and verbs in systems modelling. Analysts' search for nouns or noun phrases in the Noun Phrase Strategy. Nouns or noun phrases identify an object for which data needs to be processed and operations need to be coded. Example objects using nouns are 'invoice', 'department' or 'manager'. Analysts' search for verbs or verb phrases in the Class-Responsibility-Collaboration (CRC) strategy. Verbs or verb phrases describe what the object does, for example 'compute', 'dispense cash' or 'print statement'.

There are other object-oriented techniques that can be used to execute SSADM phases or used as part of other IS methodologies. These are: sequence diagrams, transition diagrams, event flow diagrams and collaboration diagrams.

Object-oriented tools

Object-oriented tools are fundamentally different because they need to reflect a richer ontological knowledge of the problem domain. Compared with structured tools, object-oriented tools need to be capable of modelling semantics in the problem domain. Object-oriented concepts such as object, attributes and services enable detailed semantics to be captured, so tools need to be capable of facilitating it.

There are horizontal integration and vertical integration CASE tools for object-oriented analysis. They range from tools for browsers, debugging, configuration management and GUI builders. Comments about CASE tools made in the section on structured CASE tools also apply to object-orientation. The utility of all tools is reduced after a certain point.

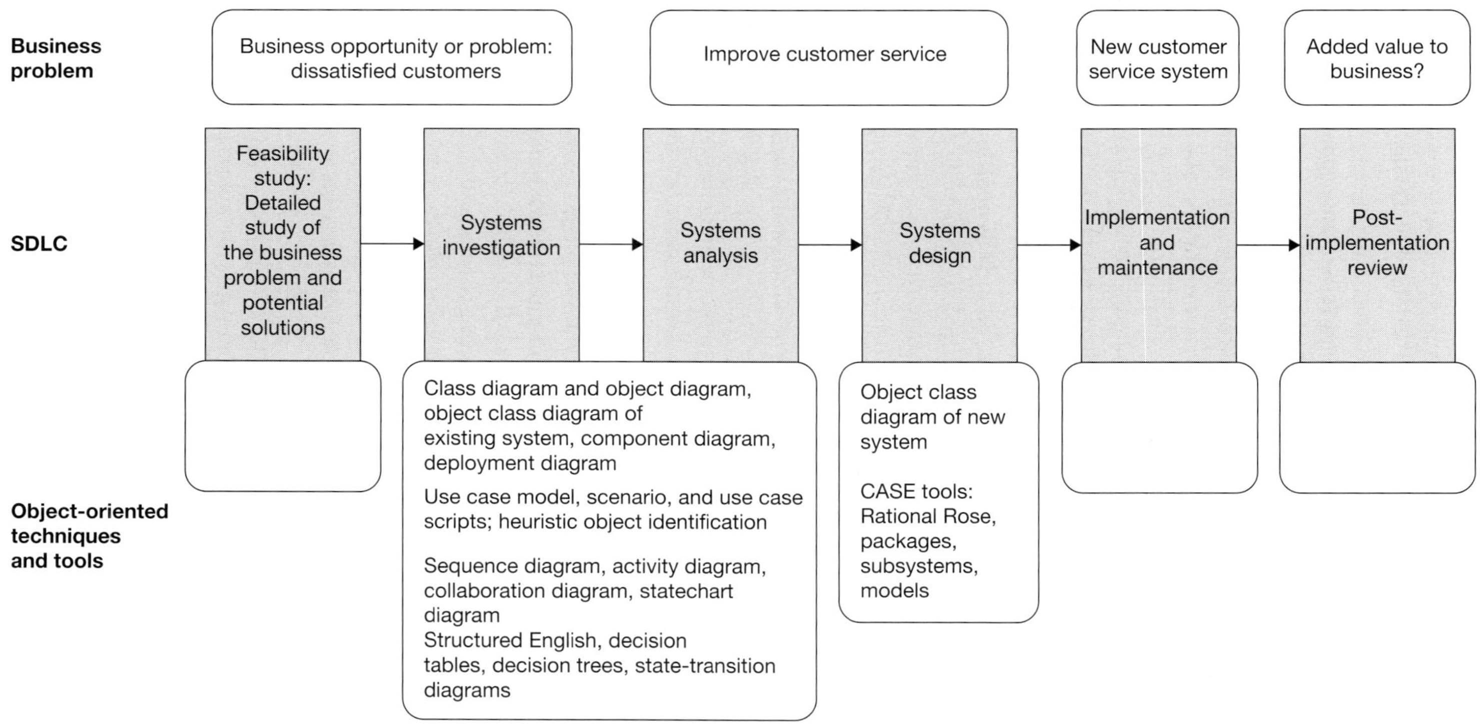

Figure 3.4 Object-oriented techniques within SDLC phases

Rational Rose is a CASE tool to support object systems modelling. It is used within UML.

CASE tools do not necessarily produce good systems models. The tools do improve the accuracy and validity of systems models produced. Good systems models depend on the thoroughness of the problem domain investigation.

3.8 IS methodologies

An IS methodology enables the articulation of an IS problem and provides a staged process for IS development. A basic premise of methodologies is that by following the stipulated stages the problem domain system requirements can be fulfilled. There are numerous methodologies available all seeking to achieve this goal. The authors of methodologies are the government or its agencies, practising consultants or companies who develop bespoke methodologies.

The British government's use of IT led it to sponsor the development of SSADM in conjunction with the National Computing Centre (NCC). Practising consultants developed various methodologies including Jackson System Development (JSD), Structured Analysis, Design and Implementation of Information System (STRADIS), Yourdon System Method, Information Engineering, and Coad's object-oriented methodology. STRADIS and Yourdon System Method use the functional decomposition problem-solving method, the latter focuses on data structures.

A methodology is underpinned by certain beliefs and knowledge held by the methodology authors about human problems and systems ontology. A consideration of these beliefs is important for criticality and practical use of a methodology. Project managers need to understand the systems ontology of a methodology and decide whether it is appropriate for the project. Analysts need to evaluate whether they agree with the assumptions made and the consequent systems ontology, and whether it is appropriate for undertaking analysis.

A methodology normally contains all the SDLC phases, though some methodologies omit the implementation phase and others do not make stringent demarcations. The characteristics of a methodology are similar to a plan, hence methodologies are referred to as detailed plans for developing IS. The significant formal part of a methodology is that it contains systematic, often sequential, phases and that each phase delivers specific documents for the next phase and the overall successful development of a new IS. The result of each phase or output is referred to as a **deliverable**.

All methodologies have a systems analysis and systems design element. This may be a concrete phase such as in SSADM or a quick process integrated with actual coding of computer programs as in ASD. In practical terms a methodology is a collection or package of techniques and tools for systems analysis and systems design, and it also prescribes systems implementation.

Methodologies may be categorized as data-centric, process-centric, or object-oriented. Some methodologies focus systems analysis on data. Other methodologies focus systems analysis on processes and information. Object-oriented methodologies focus systems analysis on identifying objects and their attributes, especially relationships. Systems models in data-centric and process-centric methodologies need to be checked and integrated to provide a consistent systems model. The systems class model in object-oriented methodologies is inherently integrated.

Methodologies are normally executed within systems projects. In business a project is the pooling of required resources to achieve predetermined objectives within predefined financial resource and time constraints. The development of an IS is referred to as a systems development project or simply system project.

A system project provides a structure for both the work to be done and the products to be delivered. Project management techniques and tools are useful for organizing the work required. The structure in projects is the result of plans formulated to identify work to be done, provide a schedule of the work, and to monitor and control the progress of the project.

3.8.1 Structured systems analysis and design method

SSADM is the UK government's preferred methodology. It was developed as a government initiative to seek efficiencies in IS projects in the public sector, particularly to control cost overruns. Software vendors are required to use it when tendering for government contracts. SSADM focuses on data in the problem domain and the requirements of a new IS to develop data models and databases. It contains elaborate standards to which systems documentation must comply.

The SSADM life cycle is shown in Figure 3.5, techniques and tools used to produce deliverables are shown below the phases. SSADM epitomizes the structured approach. Its main phases closely follow the SDLC phases and make use of structured techniques and tools for systems analysis and design. Both contain the feasibility study, systems analysis and systems design phases, but SSADM does not have the implementation of the designed system and its post-implementation evaluation phases.

SSADM is based on the structure concept. It stipulates tight rules and guidelines to which project managers have to adhere. It does not provide systems project management techniques, so project managers use techniques drawn from the Project Management In Changing Environments (**PRINCE**) method.

SSADM concentrates on systems analysis and design. It focuses on the analysts' activities required to complete a feasibility study, requirements analysis, requirements specification and logical systems design. Systems analysis and design in SSADM is achieved with the aid of formalism and CASE tools.

A thorough or rigorous systems analysis and design is achieved on the basis of the 'three-views' of the system propounded in SSADM. The three views are: (a) the data in the system, (b) the business events to which the system must respond and (c) user defined functions of the system. Systems analysts are required to cross-reference analysis and design between these three views of the system to check the integrity of systems models. A significant criticism of SSADM is that it produces much documentation, and configuration management of documents becomes problematic.

3.8.2 Jackson System Development

Many companies use JSD. It is a structured systems analysis and design methodology that complements Jackson Structured Programming (**JSP**). JSP is widely used in practice too. The problem-solving strategy used in JSD is the top-down decomposition of the problem into sub-problems that are individually and separately solvable. Companies have adopted versions of JSD that include the main phases of the SDLC.

The notion of an entity is important in JSD. An entity is something that exists in the real world, which needs to be reflected in a new IS. In JSD an entity is defined as something that is unique and performs an action or it is affected by action over time. For example, a customer is an entity that places an order at a certain time, or an invoice is generated when an order has been fulfilled. JSD contain six steps:

- entity action
- entity structure

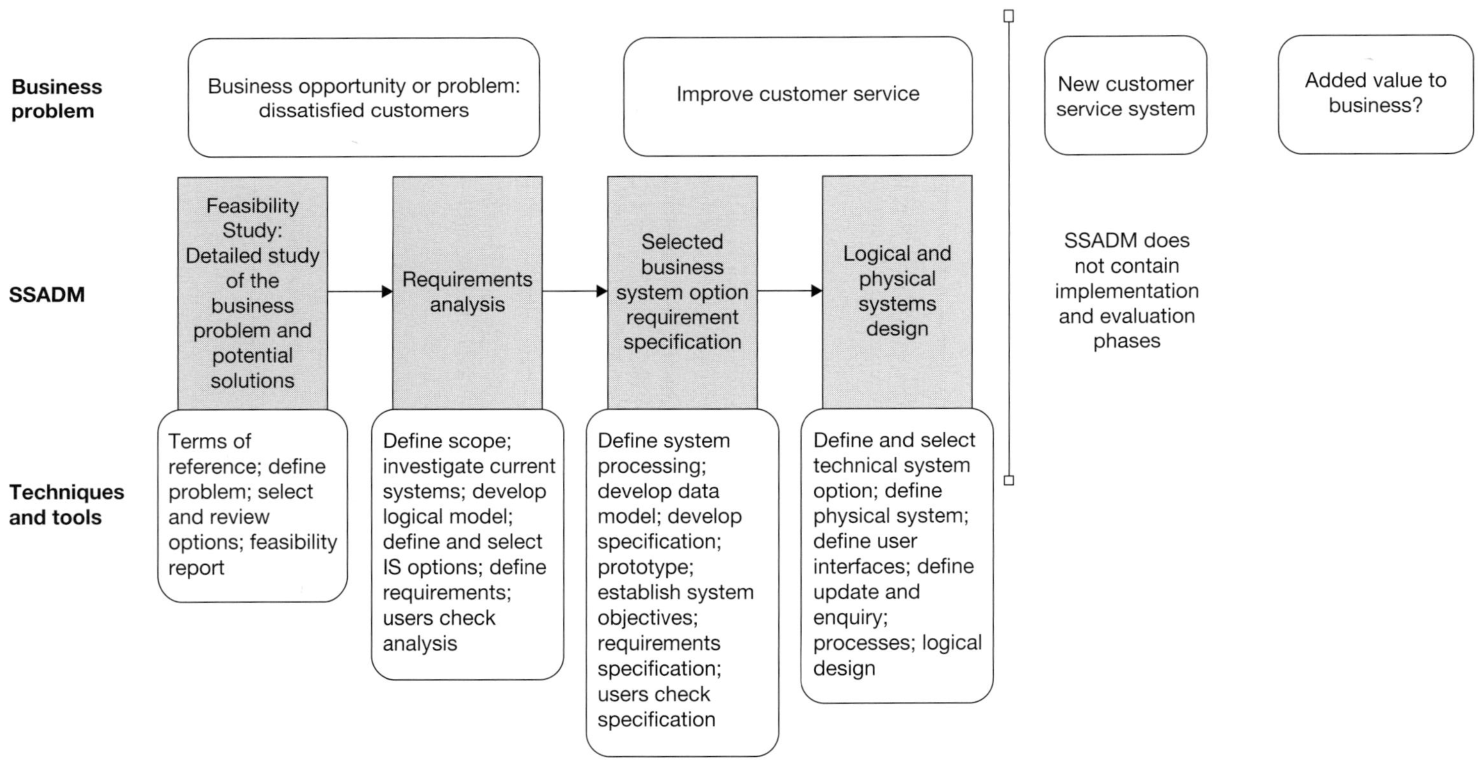

Figure 3.5 SSADM phases, techniques and tools, and deliverables

- initial model
- function
- system timing
- implementation.

The development of systems models is significant in JSD. Entities are identified and modelled to show how they affect other entities or how they themselves are affected. The focus on entities is claimed to produce a flexible IS that can be readily enhanced or amended to meet changing business needs. Systems analysis and design is regarded as an extension of software development that leads onto JSP.

Project managers do not have a significant role in JSD because of its emphasis on system level analysis and design. The identification of entities and development of models is undertaken by analysts who use formal structured diagramming techniques to develop systems models. The diagramming techniques are: JSD structure diagram, JSD system specification diagram, JSD structure text and systems implementation diagram.

3.8.3 Open Source Software

OSS is not a methodology like SSADM or JSD. It is included here because it is being recognized as an acceptable method for developing software, both for government and commercial companies.

OSS is software whose source code is open to the public. It is often developed by voluntary efforts and is usually available at no charge. It is used under a licence defined by the Open Source Initiative (OSI), which prevents the open source produced software from being redistributed under licenses that contravene the OSI ethos. OSS is now a significant source for industrial-strength software for many kinds of software applications ranging from the Linus operating system to the Oracle and Informix database systems, and Word Perfect and Corel word-processing package and office suite.

It is not clear whether OSS is a methodology. The importance of OSS is recognized by the UK government that initially encouraged the development of SSADM, which subsequently become a benchmark. The UK government's e-Government Interoperability Framework (e-GIF) mandates open standards and specifications. The UK government's policy on OSS includes the use of OSS solutions in IT procurements. It will only use products for interoperable systems that support the OSI in future systems applications. Government supported research and development will also make use of OSS as the default software. OSS is a fundamental change in the way software is developed and the UK government is developing policies to recognize its importance in the marketplace and for government contracts. The European Commission (EC) too promotes the use of OSS in the public sector and e-government best practice.

The role of project managers and systems analysts in OSS is unclear. Though the roles of project managers and systems analysts in OSS will not be redundant, it is unclear what precisely they would do. Software programmers develop most open source software. They become intensely interested in the actual programming problem, sometimes because they themselves make use of the software product and would like to make improvements to it for personal use. The use of OSS in an organization would certainly require analysis of the problem domain.

3.8.4 Coad's object-oriented methodology

Coad's object-oriented methodology focuses on data, behaviour and functions of objects. The aim is to develop one integrated class systems model of function, information and behaviour. A single integrated class systems model does not require 'balancing' or other elaborate checking across models, as required in structured analysis. The purpose of the integrated class model is to develop consistent and accurate systems models. Coad's object-oriented methodology consists of four activities:

- Identify system purpose and features.
- Identify model component's objects and patterns, for each of the model components:
  - problem domain;
  - human interaction;
  - data management;
  - system interaction.
- Establish object responsibilities.
- Define service scenarios.

The purpose of identifying objects and patterns in the problem domain and the system is to find data objects and define the behaviour of the objects. The purpose of the third activity is to establish object responsibilities by identifying the functions required of the system and the data objects that need to be captured, processed or stored, and the behaviour needed to complete the responsibilities and services of the identified objects. The purpose of the fourth activity is to determine the services to be provided by the objects.

Project managers have flexibility to determine how the methodology should be practiced. It allows parallelism, substitution and omission. A project manager can perform the activities in parallel if the problem domain permits. The sequence in which activities are performed may be changed or substituted by the project manager when the problem domain permits. The project manager may decide to omit one or more of the activities if the problem domain permits.

Analysts' task is to execute the four activities. Whether the activities are completed in the sequence stated in the methodology depends on the project manager's decisions concerning parallelism, substitution and omission. An analyst using an object-oriented methodology needs to be able to understand how to identify objects and the role of the one integrated class systems model.

3.8.5 Prototyping

Prototyping is a popular Rapid Application Development (**RAD**) approach. Prototyping is a methodology but it does not make stringent demarcations in the process of IS development. It does not divide the process into discrete phases based on the SDLC. The aim of prototyping is to reduce the phases and the time it takes to develop an IS, and to involve closely clients in development. By not demarcating the process into phases prototyping seeks to reduce the time it takes to develop a system.

Prototyping actively involves clients in the problem domain to elicit system requirements. A system is developed quickly and installed for use in the problem domain. People's usage of the system generates evaluation and feedback on the system's functionality, user interfaces and information for developers. Developers use this feedback to enhance the prototype system.

Prototyping varies the SDLC phases moving from analysis, to building, and seeking clients comments on developed prototypes. There are a variety of prototyping strategies:

- Incremental prototyping where the prototype eventually becomes the operational system.
- Throwaway prototyping where the system is a mock-up to obtain feedback.
- Cooperative prototyping involves clients actively in the design of the prototype.

The role of systems analysts is to determine an initial set of system requirements and work with software programmers to develop a prototype. Clients are not usually consulted on the initial requirements and are not involved in the design of the initial prototype. Analysts then work with clients to evaluate the system by performing certain tasks with the prototype system. The results of the evaluation are recorded as part of the requirements document and used to enhance the prototype.

The role of project managers in prototyping varies depending on the size of the prototype and type of prototyping strategy adopted. They perform the normal systems project management tasks if the prototype is organization-wide.

3.9 Interpreting concept and action in the Critical Framework

IS knowledge and IS development knowledge are continuously progressing. There is no formal theoretical knowledge on IS development. The SDLC is the fundamental conceptual and formal body of knowledge. It prescribes how to develop an IS. It constitutes a formalistic approach to IS development. Notions of optimality and the universal application of instruments underpin its systems ontology. The SLDC is the basis for most research and practice in IS development. SSADM, and other methodologies, are based on the SDLC.

The Critical Framework can be used to analyse critically concepts and practice in systems analysis and design. Fundamental concepts like the SDLC, structured analysis and object-orientation, and methodologies, constitute knowledge. The actual practice or implementation of this knowledge can be analytically evaluated with the Critical Framework.

Figure 3.6 is the Critical Framework populated with critical reflection on the conceptual basis and practice of systems analysis and design. As the bottom layer shows, many questions concerning the four themes of criticality arise from assumptions of objective and rational human behaviour in systems ontology. These questions relate to transformatory critique, refashioning of traditions, reflexivity and critical skills.

For example, in terms of reflexivity, analysts use the term 'users' or 'user organization'. The term has implications for knowledge and practice. Some researchers do not consider it to be an appropriate term to describe people in organizations. For many decades the term foreclosed 'users' to participate in IS development. It also had to be qualified because of technological development in 'end-user computing' that enables people in organizations to develop IS themselves. Patel (2003) has coined the term 'action developers' to describe people in organizations capable of tailoring IS to their particular needs.

The SDLC seeks optimal solutions. Critical evaluation of the SDLC systems ontology reveals that optimality assumes perfect information will be available to make decisions about the human problem. Such information is rarely available in practice. Similarly, instruments are underpinned by ontological assumptions. The SDLC assumption of rational organization and humans leads to instruments that produce systems models that are optimal and measurable. An optimum situation is where the best possible results are achieved, which is usually deflected by other factors in the 'messy world'.

To internalize knowledge and develop practice of the SDLC, analysts would need to accept certain kinds of personal constructs. Based on knowledge from the systems ontology theme, they need personal constructs that fit the objectivist ontology of the SDLC. For example, an analyst would need to develop an 'objective' predisposition given the SDLC's assumption that the world is objective. So, an analyst would need to be detached and impartial, and analyse an actual situation with detachment and impartiality.

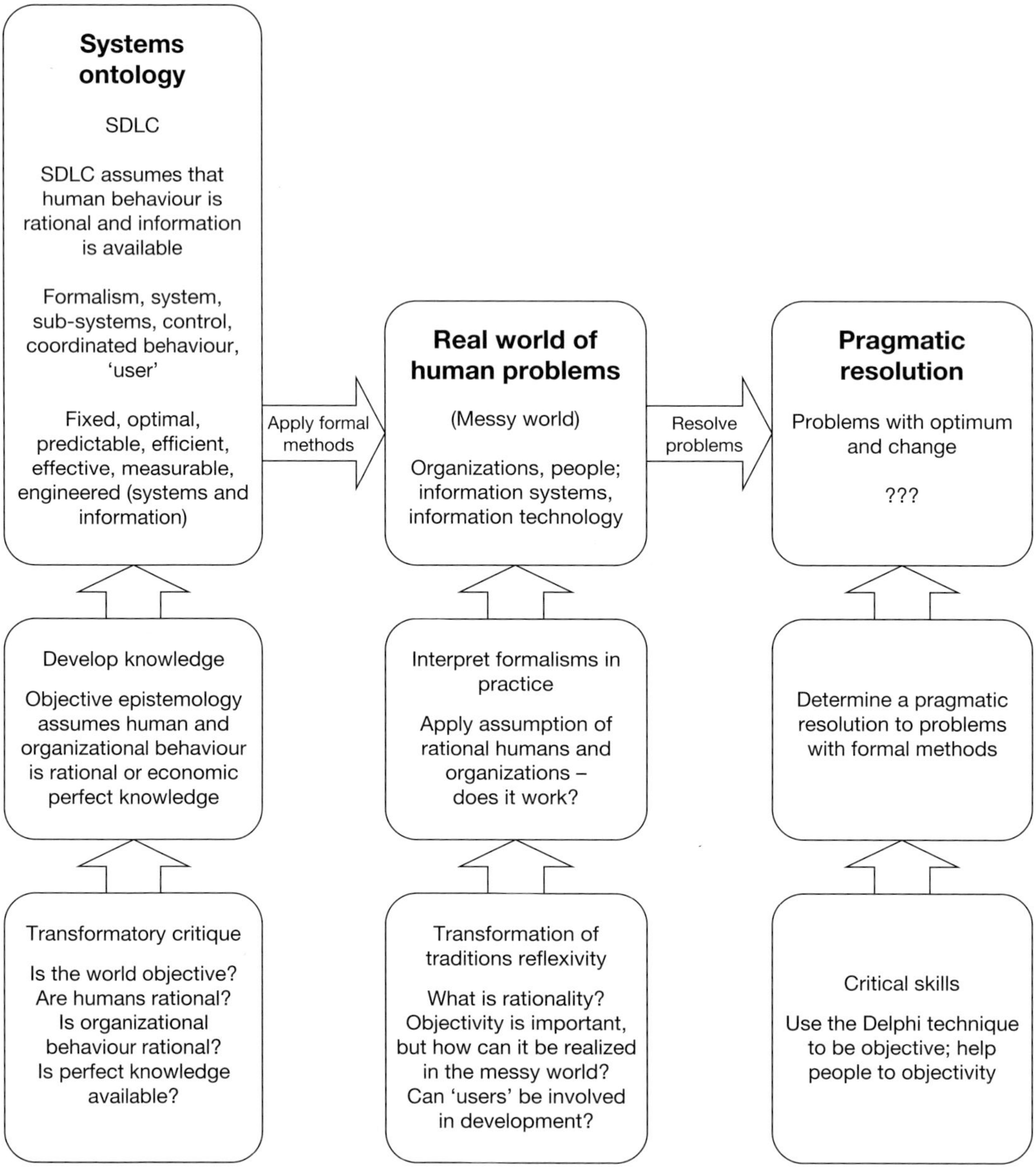

Figure 3.6 Critical framework: the ontology of the SDLC

The actual inclusion of the objectivity personal construct in a PCF should be based on critically evaluating its relevance for practice. The actual situation will reveal that people have self-interest and egos that will make them behave in a subjective and partisan manner. Analysts too need to question their self-interests and biases. A series of questions will arise that will lead to evaluating critically the objectivity personal construct and its validity in real situations. These may include: How is the analyst to be objective in such a situation? If a systems analyst makes a design decision that is not favoured by some people, will they lose interest in the IS development? The analyst would then evaluate whether to include the objectivity personal construct in a PCF, or whether it should be practiced selectively.

3.9.1 Considering sub-themes

The SDLC can be further analysed and evaluated in terms of planned action and situated action sub-themes. Objectification of personal constructs can be made clearer by considering these sub-themes to understand the SDLC systems ontology and its relevance in real situations. The analyst can then consider whether to develop personal constructs based on planned action or situated action.

Systems ontology knowledge is often couched in planning terminology in IS development. Knowledge about IS development problems and of how to overcome them is devised as appropriate instruments deployed in the context of plans or methodologies. The SDLC may be characterized as planned action, since it has systematic phases: requirements analysis, data and process modelling, and data and interface designs, determining system functionality, and actually implementing the designs using IT.

The requirement for an analyst to be objective necessitates planning, for example when eliciting requirements. The analyst's plan of activities may not be possible to implement in an actual situation. There may be limits on how much time people can give or required documents may not be available. The analyst's plan would then need to be changed, however re-planning may result in a similar situation. Alternatively, the analyst could choose to take situated action. People who are accessible can be interviewed and documents that are available can be studied.

3.9.2 Objectivity and modelling

The SDLC, structured and object-oriented systems ontology assumes an objective problem domain in which analysts themselves can behave rationally. Objectivity is a problematical assumption in systems ontology. It has implications for both knowledge development and praxis. The premise that humans behave rationally leads to the expectation that analysts are capable of applying rational instruments in an objective and detached manner.

Analysts expect people in organizations to behave rationally because of their training in structured or object-oriented methods. They often find that people do not meet their expectation, which ironically leads them to seek shortcomings in the people or application domain rather than reflect on and evaluate their assumed knowledge.

Analysts can analytically evaluate the SDLC systems ontology to develop personal constructs on objectivity, and the design of objective, unambiguous and complete systems models required in structured analysis. Analysts would then evaluate whether to include objective systems modelling personal construct in their PCF.

A model is an abstract representation to describe and explain reality – organization, people, IS and IT. It is used to understand actual human problem situations. The development of systems models is central in the SDLC, in which the tenet of objectivity affects the type of systems models developed. Its objective systems ontology assumes that an analyst can create objective, unambiguous and complete systems models. Analysts are expected to remain detached and impartial in the process of designing systems models.

The fundamental premise of the SDLC, structured, and object-oriented analyses is that human behaviour in organizations is economic or rational. This basic premise encounters problems when it is applied in real human problems.

Human behaviour normally, but especially in an organizational setting, is influenced by social and political factors that question the premise of rational behaviour. Such analytical evaluation and interpretation of knowledge and practice in terms of the Critical Framework will enable an appreciation that the *actual* practice of IS development is not a wholly rational, objective phenomenon.

3.9.3 Pragmatism

During the 1990s the new IS conceptions were based on Business Process Re-engineering (**BPR**). Analysts in consultation with people developed systems models of existing organization, processes and workflows. They were then analysed to create alternative, radically different process models that took advantage of IT. Analysts and business analysts questioned and analysed business processes to seek out inefficiencies and improve effectiveness. The new systems models aimed to redesign or 're-engineer' organization and improve productivity.

Systems models are created with formal notations and diagramming techniques. The assertion that notations provide a common language with which analysts can communicate with each other, and with project managers, software programmers or database administrators. The models created with formal notations are also used to communicate with people. These assertions can be evaluated. In structured systems analysis the models are of data and processes, and in object-oriented analysis the models depict classes and relations. Complex systems models are difficult for users to understand, and even analysts find them difficult to interpret. Such confused models are subsequently implemented using IT.

Authors of methodologies do not attempt to explain and elaborate assumed systems ontology or epistemology. Analysts can learn much about praxis from them, but little on how to develop criticality. The questions that arise in developing a new IS require critical thinking. The SDLC, structured methods, and object-oriented methods and IS methodologies are some responses to these questions but they are mostly prescriptive. Though the knowledge may be technically sound, its usefulness in practice depends on its application limits in real human problem situations, where organizations and people matter the most.

There is a close relation between writing computer programs and performing systems analysis and design. Problems in software development have led to solutions that have found relevance in systems ontology. Knowledge on how to develop computer programs has influenced knowledge on how to conduct systems analysis and design. Structured programming know-how resulted in structured analysis and design. The emergence of object-oriented programming has led to object-oriented systems analysis and design. The recent interest in agile or eXtreme programming is driving a methodological systems analysis and design. This synergy may be regarded as inherent because systems analysis and design is conducted to inform software development.

Structured methods, techniques and tools arose from the problems that the US military and space exploration organizations encountered in applying digital computers to achieve their aims. Military and space software developers looked to engineers to learn how to produce robust software, of good quality and reliability and acceptable to clients. Adopting the engineering analogy lead to the term **software engineering** to describe the disciplined development of software. The engineering analogy has persisted ever since in software development and was extended to developers of commercial software applications.

Knowledge of systems analysis and design has to be enacted in real situations. Despite the progress in IS concepts, instruments and methodologies, the perennial IS development problems of unmet requirements and cost overruns still persist. Reflective analysts need to explain why the problems still exist. Arguments for further progress or better adherence to standards do not resolve or explain the problems. Better knowledge and understanding of organization and people, IT and IS, and how these relate in the application domain is needed. Better elaboration of systems ontology is needed. An explanation of how these relate to ontological understanding of organizations and people is required. Inclusive epistemologies, capable of investigating organization and people, need to be accepted.

Mechanistic conceptions and definitions of data and information are simplistic for developing appropriate systems ontology for IS. Human experience of data and information is phenomenological, which IT is presently unable to reflect. Organizational knowledge and knowledge management in particular pose significant problems. Knowledge management systems based on algorithmic conceptions of knowledge have been unable to facilitate or improve knowledge generation and sharing. The notion of semantic data and information, particularly reflected in XML for the web, provide the necessary human dimension required to contextualize information. Similar conceptions are necessary in systems analysis and design.

3.9.4 Planned action and situated action

The planned action and situated action subthemes are particularly relevant for critically evaluating concepts and practice. It is informative to consider analytically the SDLC, structured and object-oriented systems ontology, methodologies and problem-solving in terms of planned action and situated action.

Figure 3.6 illustrates the practical problems when objectivity underpins systems ontology. Objectivity and rationality suggests a planned and predictable reality. Planned action is the basis of structured systems analysis and design and in some versions of object-oriented systems analysis and design. Entities in structured analysis and objects in object-oriented analysis are presumed to exist, and analysts' task is to discover them to develop systems models. Characterizing systems analysis and design as such objective, planned action is questionable because it negates contexts, and it negates emergent organization – things organizations and people have to respond to that cannot be planned or predicted.

Project managers deploy methodologies as project plans that detail the resources required and the work to be done. So it is possible to label IS development as planned action. To develop a critical perspective planned action and situated action will be differentiated in the context of knowledge and practice.

The SDLC

The SDLC is structured IS development. It consists of conceptions of IS and instruments for systems modelling. Its basic premise is that humans and organizations behave rationally. This premise translates into a planning perspective, with systematic phases in the SDLC characterized as a form of planned action. Project managers develop sophisticated plans detailing systems analysis, design and programming work packages and the resources required to complete them.

Though the SDLC is shown in Figure 3.2 as sequential, phased activities, it has various versions. These versions have resulted from addressing gaps in a prevalent version. The 'waterfall model' version was added because in practice developers need to go back to previous phases. An 'iterative' version was developed to enable developers to work across two or more phases to fill informational gaps in previous phases. A 'spiral' version was developed to diffuse risk analysis and enable verification and testing throughout the phases. In previous versions risk was not considered or verification and testing were carried out too late to contribute to successful completion.

Analysts are expected to deploy the SDLC objectively and systematically within project plans. The premise of objectivity has led to the design of instruments that fail to recognize the non-rational aspects of human and organizational behaviour – namely, politics, culture, limits on knowledge and communication gaps. The actual practice of systems analysis requires analysts to consider situated factors. Analysts have to account for the limitations of people's ability to state system requirements. They have to consider whether people are capable of envisioning a new IS or how it might relate to the work they do. These are situated factors that cannot be accounted for in plans.

Techniques and tools

It is assumed in structured and object-oriented systems ontology that instruments enable the development of *correct systems models*. Proponents of structured techniques and tools make various claims. For example, that instruments provide better project control and communication, they meet people's requirements and can cater for changing requirements.

CASE tools are used to document systems analysis and design activity and function as communication aids. They are used to communicate systems models between analysts and programmers and project managers. Contentiously, it is also claimed that CASE tools are useful to communicate between analysts and user too. There is no verification of this claim.

Though CASE tools are meant to achieve correctness of systems models, the consistency and accuracy of integrated models has not been achieved. Use of vertical CASE tools in structured methodologies that separate data and function modelling has not led to successful technical validation of integrated models. Object-oriented CASE tools have not matched developments in structured methods and techniques.

Instruments alone are not sufficient for developing appropriate systems models. The formalism in structured analysis and design encourages people in the problem domain to be acknowledged in systems analysis. They are encouraged to participate in the development process but they have severe technical limitations. Prime among them is lack of knowledge to read technical systems diagrams or systems models.

Appropriate systems design is the result of thorough investigation and the validity of the assumptions made of the problem domain. Limitations can be determined by examining ontological assumptions. Structured and object-oriented analyses have techniques and tools that assume an objective problem domain that can be impartially investigated. Often analysts find people have interests and attach subjective meaning to the work they do which makes the instruments ineffective.

Methodologies

A methodology is normally more than simple packaging of instruments for developing IS. Critical analysis reveals that methodologies contain philosophical assumptions and ontological beliefs that influence the type of IS and instruments developed. Methodologies harbour assumptions about the nature of systems, instruments, IS, people and organizations that heavily influences conceptions of IS and how IS are developed. They prescribe the role of potential users. Significantly, they embody hidden epistemological method which affects the contributions the IS makes to the achievement of human and organizational purpose.

Structured and object-oriented methodologies assume 'things of interest' pre-exist. Analysts have to simply identify them for systems modelling. JSD and Coad's object-oriented methodology presume that entities and objects exist in human problems. The analysts' task is to establish objectively what these entities are and develop systems models. They are required to identify entities and objects from people, business processes, and documents and to establish their importance and relevance for systems modelling. The actual process though is more involved. Analysts have to rely on people, who have subjective interests, in the organization to name entities and objects and to establish their importance and relevance. This reliance reveals the social dimension of IS, and necessitates consideration of power relations in organizations, and in IS development.

A methodology is packaged as a set of instructions for practitioners. JSD or Coad's object-orientation methods prescribe to IS developers how to develop IS. The phases of SSADM prescribe what project managers and systems analysts should do. Significantly, methodological prescriptions cause analysts to expect organizations and people to behave in prescribed ways. Actual implementation of methodological prescriptions tests the objectivity and planning assumptions and the prescribed expectations. Consequently, most practitioners do not deploy a methodology because it binds them to planned action that contradicts the problem domain. They find that the actual behaviour of people and events in the organization are not susceptible to objective and systematic analysis, and the expectations raised by methodology remain unfulfilled.

At the core of a methodology is the belief that the required or intended future can be created. Systems analysis and design is concerned with creating a socio-technological future, where work and organization depend on or are integrated with IT and IS. The future is often concerned with strategic thinking. Consequently, systems analysis has strategic significance. Most thinking and action on the future is based on planning.

Planning itself is problematical, but the assumptions underpinning development of a future with plans are more problematical to reconcile with actual situations – human problems. Planning assumes it is possible to predict the future in that the planned future can be realized. There are obvious limitations to human capability to predict. Another assumption in planning is that the actual conditions under which the plan is to be executed will remain constant, or as explicitly or implicitly assumed in the plan, to be successfully implemented. Actual conditions invariably do not comply with assumptions of the plan. So project managers have to continually adjust budgets, resources and timescales accordingly.

Some methodologists propose methodology as a form of inquiry or epistemology. For them a methodology is a method for investigating the problem domain to discover knowledge. The techniques and tools used for modelling act as instruments for investigating the problem domain and recording findings. Analysts should critically evaluate epistemological assumptions in systems ontology to understand their effect on knowledge and practice. Structured analysis and design affects organization by delivering IS based on objectivity. A new IS affects the very organization for which it is developed. The future and the present organization are intricately tied. This effect is most prominent in BPR and eCommerce.

Problem-solving

Both structured and object-oriented systems ontology assume that IS development problems can be addressed objectively. Consequently, they characterize problem-solving as reductionism – the decomposition of a problem into smaller, manageable elements. The decomposition of a problem leads to compartmentalization of organization and people's work, whereas organization is designed to function as a whole.

Objective systems ontology separates out people from the actual problem domain. Meanings that people attach to their actions are not modelled in structured or object-oriented system ontology. In structured systems ontology detached analysts carry out the identification of entities, and though entities can be people, as in customer or employee, the actual customer or employee is not consulted for their subjective perception. Similarly, in object-oriented analysis objects can be customers or employees but the actual people are not consulted for their views. The development of use case models in object-oriented analysis, though modelling the users, does not involve the actual people, or consider the meanings they attach to information.

Problem-solving in IS development largely ignores the social context. IS may be defined as a medium of social communication in organization. People rely on information to attach importance to their action. Trial and error type of problem-solving, also called 'exponential learning', is often used in social contexts. It is supported by prior formal analysis. Structured and object-oriented systems ontology does not recognize such social contexts and problem-solving strategies.

Problem-solving cannot be de-contextualized from politics. Project managers, analysts, programmers and users' roles depicted in structured systems ontology are simplistic. Stakeholder research in organizational initiatives and innovations reveals political factors in IS development. Stakeholders are individuals or groups who have a particular interest in what happens in the organization. Structured and object-oriented systems ontology do not explicitly recognize stakeholder politics in IS development. Critical theory assumes politics and conflict in systems ontology, but it is not recognized in dominant problem-solving strategies used in practice.

When the above issues are factored into IS development, objective problem-solving or planned action becomes problematical in practice. Structured and object-oriented systems ontology make assumptions about organization and people that need to be critically evaluated. In structured systems ontology terminology, 'users' are involved so that systems analysts can establish knowledge about the 'problem domain'. Users are needed to define clear and objective goals for a new IS. They assume that a clear communication channel is open between users and IS developers. Assumptions on users capability to be objective, have complete knowledge of their situations, and be able to articulate it are not tested.

Structured and object-oriented systems ontology assume that users know and can communicate details of the business problem. In structured systems ontology, users are encouraged to take an active interest in changing the current system and are assumed to be open to change. They are also assumed to be accessible and available to systems analysts for interviews and discussions on developed systems models. Peoples' interest, resistance or fear is not explicitly considered in structured or object-oriented systems modelling.

3.9.5 Information Systems and the application domain

System ontological knowledge lacks both understanding and formal knowledge on how to deal with complex application domains. Analysts need personal constructs and techniques to understand the application domain. Such knowledge will supplement analysts' technical knowledge, design relevant IS, and devise effective instruments.

A system-centric perspective on systems analysis and design fails to acknowledge organization and people – the application domain. Its systems ontology focuses on the 'system' to be developed, rather than the human, social, political and organizational factors. Consequently, its instruments are designed to model systemic knowledge only. Structured and object-oriented systems ontology are both system-centric because prescribed instruments focus on the functionality of a new IS.

Analysts need to consider personal constructs to account for socially constructed realities and evaluate whether structured and object-oriented systems ontology are capable of dealing with it. Structured and object-oriented analyses define IS in systemic, technical terms. They assumed that such definitions are not only technically *correct*, but presume that they are socially *relevant*. These assumptions are only internally validated in systems ontology because of the assumption of an objective reality. Technical definitions based on an assumed objective reality do not consider what an IS means to people and organizations from a social or individual perspective. There is no attempt at externally validating correctness or relevance.

A technical definition is different from a social perspective. Objective technical definitions do not admit a socially constructed reality. They do not allow that reality, or IS, can be constructed subjectively by people. In socially constructed realities, definitions of data, information and knowledge depend on what people make of them. Structured systems ontology does not allow for subjective meaning in IS. Object systems ontology implicitly is capable of enabling subjectively constructed realities, but methodologies like Coad's object-oriented method do not provide explicit instruments to develop systems models of socially constructed realities.

The systems ontology for both structured and object-oriented analyses does not sufficiently account for the multifarious application domain. Its assumptions need to be analytically evaluated. The term 'problem domain' itself indicates a system-centric perspective of IS. It is the label to describe individuals, groups, human collaboration and communication, and information, organizational knowledge, and much more, where IT is to be applied. A system-centric perspective neglects the relation between the computer-based IS and the human organization. Many of the limitations and problems with structured systems ontology arise because the rich application domain is not sufficiently reflected in the available systems analysis instruments.

The application domain is composed of many factors that make it difficult for analysts to deploy systemic instruments. Table 3.2 shows some of the constituents of the application domain. Each of these factors, and others, pose problems for system-centric analysis and design.

Understanding the relationship between a new IS and its environment is critical for its successful development. It is more critical for IS usage. The complex application domain depicted in Table 3.2 is not adequately factored into systems project management, the software development process, the project plan and its execution or the type of system required. Analysts should conduct an analytical assessment of why IS may lack relevance for 'users' in terms of systems ontology and its application in real situations.

People in organizations work and communicate with each other in a social context, in which information is exchanged to enable organizational tasks to be completed. Politics, power and informal dealings are characteristic of the work and communication. These factors are not accounted for in structured and

Table 3.2 Constituents of the application domain

Constituents of the application domain	*Description*	*Effect on Information System*
Individuals	People in the organization employed to work to achieve business objectives.	Individuals use IS to complete work tasks and organizational processes.
Groups	People in the organization who work together to achieve business objectives.	Groups use IS to coordinate collaborative work and to communicate information.
Business decision-making	Individuals or groups who make decisions concerning strategy or resources.	Demands on IS for information and knowledge to enable decision-making.
Business processes	Sets of activities that combine individuals, groups and IS designed to achieve business objectives.	IS transform, enable and support business processes.
Organization	Individuals, groups, goals and boundary within which business processes are designed to achieve business objectives.	There is a bi-directional relationship between organization and IS. Organization affects types of IS and their functionality and IS affect organization design.
Social context	Social, cultural and political factors in the organization. The human element of IS.	IS (system) is expected to function in the social context.
Business competition	The postures and moves of rival firms to gain larger share of the market.	New demands on information and knowledge, and IT support for business processes and radical strategic change.
Stakeholders	Individuals or groups with an interest in a new IS development project.	Powerful stakeholders can influence IS development. If neglected they can reduce its chances of success.
Systems analysts	Individuals and groups who investigate all the above to design IS that add value.	Analysts determine the scope and functionality of IS.
IT infrastructure (computer systems, internet)	Other IS that process data, information and knowledge.	Necessitate interfaces and interoperability with other IS.

objected-oriented systems ontology. Power, politics and informal dealings are not considered in the planning phase of systems project management, which assumes a sanitized problem domain. The application domain is composed of stakeholders who have an interest in a new IS. They range from high-ranking senior managers to departmental managers whose processes are affected by a new IS development.

An analyst is also a constituent of the application domain. The premise of analysts' rational behaviour in objective systems ontology becomes weaker when it is realized that systems analysts' have a powerful role relative to the role of potential users in IS development based on the SDLC and related methodologies. Their power stems from their technical knowledge. Potential users of IS are not capable of understanding technical knowledge and cannot enter into a meaningful dialogue, perhaps arising from conflict with analysts.

There is no attempt at formal modelling in structured systems analysis of the application domain as discussed above. It focuses on data and processes. In object-oriented analysis, modelling notation is available to model potential users of a new IS. Use case models are developed to show potential users and their functional interaction with the IS, but the social depth of such modelling is superficial.

3.10 Personal Critical Framework development

3.10.1 Personal constructs for systems

Activity A

Table 3.3 is a sample repertory grid for systems ontology. Reproduce the grid on a spreadsheet and add further columns and their polar opposites that you consider relevant. To objectify personal constructs in systems, complete the grid by following the details on how to use a repertory grid in section 1.10.1.

3.10.2 Planned and situated action

Questions

1 Discuss whether the SDLC can be characterized as planned action.
2 Evaluate the value that the situated action concept can add to systems analysis and design.

3.10.3 IS development

Questions

1 Appraise the role of standards in IS development. What practical relevance do standards have, as provided in SSADM for example?
2 Critically discuss whether adding 'structure' to the IS development process bridges the knowledge and communications gap between analysts and users.
3 Select a methodology and critically evaluate its future enhancement strategy – how it can be adapted for future needs. Do systems models based on the methodology provide for flexibility once implemented?

Table 3.3 Personal constructs for systems

Pole 1	*Method*	*Technique*	*Tools*	*Methodology*	*Problem domain*	*Problem-solving*	*Pole 2*
Formal							Informal
Boundary							No boundary
Control							No control
Structure							Unstructured
Problem							Solution
Practical							Unpractical

Activity A

- Consider the systems ontology and real world of human problems components of the Critical Framework.
- Identify an IS familiar to you and examine its systems ontology.
- Describe the actual situation in which it is used. What aspects of the actual situation are not reflected in the systems ontology?
- How is the effectiveness of the system affected by lacking factors you identified in the actual situation?

3.10.4 Conceptual basis of systems analysis and design

Questions

1 Discuss whether the assertion that structured techniques tend to lack context is valid. Detail reasons why you think they are prevalent in practice.
2 Evaluate the relevance for practitioners of alternative methodologies, for example ETHICS, that cater for the social context of IS.
3 Appraise the relevance of the concept of 'structure' in SSADM and 'object' in Coad's object-oriented methodology for real human problem situations.

Activity A

Work with your peers to compare the traditional, structured and object-orientation concepts for IS development. Discuss which approach you would include in your PCF. What assumptions of organization, people, and IS developers, including analysts, does the approach you select make?

3.10.5 Systems ontology

Question

You may want to read Chapter 9. Compare structured system and object-oriented techniques and tools and analyse how they:

- characterize the problem domain;
- differ on what they consider of interest in the problem domain;
- apply, effectively, the techniques;
- explain which set of techniques you would use.

Activity A

Structured systems ontology. In structured systems analysis and design the problem domain is assumed to be composed of business transactions data, dataflows, relations between data, and processes. These form the basic units of analysis. Either individually or in groups:

- Select a system that is familiar to you – a student records system or a mobile-phone billing system.
- Make a list of factors in the application domain of the system you have chosen that would not be accounted for by structured systems ontology.
- In developing systems models how would you cater for these factors?

Activity B

Object-oriented systems ontology. In object-oriented systems analysis and design, an analyst's task is to identify relevant objects, patterns, responsibilities and scenarios for a new IS. Objects can be transactions data, things and people for which data needs to be processed. Individually or in groups:

- Select a system that is familiar to you – a bank personal current account system or online shopping system.
- Make a list of factors in the application domain of the system that would not be accounted for by object-oriented analysis.
- How would you account for these factors in a class systems model?

3.10.6 Notation languages

For this sub-section you may want to refer to Chapters 7, 8 and 9.

Questions

1 Detail three critical flaws that you think exist in a structured notation language of your choice. How would you overcome these when developing systems models?
2 Detail three critical flaws that you think exist in an object-oriented notation language of your choice. How would you overcome these when developing systems models?

Activity A

Choose either an object-oriented or structured notation language you would use to conduct systems analysis. How did you decide which notation language to use? Make a list with two columns, one for aspects of the notation language that are sufficient for describing the problem domain and the other for aspects of the problem domain that the notation language does not cater. Identify and discuss the assumptions of the notation language in terms of its systems ontology.

Activity B

Individually or in groups, choose a structured or object-oriented notation language (you may use the selection from Activity A). Discuss among yourselves whether the notation language is sufficient to describe a problem domain. Make a list of its limitations and discuss how you would overcome them in practice.

Activity C

Notation languages are problem-solving devices. How important is it to enable refinement of the notation language for effective systems analysis and design? Individually or in groups, choose a structured or object-oriented notation language (you may use the selection from Activity A) and recommend at least three refinements to make it more effective in practice.

3.10.7 IS methodologies

Questions

Choose one or more from the selection below and discuss:

1 JSD is oriented towards software systems and not organization and people.
2 The author of JSD regards systems analysis and design as an extension of JSP.
3 JSD is highly structured which means that organization and people issues are not considered.
4 JSD does not consider project selection, cost justification, requirements analysis, project management, user interface and procedure design or user participation.

Activity A

Choose a methodology to include in your PCF. Make a list with two columns, one for aspects of the methodology that are sufficient for developing an IS and the other for aspects of the problem domain that the methodology does not cater. Discuss the systems ontology it assumes.

Activity B

A methodology may be regarded as a tool for gathering knowledge for IS development. Methodologies like JSD and Coad's object-oriented methodology assume objective knowledge. In groups, choose one methodology and discuss how it develops knowledge about IS and the problem domain that analysts can use to develop systems models. Discuss whether it is necessary to make the distinction between objective and subjective system knowledge.

3.10.8 Internet sources

Details of the object-oriented CASE tool Rational Rose can be found at http://www.rational.com.

Search the web for CASE tools for structured and object-oriented analyses. Choose CASE tools for systems analysis and justify your choice in terms of IS development, systems ontology, problem-solving and fit with methodologies. The Oracle site at http://www.oracle.com is a good source for information on CASE tools.

3.10.9 Further reading

For seminal work of socio-technical approaches to Information System development see: Mumford, E. (1983) *Designing Human Information Computer Systems for New Technology, The ETHICS Method*, Manchester: Manchester Business School.

A revised and updated edition is: Mumford, E. (1993) 'The Participation of User in IS Design: an Account of the Origin, Evolution, and Use of the ETHICS Method', in Douglas, S. and Namoka, A. (eds) *Participatory Design, Principles and Practice*, London: Lawrence Erlbaum Associates, pp. 257–270.

Eva, M. (1994) *SSADM Version 4: a User's Guide* (2nd edn), London: MacGrawHill.

For the original work on combining Soft System Methodology and the SDLC to account for the application domain see: Wood-Harper, A. T. and Avison, D. E. (1990) *Multiview: an Exploration in Information System Development*, Oxford: Blackwell Scientific.

Chapter 4

Systems project management

4.1 Learning outcomes

After completing this chapter you should be able to:

- Explain the key elements of a system project and the ways in which projects can be organized to relate to business objectives and strategy.
- Evaluate and select appropriate systems project management techniques for each phase of a project, justify their appropriateness, and apply them to practical situations.
- Evaluate and critique the limitations of traditional systems project management in the context of modern business organization and emerging approaches to IS development.
- Critique the major assumptions of systems project management and explain (a) why they may be inappropriate in the context of IS development and (b) how the state-of-the-art might be improved.
- Formulate a business case for a new IS development project.

These learning outcomes will provide knowledge of systems project management and enable you to be critical of system projects and project management. Practitioners develop an IS as systems project management in business contexts, in which systems analysis, design and systems analysts have a significant role.

4.2 Introduction

Systems project management is the systematic management of a process of technological and organizational change. It is central in IS development because it brings together the people, process and technology required for IS development. A system project has predetermined objectives, limited monetary budget, finite human resources and predetermined completion date. Consequently, project managers require broad knowledge in project management techniques and tools, the psychology of teams and organizational change management.

This chapter will cover the phases and basic techniques of project management. It will present knowledge of project management and discuss the issues and challenges that project managers encounter in practice. The management aspect covers planning, monitoring and controlling the process of software development. An understanding of traditional systems project management provides basic knowledge of the phases of project management. Whether projects should form the foundation of IS development will be critically considered in terms of the Critical Framework. Assumptions on planning that underpin systems project management will be critically considered to contribute to PCF development.

4.3 IS development project

A business project is a programme of business or organizational change management. It is a device used to innovate a product or service, or make significant change to organizational structure and processes. A project is used to organize and coordinate human effort to achieve specific objectives that the business thinks will result in a superior product or service to compete with other firms. The innovation and introduction of a new product to the market, the merger of two companies and restructuring a company are examples of business projects. Businesses often need to change strategic direction or operational organization, usually in response to markets or competitors. This involves large expenditure of time and human and monetary resources that need to be managed efficiently to ensure successful completion.

A system project encompasses the conceptualization, analysis, design, implementation and delivery of an IS. It assumes an objective reality and rational human behaviour. An IS is normally developed as a business-cum-software development project, drawing on business project management knowledge. Software developers use the business project concept, its techniques and tools, to develop IS because the process of developing software can be disciplined and standardized. System projects provide the discipline required to conceive and manage IS development within financial and temporal limits, and usually as part of a business change programme.

A system project is used to define and manage business and organizational change involving IT and IS. A system project has specific objectives that define the required system outcome. A timescale is determined in which the specific objectives need to be completed, this is known as the start date and end date of the project. Project managers, systems analysts and programmer resources are allocated to make the planned change happen. The elements and characteristics of a system project are:

- Setting of specific objectives to define a project.
- A specific timescale that defines start date and end date for a project.
- Specific and finite resources allocated to achieve project objectives.
- No significant prior experience of similar projects.
- Management of allocated resources to achieve project objectives.

A system project enables a business through disciplined management to complete major organizational change programmes involving IT and development of IS within allocated time and resource constraints. System projects are used to plan, monitor and control systematically the activities of analysts, programmers and others involved. Time and monetary resource constraints, including human resources, can be organized better and managed as a system project. System projects can be small with one to two people and lasting less than one person year, medium with four to ten people and lasting one to twenty person years, anything more than that is a large system project.

As a significant element of an IS is software, its quality determines the efficiency and effectiveness of the delivered IS. System projects are used to meet and improve quality standards. Software needs to be of a certain standard to ensure it is fit for purpose and meets the requirements of its users. Standards are normally set by formal bodies to ensure superior quality is achieved. Examples of standards and standards bodies for software development are Constructive Cost Model (CoCoMo), Capability Maturity Model (**CMM**) and SPICE. These standards are designed to improve the software development process and consequently the resultant software system.

Systems project managers need sophisticated technical and management skills. Software development skills need to be complemented with knowledge and the skilful practice of project management. They need to put into practice interpersonal and communication skills in the context of organization, organization culture, stakeholders and politics to realize project objectives successfully. Systems project managers are required to produce systematic and detailed plans of how software will be produced. They are responsible for monitoring the successful completion of planned activities, and taking appropriate action to control any deviances from the project plan. These are management activities in the project and constitute a significant part of project managers' work involving the allocation of financial and human resources.

4.3.1 Plans and projects

In systems project management, planning is the formal basis for decision-making, organizing and allocating resources. The resources used in IS development are systematically planned and allocated to predetermined tasks or work packages designed to achieve project objectives. Techniques are available to plan the work to be done, estimate and schedule budgets and work, monitor and control the work in progress, assess and manage risk and ensure quality standards are met.

Objective systems ontology, for example the SDLC and SSADM, has a natural synergy with projects and rational planning. The phases of the SDLC or SSADM are executed within the bounds of a project's time and resources limitations. Project management techniques and tools enable actual planning of tasks to determine software requirements and development. Start-to-end or end-to-end work planning techniques are used to organize and manage available resources.

4.4 The business case

Monetary investment in IT and IS is determined on the basis of the value it will add to a company. A proposal for a new IS is normally assessed as a **business case**, focusing on its contribution to achieving business strategy and objectives. Senior management are interested in determining benefits of investing. Some considerations include reducing operating costs, improving quality of goods or services, better customer relationship management and improving information and knowledge for operational and decision-making purposes. These considerations and others are shown in Table 4.1.

In the SDLC a business case for an IS is made in the feasibility phase. Analysts are involved in identifying and selecting IS to develop. They have a significant role in formulating a business case and work with project managers to identify business areas that would benefit from the application of IT. They determine strategies

Table 4.1 Drivers for IT/IS investment

Reasons for IT/IS investment
Reduce operating costs
Improve operational efficiency and productivity
Modernize or innovate operations
Improve competitiveness
Provide superior quality information for operational and executive decision making
Introduce a new business model, for example the internet and eCommerce
Improve knowledge management

and frameworks for identifying candidate system projects and evaluate them by selecting appropriate techniques.

4.4.1 Types of Information Systems

The type of system project is determined by the type of IS to be developed. Analysts work on different types of IS development projects. The types are based on what they contribute to improvements in managerial decisions, automation of operations and procedures, or definition of new business models. The types in business organizations include:

- eCommerce systems;
- Knowledge Management systems;
- Enterprise Resource Planning systems;
- Supply Chain Management systems;
- Customer Relationship Management systems.
- Decisions Support systems (including Executive IS);
- Accounting and Financial Management systems;
- Expert systems.

Analysts involvement in formulating a business case will normally depend on particular types of IS. They may work alone or in groups to make a business case for decision support systems or functional accounting systems. They would normally work with managers and senior executives to make a business case for customer relationship management systems or knowledge management systems.

eCommerce systems require a special business case, which is normally supported with a business model. Business analysts, executives and analysts work together to define a business model, often radically different from existing operations and business processes. Business analysts contribute knowledge of business and operational issues and systems analysts contribute knowledge of IT to the business modelling process.

4.5 Project management

Project management is the planning and organization of resources to achieve project objectives. A typical system project requires the management of scope, timescales and resources. They are related because an increase in work – scope – necessitates more resources and time. Project managers plan and organize resources and work into distinct phases. Each phase makes a particular contribution to the completion of a project.

4.5.1 Prime project roles and responsibilities

A project has roles and responsibilities, examples are shown in Table 4.2. A project manager is responsible for all the resources available to the project, including human and financial resources, and works closely with senior systems analysts and senior systems designers to determine allocation of budget, systems analysts, designers and software programmers.

The role of users and stakeholders is significant because it affects the success of a new IS. They are not responsible for determining system requirements, but for analysts they are the prime source for gathering system requirements. Their representatives on a system project form part of the user groups that work closely with the project manager and analysts.

Project managers need a variety of skills. The ability to plan, monitor and control a project are basic necessary skills. Beyond basic skills, a project manager needs to be aware of the context of the project and how people and events in it affect the project. For example, a contextual factor like additional system requirements will have an impact on the available resources. A skilled project manager develops an awareness of how the context affects a project and is capable of negotiating and managing resources accordingly.

Additional skills are required dependent on the uniqueness or originality of the project, its size, complexity and how it relates to the business and the market in which the business operates. An IS that has not been developed before will necessarily be more difficult to develop because it is not possible to rely on prior knowledge. A large IS compared with a small one will be more difficult to develop because the planned change will be more widespread, and it will require the skilful management of greater resources. The type of IS will determine how complex it is to develop. IS that are integrated into the business will require additional systems design features. Each of these factors also has a significant impact on whether a project is regarded a success or a failure.

Table 4.2 System project roles and responsibilities

Project role	*Responsibility*
Sponsor or project champion	The person or group who want the project to proceed. They provide the project aims and objectives, and organize other necessary organizational support.
Project manager	The person or group responsible for managing the available project resources.
Potential users and stakeholders	People who will make use of the developed system. They will be involved in determining what the system is required to do and in testing modules of the system.
Senior systems analyst	An experienced systems analyst who works closely with the project manager to ensure the systems analysis work is completed to the required quality standard.
Senior systems designer	An experienced software designer who works closely with the project manager and senior systems analyst to ensure that systems designs are implemented.

4.5.2 Project stakeholders and stakeholder analysis

A stakeholder is any person, group or organization, internal or external to the organization, that has an interest in the IS to be developed. Example stakeholders are the project sponsor, banks providing capital, suppliers and contract organizations, the client organization, analysis and design groups and individuals.

Obvious stakeholders can be identified relatively easily depending on the type of IS to be developed. A project sponsor or groups whose work will be affected by the system are obvious stakeholders. For more complex IS other stakeholders may not be so obvious. A project manager needs to conduct careful and detailed stakeholder analysis to ensure success. Stakeholder analysis is particularly significant for projects that are important for achieving business strategy and objectives. It involves the identification of stakeholders and stakeholder mapping.

To identify stakeholders, who may be critical for the success of a project, a project manager needs to conduct a formal stakeholder analysis. It will reveal the relative importance, influence, power and impact of stakeholders. This will involve examining values, beliefs and assumptions held by various stakeholders and how they attempt to influence each other to determine project outcomes.

Formal matrices for stakeholder analysis are available. Some of these metrics measure variables such as the power, predictability or interest of stakeholders. A project manager would use the metrics to understand key stakeholders and assess their importance for the success of a project. This can help project managers to determine what management approach to deploy and decide where to focus political effort.

People whose work will be affected by a new IS will be interested in knowing how their jobs will change. They may be fearful that a new IS will make their jobs redundant or that job role may be made trivial or uninteresting. Managers of departments or sections will be interested in knowing how a new IS will help them achieve departmental objectives. They will want to know whether it will produce efficiency gains and cost reductions.

A project manager and other IS developers are also stakeholders. An unsuccessful project will mar a project manager's record, so project managers will be interested in achieving successful completion of a project. Software programmers will be interested in developing working computer programs from given system requirements.

Systems analysts themselves will be interested in developing relevant systems models that meet the approval of managers and people whose work will be affected. Analysts work with different project stakeholders on various aspects of a project. They work closely with managers and people whose work will be affected by a new IS to elicit requirements. They communicate developed systems models to software programmers.

4.5.3 Human factors and teams

Human factors in the project team are the project manager, senior analysts, senior designers, software programmers, systems analysts, and systems designers and other technical and managerial people. (For their roles and responsibilities see section 4.5.1.) Human factors include relationships with project sponsors and champions and their psychological characteristics. These aspects are important because they determine how individuals work in a project team and the effectiveness of teams. Appropriate individual and group or social psychological characteristics are important for an effective team. Human factors encompass leadership, motivation, and teams.

Leadership

A project manager's role as manager and leader is critical in determining success. A project manager has to manage both the resources available and lead a project team to achieve project objectives. Management involves planning, monitoring and controlling of resources, but leadership requires motivating and inspiring team members to want to complete a project successfully. The ability to influence team members and stakeholders is a critical skill. Combined with competence in management, leadership ability and political skills are factors that influence others to respect the project manager and commit themselves to success.

The action-centred leadership model and the situational model of leadership are formal models of leadership. The action-centred model proposes that a leader's effectiveness in a team depends on three factors: the need to achieve the task, the need for team maintenance, and the individual needs of team members. Rather than make the distinction between management and leadership, the action-centred model focuses on the interrelationships between the three functions of task, team coherence, and individual needs, and the project manager's ability to satisfy them.

The situational model of leadership focuses on the tasks to be completed and relationships to be formed and maintained. It is a behavioural model that integrates well the management and leadership behaviour of a project manager. Task behaviour is concerned with how well a project manager (the leader) is able to set goals and define roles for team members, and provide direction to individuals and groups. Direction is critical for success. Relationship behaviour is concerned with how well the leader communicates, listens and provides support for the team and individual members. Encouraging individuals and the team is a significant aspect of relationship behaviour. The model defines four levels of individual and team development based on the permutations of individual team members' competence and willingness to complete tasks. These permutations affect the kind of leadership styles that the leader would need to deploy, spanning over the parent, coach, developer and driver types of management.

Motivation

Motivating individual team members and the team is a significant issue in systems project management. A project manager would normally be aware of motivational issues and attempt to address deficiencies in individuals' and team motivation. An effective leader with effective management skills will be aware of how to recognize motivation levels and take action to raise motivation.

Theories of motivation explain how willing an individual or a group is to increase their effort to take part in and complete a task. They relate to organization and how they lead to improve the performance of individuals and groups. They focus on how such involvement satisfies particular needs of individuals and teams. A distinction is made between motives and needs in the theories, both of which are internal to an individual. An individual's motives affect their needs and the actions they take to achieve them.

Theories of motivation explain the level of motivation in terms of content, process or reinforcement. Maslow's hierarchy of needs and Hertzberg's dual-factor theory are content theories. They take the individual as the subject and focus on the relationship between individual needs and the reward individuals get for doing work. Their basic premise is that individuals generate needs that increase their motivation to satisfy the needs. In system project terms, a project manager then provides monetary or in-kind rewards or offers challenging tasks that either increase or decrease job satisfaction, which in turn is a motivational factor.

Process theories are concerned with how the process of motivation can be developed. It can be based on either individual's consideration of equitable return for the effort they put into work, or a consideration of the expected return for effort put into work. Equity theories are based on how individuals compare their work and reward with others. The comparison of work and the return they get determines their level of motivation based on how equitable they perceive the reward. In theoretical terms, individuals generate a ratio of work done and reward received and compare this ratio with others' ratio. An equitable ratio will lead to job satisfaction and increased motivation and an inequitable ratio will lead to dissatisfaction and demotivation.

The expectancy theory of work is based on the premise that individuals make informed and rational decisions about how much work effort they should make. The decision to do work is based on whether the individual possesses the ability to do the task and whether the expected effort will be noticed and rewarded or not.

While no single theory of motivation will suffice there are likely to be elements of truth in each one. A project manager can pick elements to design job roles, work and reward structures that lead to increased motivation in individual team members and the team.

Project team

Project teams consist of project managers, systems analysts, business analysts, software programmers, database administrators and systems professionals, and representatives of user groups. A system project is normally regarded as work done in a team. Individual members of a team need to be able to work with each other and individually. Working in a team requires interpersonal and communication skills. An individual's self perception is an important aspect of team and project success.

There are various metrics that can be used to assess the predisposition of an individual and their ability to work in teams. A questionnaire designed by Belbin is often used by project managers to determine characteristics of individual project members. It focuses on how an individual would describe himself or herself in team situations and provides a self-perception inventory analysis sheet that records individual self-perception in team situations. It also classifies the types of individual who could be part of a team according to the potential contributions. These types are:

- coordinator (chair)
- shaper
- innovator (plant)
- monitor/evaluator
- resource investigator
- implementer (company worker)
- team worker
- completer/finisher
- specialist.

A project manager may not rely solely on such an assessment, but it does help to provide one measure to base decisions on whether to include an individual in a project team.

4.5.4 Techniques and tools

A project manager relies on planning techniques and tools to determine what work needs to be done, when it should be done by and what resources will be required to do it. They enable a project manager to conduct an analysis of the work to be done and an analysis of the nature of the product that will result from a project. Objectifying the work to be done and the product are both critically important for success. Some examples of project planning techniques available to project managers are listed in Table 4.3.

The tools to support the techniques include software packages for project planning. Software support tools are available for planning precedence networks and activity-on-arrow networks, though the former are more popular and widely available. Some project planning tool examples are WBS Chart Pro for planning work breakdown structure and PERT Chart Expert.

4.5.5 Quality and risk management

Quality and risk management are necessary and important. Rather than being discrete project management activity, they are a continuous activity with overall responsibility resting with a project manager. The standard of IS developed and the process of software development are incorporated in quality assurance. Quality assurance can be based on one of many approaches and frameworks available.

Quality may be regarded as an absolute or relative measure in software system. Unlike certain consumer products, the quality of IS is difficult to define in absolute terms. A software system is considered to be of quality if it conforms to client's requirements. This measure of quality is problematic because it assumes that client requirements can be completely captured. There are problems with establishing requirements that the conformance measure of quality does not address. The concept of 'client' is vague as various stakeholders within an organization can be said to be clients too. The determination of actual requirements is also difficult in practice because clients often add new requirements.

The definition of quality is important because the agreed measure will determine

Table 4.3 Project planning techniques

Technique	*Description*
Work breakdown structure	A project planning technique to define work packages required to complete a project. Also known as the means necessary to achieve project objectives.
Product breakdown structure	A project planning technique to define the results of work done. Also known as the results or ends.
Precedence networks	Precedence networks are used to schedule work packages as activities and resources. The technique is used to manage the limited time available on a project.
Activity-on-arrow networks	Activity-on-arrow networks can also be used to schedule work packages. They depict the work packages in a different form to precedence networks.
PERT	Program evaluation and review technique (PERT) produces charts that depict task, duration and dependency information. A network is produced to show connecting nodes, lines and floats.
Gantt Chart	Gantt Charts are a visual graphical form for depicting work packages against a timescale. They show the sequence of the work packages, important milestones and floats.

whether a project is deemed successful or not. A project manager needs to be aware of differing measures. Quality defined in absolute terms, such as executable software performance or percentage downtime, will be easier to quantify and measure. However, the subjective element of IS is often a significant block in regarding it as successful or as a quality product. Clients may raise objections concerning the 'quality of information' or lack of its timeliness. Such subjective objections make it difficult to identify and measure quality in IS.

There are approaches for improving the software development process. Though CoCoMo is a technique for estimating resources required to complete work packages and cost estimation, it may also be used to improve quality of the software development process. CMM is specifically designed to improve the software process. The Software Engineering Institute (SEI) at Carnegie Mellon University originated the model with the aim of improving the software process in organizations. It is shown in Figure 4.2 section 4.6.1 with its five stages of software process improvement.

Risk management

Risk is the gap between what is the expected result of effort and the actual outcome in the future. Undesirable events that may happen in a system project which require action to be taken to prevent them from happening. Risk management is the management of such uncertainty in a system project. It is effort spent to avoid undesirable outcomes. Uncertainties need to be identified and managed to mitigate their impact on the successful completion of a project. Risk management is associated with the idea of contingency measures or planning for unforeseen or unpredictable events. Contingency planning is an integrated aspect of project planning and is derived from the main planning effort.

Critically, work breakdown structure and estimation can be used to identify and manage high-risk parts of a system project. An example of risk is unrealistic or poor estimation of resources and time required to complete work packages. The judgement on risk is based on measurable quantities and the level of confidence a project manager has of them. If confidence is low then there is a high risk, if it is high then there is low risk.

A risk management plan is drawn by identifying risks, assessing their impact on the project and its objectives, planning responses and taking action to reduce the risk from occurring. The identification of risk results in compiling a risk register for the project that shows the categories of risks. The risk register contains all the identified risks. The project objectives are compared with identified risks in the risk register to assess impact and deploy appropriate response.

A project manager needs to be able to judge the likelihood of a particular risk occurring. It is the ability to judge occurrence of risk and its impact on project objectives that is important in risk management. This assessment is then used to draw the risk management plan that contains the necessary action required to reduce the impact of the risk on a project. The risk management plan will contain action to either avoid risk or mitigate risk. For some types of risk, action can be taken to avoid it. For example, a known shortage of programming skills can be avoided by hiring the right people. Other types of risk cannot be avoided, so action is taken to mitigate its impact on the project objectives. For example, a particular systems analysis work package overrunning its allocated time.

The actual assessment of risk can be based on a basic scale, or more sophisticated and comprehensive methods. At a basic level the occurrence of risk may be assessed as simple categories of 'high', 'medium' or 'low'. The consequent impact would be 'small', 'moderate', or 'large'. This kind of measure is subjective and open to wide interpretation. Detailed metrics are more comprehensive and provide a more objective measure, including probabilities.

4.5.6 Work and product breakdown structures

A system project needs to be managed to ensure efficient and effective use of limited available resources. A project manager has to determine system requirements and organize human resources and work. The relation between different packages of work needs to be determined and scheduled. This is achieved by using the work breakdown and product breakdown planning techniques.

Work and product breakdown structures are analyses techniques to plan, monitor and control work. Breakdown structures enable detailed definition of necessary work to be done and its management in well-defined packages. In terms of resource allocation, breakdown structures enable more accurate estimation of resources required to complete work packages.

Work breakdown structure and product breakdown structure are related in terms of means–ends analysis. Work breakdown structure is an analysis of the work required or the means, and product breakdown structure is an analysis of the results or the ends. Work breakdown structure enables a project manager to define what work needs to be done and determine how the required work packages are related. Product breakdown structure enables a project manager to determine what tangible products will result from the work packages.

The relation between work packages resulting from work breakdown structure is termed dependencies. Work packages can be logically linked to determine whether a particular work package is dependent on one or more other work package. For example, during systems analysis system requirements specification depends on requirements determination and cannot begin until requirements determination is completed. When one activity can only begin when another is completed a dependency exists. There are various types of dependencies:

End-to-start exists when one activity cannot begin until one or more other activities are completed. For example, process modelling in structured systems analysis cannot begin until data modelling is completed.

End-to-end exists when one or more activities can overlap. For example, in object-oriented systems analysis investigation of classes for systems design, user interfaces for instance, can begin before systems analysis is completed. Start-to-start exists when one or more activities can overlap. The same example for end-to-start applies. Start-to-end completes the logical possibilities but is rarely evident in practice.

Project networks

Work packages, dependencies and schedule of time and work can be depicted graphically in project networks. Two examples of project networks are precedence networks (also called activity-on-node) and activity-on-arrow networks. Precedence networks are popular in software tools for planning project networks. They are also simpler than activity-on-arrow networks. An example precedence network is shown in Figure 4.1.

The precedence network is an example of requirements analysis. It shows interviews, document analysis, and current IS analysis. It also shows the development of mock-up user interfaces to help 'users' visualize the envisaged IS. Analysts would show the prototypes to users and record comments they make. These would then be used when consolidating the gathered requirements before writing the requirements

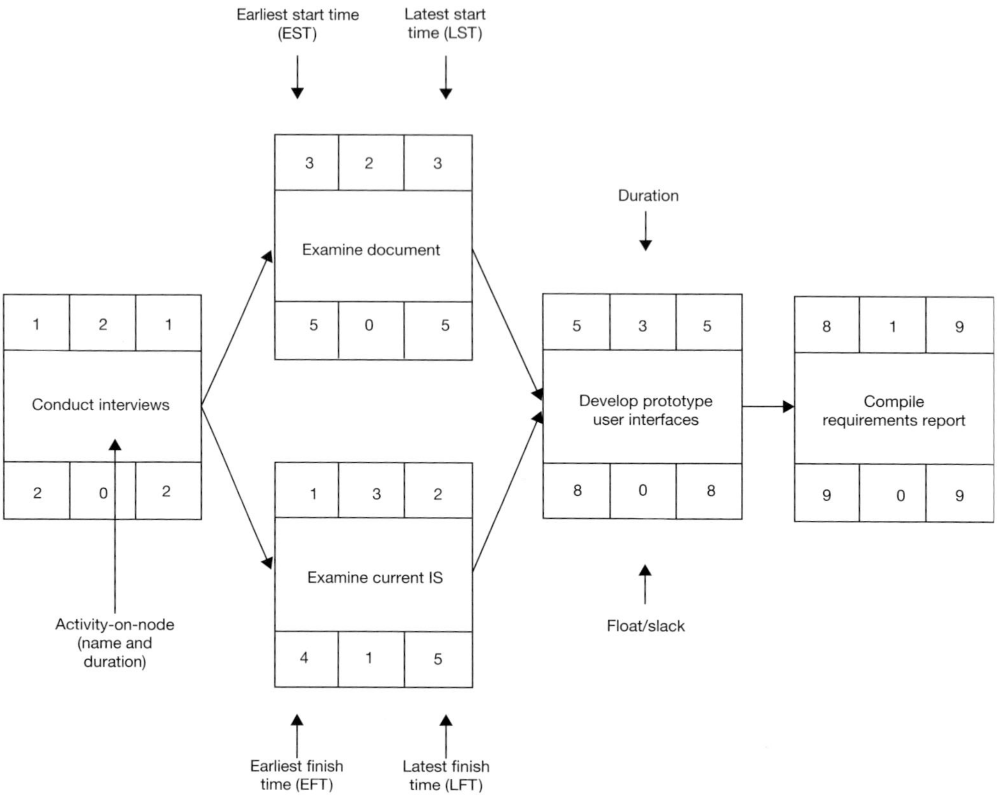

Figure 4.1 The precedence network

report. The precedence network shows several important elements for drawing a project network. These elements are listed and described in Table 4.4.

4.5.7 Estimation and schedules

A pragmatic management activity is to estimate the time and resources and to schedule the work required to meet project objectives. A project manager uses the time and resource estimates and work schedules to plan the project activities using project networks. Estimation is used to derive resource and time parameters for the project, often called 'baseline' estimates. These estimates are useful for developing work breakdown structure, ensuring quality standards, monitoring and controlling a project. Estimation is done using generic human-based or non-human techniques or techniques that have been developed specially for systems project management.

The Delphi method for estimation is based on human judgement. It is composed of expert humans in their particular sub-fields like project management, systems analysis, or programming who are brought together to form a panel. Panellists need to be open to sharing their experiences and willing to reach a consensus of opinion. A project manager would set the agenda to avoid a particular panellist, someone with higher status or more success, from dominating the proceedings. The panellists go through various rounds making judgements on

Table 4.4 Elements of a precedence network

Element	*Description*
Activities	An activity is a work package derived from analysis of work breakdown. In systems analysis, examples are user case analysis or user interface design.
Earliest start time (EST)	This the earliest time that a particular activity can begin. For example, user interface design may begin before other systems analysis tasks are completed.
Duration (D)	This is the time duration of a particular scheduled activity.
Earliest finish time (EFT)	This is the earliest time that a particular activity can finish. For example, physical database design cannot finish until requirements specification is complete.
Latest start time (LST)	This is the latest time that a particular activity can start. For example, examining current IS can be delayed but must finish before the next activity on the critical path.
Float/slack (F)	This is the spare time available before a particular activity must start. Often this time is useful when an activity overruns.
Latest finish time (LFT)	This is the latest time that a particular activity must finish. For example, in structured analysis the feasibility study must finish before systems analysis can begin, or testing must finish before compiling the final report.
Critical path	The set of activities that has the longest time path is the critical path. The activities on the critical path must all be completed on time for a project to be successfully completed.

particular project objectives and resources required to complete them. They would make written responses to such issues. These responses would then be used for making actual estimation decisions.

Other estimation techniques based on human judgement include: the analogy method, analysis effort method, programming method, and direct estimation. The analogy method is used when the experiences of a previous similar project are available to a project manager or organization. The previous project is used as an analogy for current project to determine estimates of required time and resources.

Non-human estimation techniques are generally called **estimation models**. They involve meticulous calculation using statistical techniques. CoCoMo and Function Point Analysis are examples of estimation models. In Function Point Analysis the aim is to measure the size of a computer program based on the number and complexity of inputs, outputs, queries, files and program interfaces. The number and complexity of each component of the system is recorded on a predefined worksheet, these are then used to calculate the Total Unadjusted Function Points (TUFP). Data entry screens, reports or databases are examples of a component. A project manager makes a subjective assessment of the complexity of each of these components. The TUFP is then assessed according to 14 factors that have an impact on complexity to determine the Adjusted Project Complexity (PCA). Examples of these factors are reusability, data communications and end-user efficiency. The TUFP value is then multiplied by the PCA value to arrive at the Total Adjusted Function Points (TAFP).

In the CoCoMo estimation model the aim is to measure the effort required to complete project objectives by converting the lines-of-code effort into a person–month estimate. Measuring effort requires knowledge of the size of the project and the software production rates of project staff. The formula to calculate the person–month effort is:

Effort (in person–month) = 1.4 * thousands of lines of code

For example, for a project with 20,000 lines of code the project would take 28 person–months to complete. This kind of simple model in practice will be complicated by other factors like the complexity of the algorithms, experiences of programmers and the type of software being developed. Source lines of codes can be replaced with function points at this stage.

The complexity of the software and experience of the programmers are examples of contingency that need to be incorporated in project plans. The actual complexity of software may be more than its estimated complexity. A project manager needs to think ahead and ensure that appropriately qualified and experienced development staff is included in the project.

4.5.8 Checking progress

Once a project begins, its actual management requires monitoring and controlling the project plan, with its associated work and product breakdown, estimates and schedules. A project manager has to check progress and make adjustments to time and work allocation where required. The project plan when implemented will require adjustments and actual resource and time allocations to be revised.

There are various techniques for checking progress and taking action. Project evaluation techniques focus on gathering quantitative data to monitor progress and are categorized in terms of backward-looking and forward-looking. They provide quantitative data to compare with baselines estimates. Backward-looking is a comparison of forecast or baseline costs with actual costs incurred. This focus on costs alone fails to measure how much work is actually completed. Forward-looking monitoring overcomes this problem by measuring both costs and the work done. This requires determining budgeted cost of work scheduled (BCWS), actual cost of work performed (ACWP) and budgeted cost of work performed (BCWP). These quantities enable comparison of scheduled cost of work with actual cost of work or other combinations to determine variances.

A project manager combines monitoring with control to ensure the attainment of project objectives. Monitoring provides information on variances from the project plan and control is used to take action to ensure the project plan is successful. Project managers have a choice of action they may take to check variances, which includes aborting if the project looses credibility with the sponsor and influential stakeholders. The other actions used in practice include reallocating resources to recover the project or compromise cost and time. If some project objectives can be jettisoned without compromising main required system functionality then the scope is redefined. There may be situations where taking no action is viable, where the actual cost or work variance has no significant impact on other activities.

4.6 Planned action

Planning is a significant activity in project management. Planning activities in the previous section and the consequent action required to implement it is characterized here as as planned action. This kind of planned action is reflected in many project management methods. PRINCE is one example. Planned action is crystallized in the CMM.

4.6.1 Capability Maturity Model for software

CMM is a process or process-centric model of software development, illustrated in Figure 4.2. It was conceived to improve the production of software through carefully planned activities with well-defined inputs and outputs. It is a process method of systems project management that is defined, managed and repeatable.

CMM depicts software development as a planned process which can be progressively improved. The various improved processes are defined as levels of development of the software process. Software houses and organizations may seek accreditation from CMM and be classified according to the level their particular process has reached. The process levels are:

- continuously improving process;
- predictable process;
- standard consistent process;
- disciplined process.

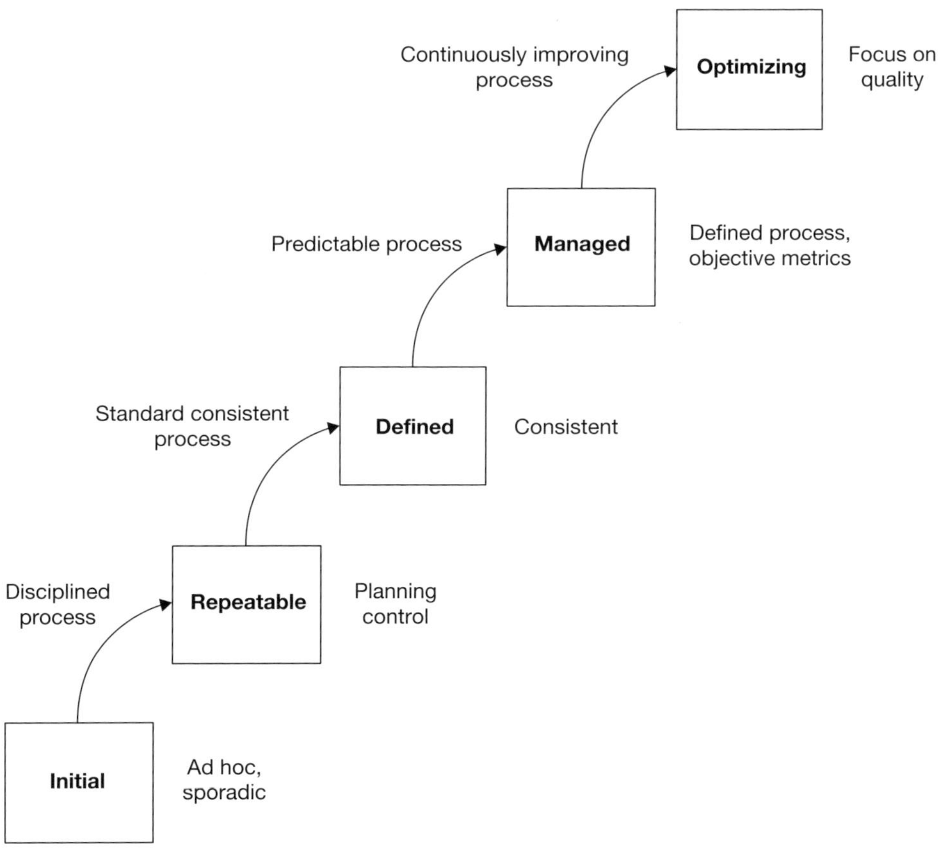

Figure 4.2 Capability Maturity Model for software

The lowest process is the disciplined process and the highest is the continuously improving process. These processes are associated with a particular hierarchy of stages or levels.

Initial level is where an organization has no clear and objective process and no effort is made to have precise work schedules and costing. The process is ad-hoc and even sporadic.

Repeatable level is reached when the organization has a disciplined process. The process is characterized as having management control, planned work schedules and costing, and careful control of changes.

Defined level is reached when the software process forms the basis for developing IS. The software process is defined and consistently implemented.

Managed level is reached when the defined process is improved by adding process metrics to measure the quality of the process. Systems project management is conducted on the basis of identified process elements and their careful analyses to inform decisions that lead to improving the quality of both the product and the process.

Optimizing level is reached when the process is developed to continuously improve and optimums are continuously sought. The optimum process is characterized with innovation in the software development process that leads to improvements in quality and productivity.

Planning is a critical feature in CMM and the highest optimizing level requires meticulous planning. CMM is important for systems project management because it sets a standard for others. In practice though, few organizations seek CMM accreditation because of the stringent process improvements required. Though the impact of CMM on systems project management is significant, its actual implementation by organizations is sparse.

4.7 Systems analysis and design

Significant components of a system project are systems analysis and systems design. They are work packages in terms of work breakdown and product breakdown structures. A project manager, senior analysts and designers, would need to describe them in detail for analysts and designers. Analysis and design activities and outcomes are identified in the work breakdown structure and product breakdown structure. As systems models produced by analysts form the basis for software programming and implementation work packages, dependencies would need to be carefully plotted on project networks.

Senior systems analysts and systems designers each manage a team of systems analysts and designers. The senior systems analyst organizes the team to conduct interviews with stakeholders and employees, analyse relevant business documents, and shadow relevant people. These activities are to gather information on the functions required of a new IS. The senior systems designer and the design team use the system requirements to produce systems designs and a system specification.

Structured systems analysis and design complements systems project management. It is used to manage effectively the software process by identifying roles, work and modelling techniques to construct systems models. Systems project management is similarly structured, as reflected in planning techniques and activities.

4.8 Project management and the Critical Framework

The Critical Framework provides a critical orientation for considering the role of systems project management in IS development and the role of systems analysis and design within system projects. It can be the basis of reasoned understanding of the uses and limits of established systems project management knowledge and the need to keep an open perspective on practice. There are various theories underpinning practice in systems project management, some are conflicting. Practitioners need to focus on the pragmatic resolution of problems that arise in the situation while benefiting from theoretical knowledge.

Figure 4.3 is the Critical Framework populated with critical reflection on systems project management and the practice of systems analysis and design within it. As the bottom layer shows, many questions concerning the four themes of criticality arise from the assumption of planned action, rational human behaviour and stable system project environment.

A critical analysis of the role and assumptions of systems project management will contribute to improving a PCF. Further enhancement of a PCF will result from a critical consideration of alternative conceptions of IS projects, techniques and measures of success. Overall, critical reflection will enable practice to be improved by considering how alternatives can be used and how they can be enhanced.

4.8.1 Introduction

The conceptual basis of systems project management and its instruments need to be analytically evaluated because the underpinning philosophical and conceptual thinking poses practical difficulties. Developing an IS using the SDLC within a project assumes formalism, an objective epistemology and rationalism. These lead to the expectation that:

1 System requirements for a new IS can be established (predicted and remain constant).
2 Formalism and formal instruments can be used to frame and solve an IS problem.
3 Project plans can be developed and effectively implemented.

These assumptions underpin systems project formalism. Systems project management in turn assume that universalistic principles operate, and that attainment of an 'optimal' is possible in every case, for example, as in CMM. The assumption that universalistic principles can be applied in all cases is based on the objectivist epistemology, which in turn leads to standardization and transfer of knowledge. Combined with rationalism and economics, it is assumed that economies of scale and specialization are possible.

Systems project management is based on the discipline of project management reflected in planned or projected activities, but there are differences between the two that a project manager needs to appreciate. The differences concern the nature of software and IS development. Unlike a physical product or other business project, the development of social software – for human informational and knowledge use – does not succumb to planned action.

An allied problem is the conceptualization of IS development as a distinct project rather than a business operational activity. This is a significant problem since the flow of information is part and parcel of organized activity. Its separation into a systems development project assumes that organized activity will be stable during the development process. Actual practice is quite different, where change is the norm. Such change affects project scope and estimates, which need to be revised to reflect actual conditions.

Managing risk requires understanding the parameters of risk, namely constraints and uncertainty in a project. Constraints may include human resources or time limits. Uncertainty can arise internally in the organization and project or externally from competitors or market forces. A prime problem in managing risk is the assumption that constraints and uncertainty can be identified and understood to enable risk management. The available techniques for risk management seem plausible, but are arguably only guesstimates.

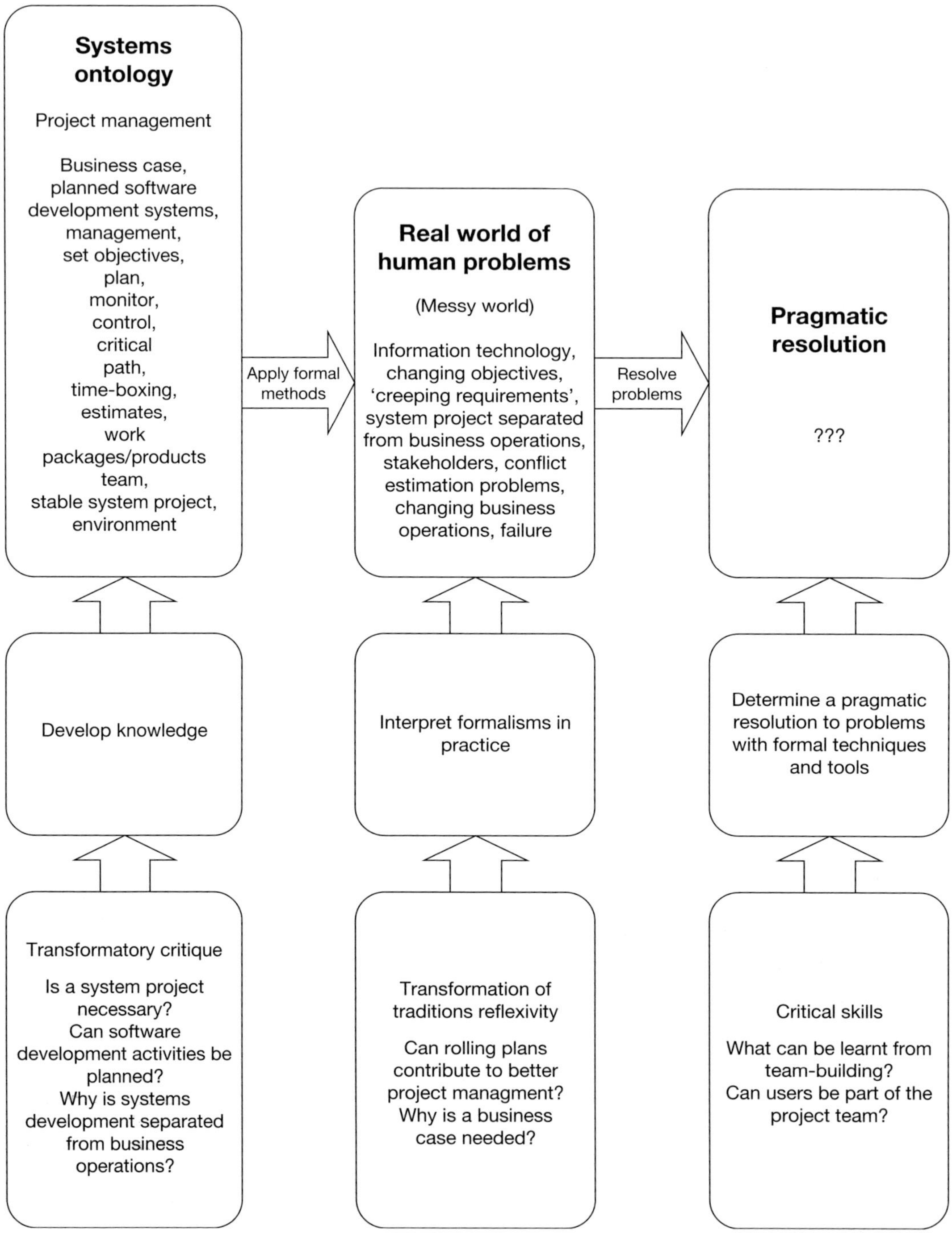

Figure 4.3 Critical framework: systems project management

There is an operational barrier between planning and the implementation of a plan. Associated with this is the idea of allocating time periods to planned activities or **time boxing**. Planning itself is directed by business strategy and can be the product of rational processes, but the actual plan in practice is difficult to implement as intended. The actual situation often does not reflect the plan and timed activities are often not realized as required.

4.8.2 Problems with rationalism and planned action

The systems ontology component of the Critical Framework in Figure 4.3 has the subtitle 'project management'. The basis of systems project management is rational thinking. The rational presumption is clear. A project is an entity that has specific and clear objectives, a beginning, duration and an end, and it is required to produce an explicit prescribed outcome. It is broken down into work packages that are known in advance, predictable and assumed to be controllable. It is assumed they can be contained within available resources. The risks associated with the project can be predicted, measured and managed, and avoided or mitigated. Above all, a project can be optimized.

IS development is normally regarded as a special planned project. The basic premise of the planning is that humans behave rationally or economically. IS projects are selected rationally for development. The rational selection assumes objectivity that is free of bias or prejudice and is normally based on financial methods such as cost-benefit analysis.

IS development in actual situations does not reflect planned action – the human problem component.

A business case is made in terms of cost savings, efficiency and production benefit. This kind of justification though has proved to be difficult to measure and evaluate, as information economists have found no specific correlation between investment in IT and business gain. The claims that the use of computers and IT is integral to business are even more difficult to quantify. Nevertheless IT and IS are significant in modern organization.

Often organizational change programmes are unique with no prior experience in an organization to manage a project. An IS project is unique in the sense that no exact same previous experience can be called upon. So even experienced project managers have to be open to learning. While they can draw on previous experience of using project management techniques, they will not have relevant experience of the particular current project. Ironically, no prior experience of similar projects reinforces the need to plan to prevent costly mistakes.

Consideration of complexity, change and the lack of requisite knowledge for the development of effective plans raise the issue of the tension between planned action and situated action. Project managers invariably need to take corrective action to ensure that the work and resources are being done according to the project plan. This type of action-in-the-situation undermines assumptions of rationality and objectivity in planned action. The situation and situated action rather than plans seem to dictate how project managers act.

Project managers' knowledge of planning techniques needs to be combined with knowledge of organizational work. The rationally derived techniques and tools of structured analysis and design and project management do not reflect adequately human aspects of software development, IS development and organizations. The effective use of planning techniques requires knowledge of organizations and human behaviour. Stakeholder analysis seeks to redress the gap, but has low profile in practice.

While planning is necessary, plans are difficult to implement in organizations. Planning and the implementation of a plan is difficult because of complexity, change and lack of knowledge. Complexity and change in particular pose practical software development problems. While change is considered in contingency planning, project managers need to take corrective action when these or other factors impinge on the plan. In doing so they need to understand the cause of the problem and what effect the corrective action will have on the project. It is necessary to understand the reasons for delay and overspend to improve the effectiveness of the project team. Managing change often involves project managers seeking support and commitment from the sponsor and relevant stakeholders.

4.8.3 Success and failure

Success and failure in IS development is evaluated and understood from different perspectives. Systems analysts believe that objectivity and rationality is possible in practice. A project mentality is expected to bring discipline and effectiveness to IS development and success, but this has largely failed to materialize. IS projects have failed incurring hefty costs to companies and public sector organizations. Executives' expectations of IT have not been fulfilled, generating a debate among researchers and practitioners on what constitutes a 'successful project'.

While a project is useful for organizing work packages, it has limitations. They have often resulted in 'disappointed users' or unused developed IS. Costly projects have tended to be unsuccessful in project terms, for instance keeping to the set budget or time. The flaws are exemplified by failures like the London Stock Exchange's Taurus system and the London Ambulance's dispatch system. These projects each had budgets of several hundred millions. The Taurus project was abandoned and the London Ambulance system had operational failures leading to tragic consequences for human life. Alternative approaches to systems development based on similar assumptions are susceptible to similar problems.

It is problematic to measure success in IS development because of the nature of information and knowledge. IT is used to process data but the actual consumption of information and knowledge is by humans individually and in organizational contexts. Developed IS remain unused because executives, managers or other employees regard the information produced as irrelevant to their needs.

From a different perspective practitioners are questioning absolute measures of project success. This perspective raises doubts on the rational assumptions in systems ontology. They argue that the measures relevant to non-systems projects like meeting project objectives, budgeted costs and time targets cannot be simply translated into IS projects. Some practitioners argue that the learning experience is more important than the actual achievement of project objectives. They reason that the experience gained from being involved in a project can be reflected upon and used in other projects.

4.8.4 Alternatives

Real situations provide contrary observations of project behaviour, organizational factors, information needs and definitions, which lead to alternative perspectives on software and IS development. There continues to be progress and development in projects and IS development process. New and emerging approaches cater for gaps in traditional project management and issues in IS development such as:

- Who is a systems developer?
- How is a system selected for development?
- How is a system developed?

- Why separate a system project from business operations?

Structured and object-oriented systems ontology assumes that a systems developer is a qualified professional. The systems ontology does not admit so-called 'users' to be regarded as developers. New approaches are breaking the traditional boundary between the 'developer' and the 'user'.

The major issue in identifying and selecting an IS project is who should do it. The SDLC and SSADM make business executives, business managers and IT managers responsible. This has been challenged for some time in End-User Computing technology, making it possible for employees of an organization to make the selection decision and themselves develop a system.

RAD and Component-Based System Development (CBD) are two alternatives to traditional systems project management. They seek effective use of limited resources. Both alternatives reduce reliance on meticulous planned action. RAD removes stringent discipline or planned action of projects. It replaces the SDLC with an alternative that is designed to be flexible in practice and true to context. This alternative is based on involving users in IS development.

CBD is the development of components of software that can be reused and composed into IS. This is possible using object-oriented technology, though components can be developed in procedural languages too. CBD aims to build software assets and generic models that can be commonly used. Software that is reusable is termed a 'component'. The task for a systems project manager developing IS based on CBD is to identify relevant components and compose them to produce the required IS configuration.

Table 4.5 Personal constructs for project management

Pole 1	*Risk management*	*Human resource skills*	*Quality*	*Time*	*Scheduling*	*Estimation*	*Actual situation*	*Pole 2*
Plan								Unplanned
Set objectives								No objectives
Control								Uncontrolled
Resourced								Under-resourced
Expected								Unexpected
Team cooperate								Team conflict
Motivated								Motiveless
Manageable								Unmanageable
Consensus								Conflict
Agreement								Disagreement

4.9 Personal Critical Framework development

4.9.1 Personal constructs for systems project management

Table 4.5 is a sample repertory grid for project management. Reproduce the grid on a spreadsheet and add further columns and their polar opposites that you consider relevant. To objectify personal constructs in project management, complete the grid by following the details on how to use a repertory grid in section 1.10.1.

4.9.2 Systems project selection

Questions

1 Evaluate whether the notion of a project is appropriate for developing IS.
2 Evaluate the utility of generic competitive strategies for selection of strategic systems projects. Some examples of generic competitive strategies are: low cost producer, product differentiation, product focus or niche. How relevant are these for your organization?
3 Systems project management provides techniques for planning and executing software development. What practical issues and problems can you identify with the use of projects and project management in IS development?
4 Discuss whether senior management or people who do the everyday work in organizations should select systems projects.

Activity A

Identify an IS in your organization. Determine how it was selected for development. Was it selected on the basis of a formal methodology similar to the SDLC using feasibility study or some other in-house method?

Activity B

The first airline reservation system is considered to be a strategic IS. Strategic IS provide competitive advantage to companies who develop them. Individually or in groups, investigate and describe the process and methods used in your organization to identify and develop strategic IS. Consider the following points:

- What assumptions can you identify?
- What systems ontology model informs the system?
- What evaluation criteria were used to select the system for development? Some examples are value chain analysis, strategic alignment, potential benefits, resource availability, project size and duration, and technical difficulty and risk.
- Did the IS development proceed according to plan? If not, explain why it did not.

Activity C

Identify an area of your organization that you think might benefit from the application of IT. It may be a business function like sales or production, or a core business process. Develop a business case for its selection as a system project. Consider the following points:

- What value or benefit will be derived from the new IS for the business?
- What monetary or other resource savings will it provide?
- Does it have the potential to be a strategic IS that provides the organization with competitive advantage?
- Why should it be selected and not some other system?

4.9.3 Planning

Question

Critically compare Agile Software Development (read section 15.5) with systems project management. Justify which method you would choose to develop IS.

Activity A

IT and IS planning are done in conjunction with business planning and strategy. Planners have to deal with actual situations that do not reflect the assumptions, ethos and techniques of project management. In groups, discuss the following issues found in real situations and suggest how you as a project management team would deal with them:

- Lack of complete knowledge and information, both internal to the organization and external regarding competitors and markets.
- The validity of the assumptions made in the plans.
- Issues surrounding the scope of the project and difficulties in defining the scope.
- Identifying and catering for contingencies in case the plan falters.
- Authority and power issues. Will people in the organization accept the planners and the plan?

Activity B

Refer to Activity C in 4.9.2. Bearing in mind that project plans are not absolute right or wrong ways of doing work:

- Construct a work breakdown structure for the identified system.
- Construct a product breakdown structure for the identified system.
- Develop a project network.

4.9.4 Project management and business models

Questions

1 Business planners use the value chain or business processes, or other business concept, to develop business models. Discuss how analyst's systems models would be affected by the need to consider business models?
2 Systems project management is used to implement business models that contain significant investment in IT and IS. Discuss whether systems project management techniques are sufficient to cater for a major organizational change involving a new business model.
3 Discuss the personal skills and competencies a project manager requires. How would you develop such skills and competencies?

Activity A

Think of a situation where you wanted to achieve something that required others' help. Briefly describe the situation in text form. Describe how you motivated others to help you. Discuss how you would motivate people in a team.

Activity B

Based on Activity C in 4.9.2, was the system developed on the basis of a business model? If so, describe the business model and how it translated into the developed system. If a business model was not used, define an appropriate business model for the application area for the system identified and discuss the expected benefits.

Activity C

Dell.com operates on a successful business model based on IT and internet technology.

Surf the company's website. Describe the business model you can adduce. Identify particular technology that enables Dell.com to be successful.

4.9.5 Stakeholder analysis

Activity A

Based on Activity C in 4.9.2, compile a list of the main stakeholders. What is the particular interest of each type of stakeholder? If the system were to be developed anew how would you as a project manager balance their particular interests and manage power relationships? You can use any method for doing the stakeholder analysis, for example see: http://www.scenarioplus.org.uk/stakeholders/stakeholders_template.doc.

4.9.6 Project failure

Activity A

The London Stock Exchange's Taurus system and the London Ambulance Service's dispatch system are highly publicized IS failures. Individually or in groups, identify an IS project that has failed in your organization. You may interpret 'failure' to suit your organization's needs. Consider the following:

- Why did the project fail?
- Is 'failure' a relative measure? Do IS developers consider it a success or something from which they have learnt valuable lessons?
- What knowledge can be learnt from the failure to ensure better project management in the future?
- Might the use of the CMM produce a different outcome?
- Define your personal construct on 'failure' and 'success' for your PCF.

Activity B

Identify an IS familiar to you:

- Determine how quality was defined during its development.
- Solicit the views of its current users to determine whether they agree with the original definition of quality.
- Determine with the current users what may be regarded as an appropriate definition of quality for the system.
- Discuss whether you agree with the users' definition. Is an objective definition appropriate?

4.9.7 Internet sources

CMM is systems project management that is highly standardized and regulated. You can explore its details at http://www.sei.cmu.edu/cmm/cmms/transition.html.

For information on team and group roles, see Belbin's work at http://www.belbin.com/.

For information on project management see www.fek.umu.se/irnop/projweb.

Information on SSADM and project management is available at www.comp.glam.ac.uk/pages/staff/dwfarthi.

PRINCE is a project management methodology. For information on PRINCE see www.ccta.gov.uk.

For risk management see http://mijuno.larc.nasa.gov/dfc/rsk.html. It contains an overview and a bibliography. Also, http://www.rspa.com has a list of references.

For information and method details for CoCoMo see http://sunset.usc.edu/COCOMOII/cocomo.html.

4.9.8 Further reading

A standard textbook on project management techniques is Yeates, D. and Cadle, J. (2001) Project Management for Information System, London: Pitman Publishing.

For a reading of alternative perspectives on software see: Truex, D., Baskerville, R. and Klein, H. (1999) 'Growing IS in Emergent Organizations', Communications of the ACM, 42(9): 117–123.

For a critique of planning read Mintzberg, H. (1994) 'Fall and Rise of Strategic Planning', Harvard Business Review, January–February 1994: 107–114.

To construct activity-on-arrow networks see: Yeates, Y. and Cadle, J. (1996) Project Management for Information Systems, Harlow: Prentice Hall. Chapter 6 provides an overview of activity-on-arrow networks and work breakdown.

For an alternative perspective on software development read Chapter 1 by Patel in: Patel, N. V. (2003) (ed.) Evolutionary Adaptive Information Systems, Hershey, PA: Idea Group Publishing.

An alternative model of IT governance is in Patel, N. V. (2002) 'Emergent Forms of IT Governance to Support Global eBusiness Models', Journal of Information Technology Theory and Applications, 4(2): 33–48.

Chapter 5

Systems analyst

5.1 Learning outcomes

After completing this chapter you should be able to:

- Analyse the organizational problems systems analysts face in applying their skills and suggest strategies they can use to overcome them.
- Identify and propose solutions to the problems of (dis)communication between analysts and their clients and stakeholders, and propose pragmatic solutions for systems modelling problems.
- Critically evaluate the problem-solving, analytical and critical skills required of a systems analyst.
- Determine appropriate personal construct for 'systems analyst' for a PCF.
- Apply transformatory critique, refashioning of traditions, reflexivity and critical skills to the 'self' as systems analyst.

5.2 Introduction

Systems analysts are arguably important members of a systems project team because of their systems analysis and systems modelling skills. The term 'analyst' emphasizes the ability to think logically, analytically, critically and creatively. An analyst is the person who communicates between the needs of the business and the capability of IT, and other related communications digital technology, to meet those needs.

Systems analysis has a long tradition in organized human activity. It is an important and vital technique to understand and develop knowledge of human problems, define human goals as problems, and to organize knowledge and understanding to resolve defined problems in systemic terms. It has been applied to military, government and business problems, and it is the prime form of problem definition and resolution for IS.

Though a specialist in IS, an analyst needs to possess other skills. Their work is vital in organizational information and knowledge management. Systems analysis and design has broader implications for how an organization defines its information and knowledge systems, and management systems for executives. In addition to contributing modelling skills, analysts need knowledge of business models, business processes, organization, management systems and decision-making processes. For these reasons, analysts can make a better contribution if they are capable of critical investigation.

5.3 Qualities and skills of systems analysts

Systems analysts need to develop various qualities to become systems modellers. An interest in how organizations work is a prime quality because analysts' work is to understand people and work and how these are organized to achieve specific goals in systemic terms. Analysts contribute knowledge of IT, IS and systems to define IS. They work in a systems project team to redesign organizations and systems radically. Table 5.1 is a summary of the non-technical qualities of an analyst.

Analysts require a broad range of knowledge and skills to investigate and model organizational work and systems. Major knowledge and skill area is systems modelling, but other areas include IT, business modelling, organization and people. They cover an understanding of organization, communicating with people, construction of systems models and the innovative application of IT.

5.3.1 Organization and people knowledge and skills

Though all organizations have common attributes like business processes and functions, each company's organization is unique. Business understanding is required to underpin the process of systems modelling and systems models. Knowledge of organization and how organizations work is necessary for analysts to undertake informed systems analysis. A genuine interest in the business, manufacturing cars or retailing food, is necessary to develop understanding. The business of the organization will determine the kind of data it produces that could be processed by IS to provide information for managers or enable business processes.

Table 5.1 Qualities of systems analysts

Analysts' qualities	*Qualities in practice*
Investigative interest	Ability to investigate organizations, organizational situations, people, workflows and business processes.
Innovative attitude	Ability to create new organization with IT and IS. Improve or replace existing organization.
Conceptual thinking	Ability to make observations and convert them into concepts and generate new concepts from them to address business and IS problems.
Critical thinking	Ability to ask searching questions based on interpreted data, facts, perceptions or opinions. Think differently and uniquely.
Diplomacy and tact	Ability to keep good relations with a project team and stakeholders. Competing and politicking groups should not be alienated in favour of one, so that modelling information can be extracted.
Objectivity	Ability to remain impartial to develop effective systems models based on all the sources of information equally. (This quality may be replaced with subjectivity, depending on analysts' predisposition.)

An important quality is diplomacy and tact. Much of analysts' work requires communicating with people who make use of the information from an IS. As analysts need to communicate with competing and politicking groups or stakeholders they need to be diplomatic and tactful. They have to keep themselves in a position to extract required information from all relevant groups and individuals. People and groups often have interests that vary from others'. Analysts need to ensure that they have access to conflicting stakeholders, and that they are impartial in how they deal with conflicting interests.

In organizational terms it is the information and knowledge in organization that is important for the technical systems design. Analysts need to consider all the sources of information impartially. Partial information will result in poor or ineffective systems models, and consequently poor usage of the resultant IS. Diplomacy and tact are needed not to alienate individuals or groups from the modelling process. Interpersonal skills are important in this process.

Analysts' knowledge of organizations is enhanced by an investigative interest and how IS can be used to improve business processes or achieve strategic objectives. Business problems can be resolved with IT and IS but the organization needs to be systematically and systemically investigated. Improvements in customer service or product quality can be obtained by applying IT and the development of appropriate IS, only if analysts' investigation is thorough and informed.

An innovative attitude is necessary to resolve business problems involving IT solutions. System models incorporate organizational aspects. Analysts redesign workflows, business processes and organization, often radically, by developing innovative systems models. During the 1990s innovative systems models were created based on BPR. Analysts were part of systems project teams that redesigned organizations radically. They combined organization and systems models to exploit the capabilities of IT.

5.3.2 Conceptual and logical skills

The precursor to innovative design is concept. Conceptual skills are important. Analysts need to be able to make abstractions from observed actual situations. Abstracting elements and details from an actual situation requires conceptual thinking. Systems models are an abstraction from actual problem situations. The notion of 'system' is itself an abstraction used to organize thinking on human problems. The radical redesign of organizations requires an ability to conceptualize an alternative organization. Experienced analysts think conceptually about organization, work and IS and IT to develop systems models. Conceptual thinking requires an ability to make acute observations and draw abstractions that can be practically applied.

Analysts require cognitive skills to frame and solve problems in systemic terms. Effective problem-solvers are people who can develop a good conceptual understanding of the problem and then begin to factor it into manageable subproblems. The skill to analyse a problem systematically and to use logical reasoning is important for analysts to develop. Abstract concepts derived from observations of how an organization works underpin systems models. Analysts need to be able to translate multifarious individual observations into concepts to develop systems models. 'Business process' is such a concept that describes organizational work that crosses functional boundaries in an organization. For example, the production process begins with the sales office taking an order from a customer, passing it onto the production department to produce the goods, and production then passing it onto logistics to make the delivery.

Notation languages enable and support logical reasoning. Analysts need the skill to identify, define and resolve business and organizational problems in terms of a notation language. Many business and organizational problems are complex and intertwined. Notation languages are used to define and resolve these problems. Analysts need to be able to use notation languages to represent symbolically actual problems.

5.3.3 Technical skills

Analysts' work requires knowledge of and skills in IT and IS modelling. Knowledge of IT is necessary to be able to communicate with software programmers, database administrators and computer network specialists. Analysts do not need to be able to write computer programs or configure systems and computer networks, but they need to understand IT in terms of the 'problem domain' and how it can be used in it. An awareness of IT capabilities is essential.

Though not essential, an analyst who has knowledge of computer programming will probably develop efficient systems models. Systems modelling skills are required to develop effective and efficient systems models. Knowledge of and the application of systems notation languages to business and organizational problems are essential to develop systems models. Systems modelling notation languages consist of technical symbols to represent and manipulate actual problems.

5.3.4 Critical skills

Analysts need critical thinking skills because many applications of IT involve radical and fundamental change in organization and business processes. Analysts need to be able to question existing organization and work, and analyse actual problems from the perspective of proposing radically different alternative organization. In Barnett's (1997) terms, criticality is the predisposition 'that the object of attention can be other than it is'.

Investigative and innovative skills are underpinned with critical thinking skills. Critical thinking involves interpretation, analysis, evaluation, inference, explanation and self-regulation (see section 1.8.1). In a particular systems modelling situation, all these skills are required to first assess the current situation and then to generate alternatives to resolve problems. To assess the current situation the observed data, facts, or perceptions have to be first interpreted and then analysed. Their credibility is then evaluated. From this position appropriate and relevant inferences can be drawn. The explanation is the justified alternative or alternatives generated.

At a more complex level, critical thinking skills are needed to design organization and systems. BPR, eCommerce and eBusiness are examples of radical business reorganization where criticality is paramount. Existing forms of organization and business models need to be questioned intelligently to produce alternatives that make efficient and effective use of information and internet technologies.

Critical skills are needed to make decisions on what systems analysis instruments to use. Undertaking analysis of a particular problem domain requires knowledge of instruments that will produce the required information to draw systems models. Though an analyst alone will not make these decisions, especially in a systems project team, an analyst needs to have the ability to make judgements on the appropriateness of instruments in a given problem domain.

5.4 Role of the systems analyst

The analyst's role is central in a system project as shown in Figure 5.1. Analysts interact with

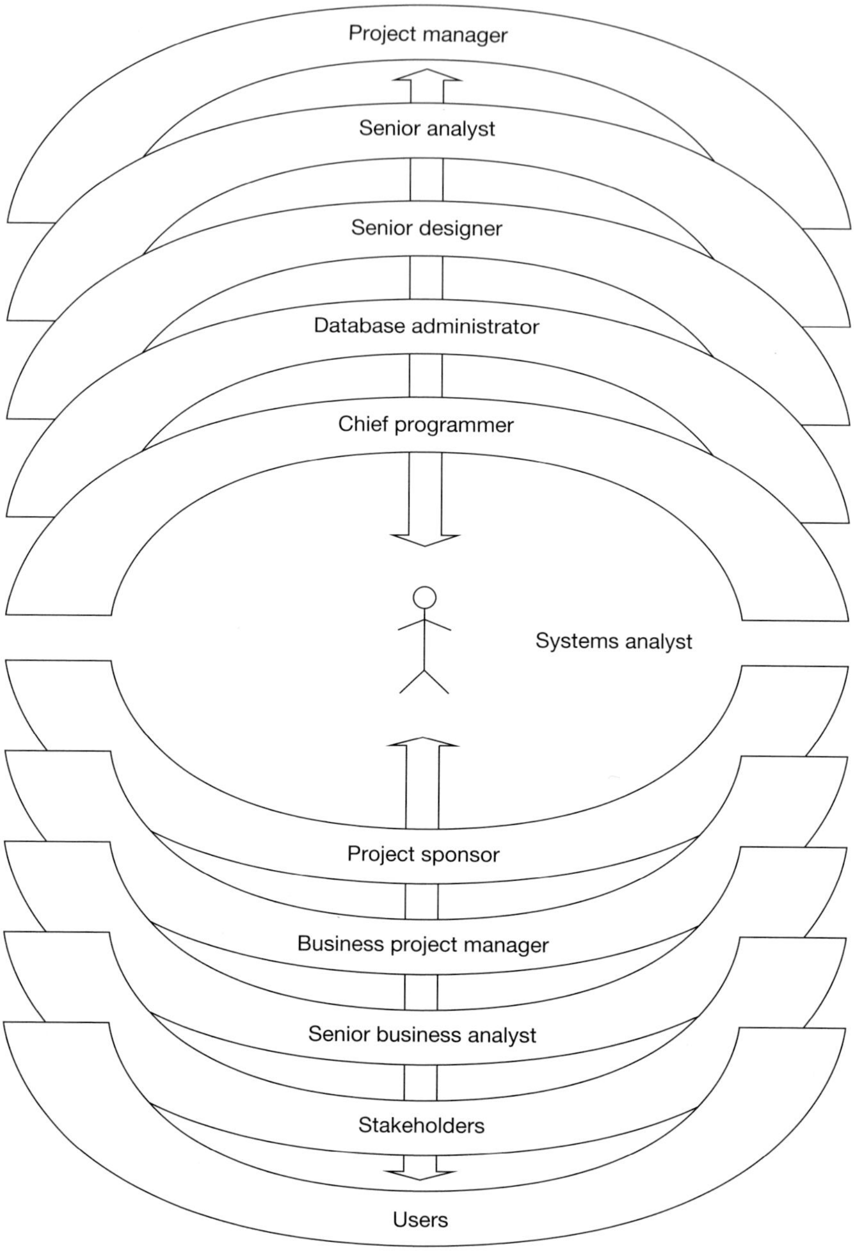

Figure 5.1 Role of the systems analyst

the project sponsor, stakeholders and potential users of a new IS on the business side, and with the project manager, senior analyst and designer, database administrator and chief programmers on the technical side. They provide systems programmers and database designers with systems models based on information gathered from business people. Analysts' tasks are to:

- Analyse and investigate the existing system, whether manual or computerized, and determine the requirements for a new IS.
- Make a judgement regarding the feasibility of developing a computer-based system for the application domain.
- Design a new IS: develop process models, data models or class models, and specify programs, hardware, data structures, and control and maintenance procedures.
- Test and oversee the installation of a new IS.
- Document a new IS and develop operating procedures for it.

Software programmers depend on a specification of the required suite of programs for a new IS. Analysts provide a system specification based on investigation of the problem domain. Computer network specialists also need to be told how to configure computers and networks to provide an integrated organization-wide IS. They depend on analysts' systems models to implement the designs with internet or intranet technology.

5.5 Systems analysis techniques

Systems analysis consists of examining the problem domain by considering its constituent parts with systems analysis and design techniques and tools – instruments. These instruments function to gather information on the actual problem and to develop systems models. Some techniques are borrowed from other forms of investigation, for example social science research, and others were developed specifically for systems modelling. Many of the requirements gathering techniques are common to structured and object-oriented systems ontology. They are summarized in Table 5.2. Other systems modelling instruments, covered in later chapters, differ between structured and object-oriented systems ontology.

5.5.1 Interview

Analysts design and conduct interviews with stakeholders and potential users. Analysts have to identify relevant people to interview and seek permission from managers. Interviews are generally preferred over other techniques because analysts can gain a deeper and thorough understanding of the problem domain. The method is used to gather general and detailed information on the current situation and requirements for a new IS.

The interviews have to be planned carefully to ensure that useful information is gathered and make efficient use of both analysts' and interviewee's time. The analyst has to prepare a set of questions and make decisions on how to record the replies to the questions. The questions need to be carefully formulated to extract required information. Interviewees' responses may be recorded as notes during an interview or captured on a systems documentation form, the normal capture method for large system projects.

Analysts need to think through the questions to ask at the interview. The set of questions will probably be agreed with senior systems analysts to ensure that the relevant information is gathered to draw systems models. The kind of questions posed will cover existing work, workflows and business processes; data generated and used; information required and communicated; other business areas affected. More detailed

Table 5.2 Systems analysis techniques

Systems analysis techniques	*Description*
Interview	Analysts meet with individual people or groups in the problem domain. Questions on roles, responsibilities and interaction are asked and responses recorded. Questions on information needs and uses of existing IS are posed and responses recorded.
Observation	Analysts observe existing workflows and processes and make a record of the events. Uses of existing IS are observed and recorded.
Questionnaire	Predetermined sets of open-ended or close-ended questions are laid out on paper and circulated to people. Responses are quantified and tabulated to facilitate analysis.
Document analysis	For manual systems, copies of manual documents are gathered to gather information. For existing IS, systems project documents are retrieved from archives to study objectives, specifications, procedures and reports.

questions will focus on the specifics of the problem, data and information needs.

The kind of questions asked will also depend on the type of interview used. Structured and semi-structured interviews can be used. A structured interview, also known as a close-ended interview, is composed of questions that are asked systematically in turn and the responses recorded. Scaling is the most frequently used form of close-ended questions. An example close-ended question, using a Likert scale, where a circle or tick is placed by the respondent around the relevant item, is:

> How accurate is the information in the management report you receive?
>
> Never accurate — Sometimes accurate — Always accurate

The granularity of the predetermined response is determined by the problem being addressed and the analyst's skill to extract the required information. Close-ended questions do not allow respondents scope to offer their own views. They have to select an answer from the provided list. For instance, in the above question, reports may be on time but not accurate or provide other information than required at a particular time, but respondents do not have the scope to make this known to analysts. Close-ended questions are asked to enable quantification of responses and facilitate subsequent analysis of the responses.

In a semi-structured interview, also known as open-ended, some initial questions are designed to elicit responses on particular topics, and the analyst uses the responses to ask one or more further relevant questions. So not all the questions have to be predetermined in a semi-structured interview. A semi-structured interview generates much data, often creative and rich in content, that needs to be analysed subsequently. Responses are usually recorded on audio to enable analysis. Analysis of gathered information can be highly structured to address specific issues and problems. Both initial structured and semi-structured interviews may be followed up with second interviews to gather further information or corroborate existing information. Examples of open-ended questions are:

> What are the problems with the existing system?
>
> What is the cause or causes of these problems?

5.5.2 Observation

Observation is used when analysts want to see what is happening in the problem domain and record the information personally. Observation affords analysts thinking space to form views based on observed events. To prepare for observation analysts need to identify the area of the problem domain and the people they want to observe. They need permission from appropriate business managers to observe people at work. A form also needs to be prepared to record observations systematically. The observation form should be structured to enable systematic observation.

If systematic observation is not required, then shadowing can be used. Shadowing is used to gather information in the problem domain as it happens. It differs from observation because in shadowing the analyst is in the background. The analyst spends time with an employee in the problem domain. The analyst spends time sitting with the employee while they work or follows the employee to meetings and other activities. The observation is recorded as notes or on standard documents.

5.5.3 Questionnaire

Questionnaires provide an efficient means for reaching many people, which is not possible with interviews or observation. The format, circulation and analysis of a questionnaire is more efficient than interviews. They are used for gathering structured information, where the information sought is known in advance. Questionnaires can be open-ended or closed-ended. Information from a large number of people that needs to be quantified and tabulated is normally gathered by close-ended questionnaire.

In a questionnaire, unlike a semi-structured interview, all the questions have to be predetermined. Close-ended questionnaires are used when the nature of the information is known. Senior analysts work with analysts to determine what information they need to understand and develop systems models of the problem domain. Questions are formulated to elicit that information from individuals or groups.

5.5.4 Document analysis

Examination of existing documents in the problem domain is a logical starting point before other methods are used. If the problem domain has no previous IS then existing manual documents are analysed. The kind of documents analysts may search for and analyse include documents that record data, or pass between people and across departments, or are compiled for managers. Where there is an existing IS, analysts will examine various systems documents, including system specification, project objectives, reports, database and program specifications, and equipment used.

To understand the business problem and workflows analysts identify and examine documents used by people. Documents, though, cannot be the sole source of information because they may be outdated. It is possible that the description of the system in the documents may not reflect the actual system and practice. Document analysis is normally supported with observation.

Standard forms are available for systems analysis techniques in IS methodologies. In SSADM interviews and observation are used in the systems investigation phase to provide insights into the problem domain, particularly the problems experienced by people. These forms become part of the systems documentation which itself becomes a source of information for project managers, software programmers and database administrators on a new IS.

5.6 Systems analysts and stakeholders

Stakeholders may make the difference between the success and failure of a new IS. Not all stakeholders will have the same influence and power. Analysts need to be aware of the influential and powerful ones. These stakeholders may have important information required for thorough and complete information gathering during systems analysis. Examples are: the sponsor of the project, auditors and departments such as accounts and finance. Potential users of a new IS are an example of influential stakeholders. Analysts interact with stakeholders through systems analysis techniques detailed earlier to elicit system requirements.

5.6.1 Requirements elicitation

The purpose of systems analysis is to establish system requirements for a new IS. Analysts need to interact with potential users and other stakeholders to elicit system requirements. They use analysis techniques like interviews and observation. Analysts need to use diplomacy and tact with stakeholders and potential users to ensure that various user groups are kept informed and satisfied. Alienating a particular user group will result in deficiencies in information gathering because the alienated group may harbour vital information from analysts.

System requirements elicitation requires analysts to apply their skills in various areas. Diplomacy and tact are prime among them. The skilful use of systems analysis technique is another. Central to both these is knowledge and understanding of organization and system, and how they need to cohere to enable business processes and objectives to be achieved.

5.7 Systems analysts and developers

Analysts are the main conduit for information to other project team members, whose interaction with people in the problem domain is limited. Systems designers, software programmers and database administrators rely on systems analysts to provide information and knowledge on the problem domain. So they depend on systems analysts to provide information to enable systems design and implementation.

Analysts are in a position to develop an overall view of the problem domain and requirements for a new IS. Their interaction with stakeholders and potential users gives them insight. They discuss systems models, design, and implementation issues in detail with database administrators and software programmers.

Analysts communicate information to other team members in various systems documents, primarily the system specification. Other documents include the feasibility study containing functional requirements of any existing system, organization charts, grid charts and records of interviews and questionnaire data collected during systems analysis. These documents will vary according to the IS development methodology used or, in the absence of a methodology, according to the practices of the organization.

5.7.1 System specification

The system specification is a collection of 'deliverables' or documents that detail the systems models for a new IS. It includes the system architecture design, data dictionary, the user and system interface design, database and file specifications, input and output definitions, and program design. Collectively these may be referred to as systems models in terms of systems ontology.

Systems analysts develop systems models in the analysis and design phases in structured systems ontology. They structure the data and information to be processed by a new IS. These system specification documents are communicated to software programmers and database administrators. In object-oriented systems ontology the class systems model is developed during analysis and design. The systems models detail what a new IS will do and how it will do it.

The system specification is important for several reasons. It will provide details to software programmers on how the system should work and database administrators will use it to design records for database implementation. It will also be used to evaluate and judge the system once it is operational. The system specification will become the benchmark for evaluating the accuracy and timeliness of information an IS provides once 'live' or in operation.

5.8 Critical Framework and systems analysts

A skilled and experienced analyst will have knowledge of IS development projects. This knowledge alone is not sufficient to act with critical insight in new situations. Reliance on skill alone is impoverishing, but practice can be enhanced through critical thinking. Figure 5.2 is the Critical Framework populated with critical reflection on the concept of systems analyst. As the bottom layer shows, many questions concerning the four themes of criticality arise from the assumption of objective and rational human behaviour. Alternative instruments and conceptions of systems developers are possible.

5.8.1 Professionalism and knowledge

The development of a PCF is a significant step in professionalism. A skilled analyst with a PCF is likely to be more effective and contribute intelligently to systems projects. Learning from knowledge and practice to develop practice further is the major function of a PCF. Knowledge, skills and experiences of previous IS projects can be enhanced and honed by reflecting critically.

Analysts need to reflect critically on knowledge, skills and experience they acquire from formal education and practice. With critical thinking analysts can think independently of formalism and draw on experience from practice. The PCF components – ethics, assumptions of reality, personal constructs, instruments, relations between components – can be drawn on to develop independence, which is an important trait for recognizing human problems and seeking resolution. A fictional anecdote illustrates the richness of critical thinking:

> This has been my experience in Washington when I had made money to give away. If I gave a contract to a designer and said, 'The doorknob to my office doesn't have much imagination, much design content. Will you design me a new doorknob?' He would say, 'Yes,' and after we establish a price he goes away. A week later he comes back and says, 'Mr Eberhard, I have been thinking about that doorknob. First we ought to ask ourselves whether a doorknob is the best way of opening and closing a door.' I say, 'Fine, I believe in imagination, go to it.' He comes back later and says, 'You know, I have been thinking about your problem, and the only reason you want a doorknob is you presume you want a door to your office. Are you sure that a door is the best way of controlling egress, exist and privacy?' 'No, I'm not sure at all.' 'Well, I want to worry about that problem.' He comes back a week later and says, 'The only reason we have to worry about the aperture problem is that you insist on having four walls

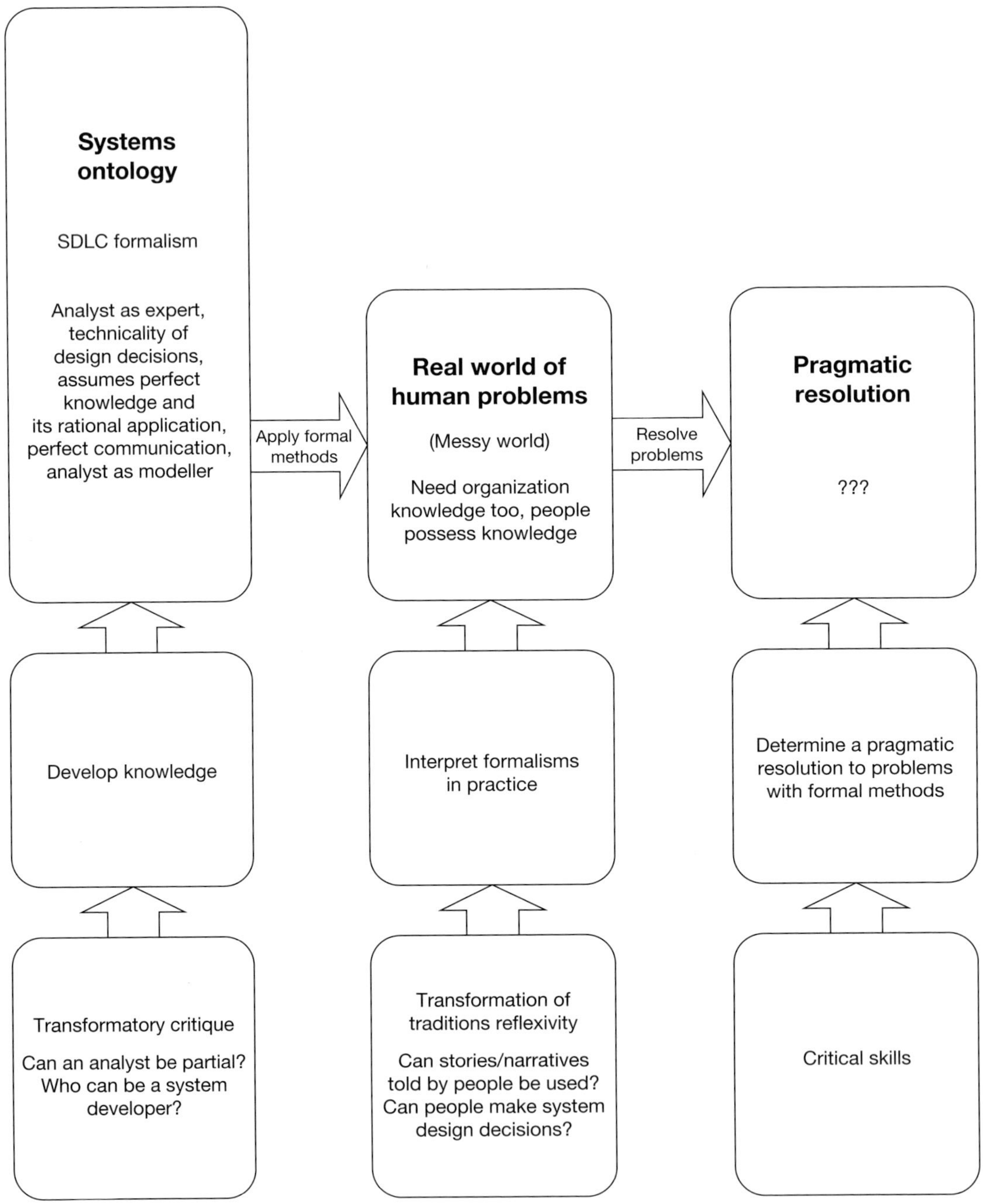

Figure 5.2 Critical framework: systems analyst and the SDLC

> around your office. Are you sure that is the best way of organizing this space for the kind of work you do as a bureaucrat?' I say, 'No, I'm not sure at all.' Well, this escalates until (and this has literally happened in two contracts, although not exactly through this process) our physical designer comes back with a very serious face. 'Mr Eberhard, we have to decide whether capitalistic democracy is the best way to organize our country before I can possibly attack your problem.'

The extent to which practitioners can rely on formalism in practice is a problem. Practitioners need to develop a PCF based on learnt knowledge and experience informed by critical thinking to become more effective. The fictional anecdote above illustrates that analysis of existing ways of working needs to be deep enough to question implicit and explicit assumptions of organization and work, yet it also needs to be realistic to address the pragmatic problem of developing an IS. The formalism of systems depicted in the Critical Framework (Chapter 3) leads to a form of *extreme* logical investigation and intervention in human activity and organizations that arguably looses realism, as in the above anecdote. The Critical Framework's other two components provide the necessary realistic perspectives.

To become effective systems analysts, analysts should invoke the Critical Framework to question learnt and experienced knowledge. It will support critical thinking on systems ontology and enable it to be contrasted with real situations to seek pragmatic resolution. Whether the requirement for analysts to be objective is valid systemic ontological knowledge or whether things in the problem domain exist independent of analysts can be analytically evaluated. The SDLC makes the assumption that analysts are an objective agent in IS development. What is objectivity and whether analysts can be objective can be understood independently by interpreting, analysing and evaluating the notion.

With reference to the Critical Framework, analysts can question whether it is possible to be objective in social situations like organizations and develop an appropriate objectivity personal construct. Analysts should assess the issue of 'ownership' of the data and information in the problem domain, as potential users often claim ownership of work. Similarly, questions about practice can be analysed. For example, are analysts' instruments effective or do they produce required information.

5.8.2 Knowledge and practice

Analysts should systematically identify other areas of knowledge and practice and question them appropriately. These include:

- Application of systems analysis and design skill in real organizations and IS development contexts.
- Clarity and difficulties of communication with potential users, stakeholders and even other team members.
- Acting as a change agent in a systems analysis and design role.
- Systems analyst as the authority or expert on systems design decisions.

Analysts' roles are determined by assumptions made in IS methodologies. In an objective methodology like SSADM, the systems analyst will need to behave impartially and make systems design decisions on the basis of evidence gathered from the problem domain. In an interpretive methodology like ETHICS or Multiview, the systems analyst will behave subjectively and make systems design decisions on the basis of interpreting the application domain. These need to be evaluated.

5.8.3 Systems ontology

Systems ontology questions can be raised through the bottom layer of the Critical Framework. Who should be regarded as systems developer? Who is best placed to make the systems design decisions? Should IS development be top-down selection of projects or bottom-up issues raised in work situations? Is timeboxing an appropriate assumption to make about IS in the real world? The answers to these questions can be explored in structured and object-oriented systems ontology.

The issue of a systems analyst as the authority on systems design decisions is illustrated in Figure 5.2. The SDLC assumes that only IS professionals should develop IS and that a systems analyst is an important intermediary in IS development. It assumes that systems analysts, and other systems professionals, should make systems design decisions. This assumption is made in structured and object-oriented systems ontology, and many other methodologies. Deficiencies in the application of such knowledge to human problems have led to the notion of **participative design**, which recognizes that people in the problem domain too can and should make contributions to systems design decisions.

The Critical Framework can be used to juxtapose systems ontology with its actual implementation in real situations, and to bring to surface its assumptions that may contradict practice. This knowledge can then be used to determine pragmatic resolutions. For example, is the systems analyst intermediary necessary? As an intermediary the analyst has to communicate with potential users, which raises many practical problems in IS development. For example, often there is confused communication. Consequently, analysts make many assumptions when making systems design decisions. It is not unreasonable to assert that many systems design decisions are based on assumptions made by analysts and designers rather than gathered requirements.

Critical thinking leads to questioning of received knowledge. Table 5.3 shows, in column one, example skills that a systems analyst

Table 5.3 Systems analyst skills

Acquired skills and techniques – systems ontology	*Practical limitations in the real world of human problems*	*Possible pragmatic resolution*
Objectivity	Social interaction, human subjectivity (even ego).	Diplomacy, power brokering.
	People often mistrust analysts' intentions and become fearful of change that a new IS brings.	Involve people in design decisions to overcome users' mistrust and fear of change.
Systems modelling techniques	Limits of technique or tool to capture rich human and organizational situations.	Seek an independent review of most appropriate technique or tool.
	Complexity of technique or tool. Real situations do not allow it to be used.	Reduce level of knowledge needed to use the tools
Technical	Pace of change. Availability of knowledge.	Use simpler methods, tools and techniques.

acquires through education and practice. It shows, in column three, how these skills need to be supplemented with skills analysts require to interact with others in organizational settings. They also question methodological assumptions made in systems ontology.

Structured systems ontology assumes perfect knowledge. Figure 5.2 shows the assumption of perfect knowledge. It applies to SDLC and structured systems ontology (and other methodologies). Analysts should examine its validity in practice. The knowledge that an analyst brings to the problem domain is limited. In actual situations it may not be possible to have all the required knowledge of the problem domain before making systems design decisions. This results in analysts and other team members making assumptions about the problem domain. As perfect knowledge and rationality are related, lack of perfect knowledge also questions analysts' ability to act in a rational manner in real situations.

Similarly, analysts' role in software engineering needs deeper reflection. Software engineers rely on analysts to specify what a new IS is required to do. The engineering metaphor may be intellectually challenged when explaining the role of systems analysts. Though the term software engineering is common when describing the development of software, the person who investigates the problem domain is called a 'systems analyst'. Analysis is a better description of investigating requirements of a new IS, because it caters for human factors in IS development.

There is a real and fundamental problem in developing systems models. The real or ontology of a thing is understood through someone's interpretation of it (unless objectivism is accepted). An analyst interprets the problem domain. An interpreted systems model will seem incredulous to people in the problem domain because their own interpretations vary from analysts'. They are often critical of such interpreted systems models.

A systems investigation relies on analysts' interpretation. Adequacy of analysis depends on the competencies, skills and experience of analysts. It is possible that analysts may not interview the right people, those who are the most powerful and relevant stakeholders. Research reveals analysts have interviewed the wrong people at the wrong times. They have posed the wrong questions resulting in wrong answers. Such practice results in bad feeling among professional developers and potential users and further weakens the assumption of rational behaviour and the engineering metaphor.

Systems models developed by analysts are interpretations of real situations. They are not the systems models that potential users would necessarily build. Even among different analysts or groups of analysts modelling the same problem domain can result in different systems models. Analysts emphasize certain aspects of a new IS and de-emphasize other aspects. Leading advocates of structured systems ontology argue that analysts do this to facilitate communication with potential users. These increase the risk that analysts may inadvertently exclude critical features in their interpretations, and develop a system that has meaning for them rather than for 'users'.

As interpretive modelling challenges the objectivity assumption of structured and object-oriented systems ontology, it also weakens the assumption of logic in problem-solving. Studies of highly skilled analysts reveal that they rely on human intuition to help them solve problems. Although analysts may be trained in formal methods of systems analysis that emphasize rationality and logic, the actual practice of systems analysis draws on deep human characteristics like intuition and prior experience. Though analytical skills are important, systems analysts also need creativity skills. Transforming organizations with IT is not a simple matter of mechanistic thinking, it requires creativity and originality. Analysts' role is potentially much broader than defined in the SDLC systems ontology or others.

5.9 Personal Critical Framework development

The purpose of systems analysis is to describe a new IS through systems modelling. In developing your PCF you should reflect critically on how you would undertake systems analysis, in particular reflect on how you would define your role.

5.9.1 Personal constructs for systems analysts

Activity A

Table 5.4 is a sample repertory grid for systems analysts. Reproduce the grid on a spreadsheet and add further columns and their polar opposites that you consider relevant. To objectify personal constructs descriptive of systems analysts, complete the grid by following the details on how to use a repertory grid in section 1.10.1.

5.9.2 Role of systems analysts and method of knowing

Questions

1 List the skills required by a systems analyst. What difficulties might an analyst experience in practising them in an organization?
2 Critically discuss whether the systems analyst's role as an intermediary is required in IS development.
3 Practising analysts may not be aware of assumptions they make during systems modelling. Discuss whether objectivity can result in factual systems modelling.

Table 5.4 Personal constructs for systems analysts

Pole 1	*Systems analyst*	*Organization*	*Employees*	*Project team*	*Information System*	*Objectivity*	*Perfect knowledge*	*Systems models*	*Plans*	*Situations*	*Subjectivity*	*Context*	*Problem solving*	*Creativity*	*Rationale*	*Stakeholders*	*Power*	*Requirements*	*Questionnaire*	*Observation*	*Interviewing*	*Pole 2*
Rational																						Emergent
Articulated																						Hidden
Objective																						Subjective
Complete																						Incomplete
Ambiguous																						Unambiguous
Static																						Changeable
Perfect																						Imperfect
Expert																						Non-expert
Reason/logic																						Intuition

4 If the assertion that 'users' know the problem domain better than analysts is true, evaluate the effectiveness of system requirement analysis techniques to gather requirements.

5.9.3 Rationality and logic in practice

Questions

1 Structured systems ontology requires analysts to be rational and logical. What problems can you identify with these assumptions in practice?
2 What difficulties might arise in applying systems modelling notation languages in practice? How would you resolve them pragmatically to develop systems models?

Activity A

A commercial company is not a clean whiteboard onto which a systems analyst can apply learnt skills. It consists of people, departments, global divisions and political and power relationships among employees. These aspects of a company (organization) pose barriers for analysts to apply their skills rationally and logically.

- Assume that an enterprise-wide system is to be developed in your organization.
- Identify the kinds of problems you would encounter in practising rational and logical systems modelling.
- How would you overcome the difficulties? Think about innovative ways of interacting with potential users to achieve your aims.

Activity B

A systems analyst has to operate in an environment of diversity, fluidity and ambiguity, yet system project objectives have to be met.

- Identify departments or sections in your organization that have diverse needs and make a list of them.
- Make a list of the changes that happen in your organization.
- Taking particular departments or sections, make a list of things that are ambiguous.
- Discussion points: How would you cope with such an application domain? How would you reduce diversity, fluidity, and ambiguity? Is it necessary to do so? Is it possible to do so?

5.9.4 Planned action and situated action

Questions

1 Discuss the quality of system requirements gathered by a set of planned activities involving interviewing and observations.
2 What kind of system requirements might arise in actual situations compared to planned elicitation events like interviewing or observation. Evaluate the effectiveness of the shadowing technique to capture such requirements.

Activity A

This activity will evaluate your PCF by categorizing personal constructs into planned action and situated action. Make two lists. One of things you would do to develop systems models that are planned, label this list 'planned action'. The other of things you think you would do when required (as opposed to planned), label this 'situated action'.

If your PCF has only planned action, explain why you have not got situated action items. If your PCF has a dominance of planned action or situated action personal constructs explain why it is so. Compare your list of planned and situated actions with a trusted peer. Discuss the differences.

5.9.5 Your analysis techniques and tools

Questions

1 Discuss the effectiveness of one system requirements elicitation technique of your choice.
2 Explain how you would decide to use particular requirements elicitations techniques.

Activity A

Identify a peer or department member to conduct a systems analysis interview. Think about how you would prepare to use the interview technique with the person identified.

- What information do you expect to get?
- Prepare a set of questions to ask them, and explain the reason for asking each question.
- Conduct the interview and record the responses.
- Analyse the information to determine its usefulness.
- What changes would you make if you were to re-conduct the interview?

5.9.6 Internet sources

There are not many internet sources on systems analysts. An account by a working analyst is given at http://www.itcareers.acs.org.au/people/shekar.htm/. You could also surf the internet searching for professional institutes of systems analysts to learn more about their work.

The Institute of Systems Analysts at http://www.iap.org.uk/ is a useful site for practise in systems analysis.

Part III

Systems analysis

In the previous parts problems with systems ontology have been examined from the perspective of the Critical Framework. Particular and overall criticisms of IS development have been discussed and PCF developed. Part III covers systems analysis. Systems analysis is a set of analysts' activities to describe *what* a new IS will do. It results in systems models that describe what the system will do and how its parts are related.

Techniques and tools – instruments – to develop systems models in structured and object-oriented systems ontology are covered. Analysts' use instruments in the problem domain to determine the data and functions of a new IS. Instruments are used to investigate, describe and model existing workflows and business processes and to develop systems models of alternatives integrated with or supported by IS.

Instruments will first be introduced and then critically analysed. Instruments will be appraised from the perspective of the Critical Framework to engender critical thought and practical relevance. Instruments are limited in the kind of knowledge that can be gathered using them because of ontological assumptions they make. They will be critically considered to develop critical awareness.

The foundation of an IS is determining what it is required to do. This is termed 'requirements analysis'. In structured systems ontology a system is modelled in terms of data and process, and the logic of processes. Chapter 6 focuses on how the requirements for a new IS are determined. System requirements are determined by developing systems models of data and processes. Chapter 7 covers how structured data systems models are developed and Chapter 8 examines how structured process systems models are developed.

Though structured data and process modelling are covered in two chapters they are logically related. Data systems models are required to identify corporate data that may be used in more than one process systems model. The aim of separating the two types of modelling is to reduce data redundancy in the system. Analysts using structured analysis have to focus on both data and process systems models.

In object-oriented systems ontology a system is modelled in terms of classes and objects. Establishing system requirements and developing systems class models in object-oriented analysis and design is covered in Chapter 9. The class systems model consists of four components: problem domain class model, human interaction class model, data management class model, and system interaction class model. Together they compose an IS design. Object-oriented analysis focuses on the problem domain class model. The other three components are discussed in Part IV, Systems Design.

In terms of the Critical Framework, as well as having knowledge of instruments, analysts need to be aware of their limitations and how they can be improved. PCF can be enhanced and improved by examining them critically, especially by considering the value they add to understanding particular systems analysis and design problems. In terms of criticality, the chapters in this part cover critical skills developed during praxis. The Critical Framework will be invoked in each chapter to show how systems modelling instruments are based on systems ontology. The assumptions underpinning particular systems ontology will be examined to facilitate the development of a PCF.

Chapter 6

Requirements

The system to be (or not)

6.1 Learning outcomes

After completing this chapter you should be able to:

- Assess the effectiveness of techniques to determine system requirements.
- Critically evaluate system requirements analysis in the context of humans and organization.
- Apply critical skills to knowledge and practice of system requirements analysis.

Techniques for requirements analysis in structured and objected oriented systems ontology will be examined. The problem of determining the requirements for a new IS is common to both. Developing knowledge of requirements analysis and applying learnt techniques to the problem domain are vital aspects of analysts' work.

6.2 Introduction

What is system requirements analysis, how does it relate to humans and technology, and how does it determine the outcomes of a system project, are all problematic issues in IS development. This problem is reflected in the title of the chapter and is critically discussed in the section on system requirements and the Critical Framework.

The initial stage in systems analysis is determining what a new IS is required to do. It is the most critical feature of systems analysis and design and the most challenging for analysts. It is erroneous to assume that the simple application of requirements analysis techniques, for instance from SSADM, results in the required information.

The difficulties in determining what a system should do arise when analysis techniques are applied in the problem domain – actual situations. Analysts have to overcome many issues and problems that stem from the human and organizational activity. They need to communicate with people and understand what the system should do in terms of human and organization purpose and work.

6.3 Application domains and system requirements

When a new IS is developed or an existing one enhanced it is necessary to know what functions it should perform. This is termed the system requirements. System requirements are elicited from the application domain and are recorded formally to provide system specification documents to software programmers. They use the documentation to design and code computer programs.

Knowledge of what a new IS should do is contained in system requirements documents. Documented system requirements feature prominently in a system project and different methodologies have varying formats for recording system requirements. In systemic terms, system requirements is a statement, consisting of various types of documents, of the services that a new IS should provide. Though the term 'service' is object-oriented terminology, it is appropriate here to communicate the service nature of IT and its application.

Determining system requirements in systemic terms is the process of identifying the functions a new IS will perform, determining the data for the functions to process, and defining the processes that the data will be put through to deliver required information. In an object-oriented system, it is the process of identifying classes, objects, data, operations and relationships in a class model.

Analysts elicit system requirements from the application domain. They interact with people to develop knowledge of the work they do, the documents they use, and the data and information they need to do their work. For IS restricted to departments or sections, analysts examine workflows, the sequence of events required to complete particular tasks, to gather information about work, data and information. For larger IS, system requirements focus on organization-wide business processes and encompass investigation of divisions, departments and sections of an organization.

Clients, project managers, analysts, software programmers and stakeholders use the requirements document as a repository of information on a new IS. They all use it as the basis for understanding what a new system is required to do and for negotiating what they want it to do. Analysts use it to design systems models. Software programmers use it, and the systems models developed by analysts, to write the required suites of software programs.

6.3.1 Formal definitions of system requirements

Professional bodies provide formal definitions of system requirements. The Institute for Electrical and Electronic Engineers (IEEE) Standard 610 defines requirements as:

1 A condition or capability needed by a user to solve a problem or achieve an objective.
2 A condition or capacity that must be met or possessed by a system or system component to satisfy a contract, standard, specification or other formally imposed document.
3 A documented representation of a condition or capability as in 1 or 3.

There are different types of system requirements:

- general requirements
- functional requirements
- implementation requirements
- usability requirements.

In the SDLC requirements analysis occurs once the feasibility study is completed and the decision to proceed with a new IS is made by management. If there are existing IS, it consists of determining their requirements to ascertain whether they are being met or not. This is called the 'as-is' system. The requirements for a new IS are then determined. This is called the 'to-be' system.

6.3.2 Systems analysts

Systems analysts are central to the process of determining system requirements. Establishing system requirements is a significant component of their work. They work with people who need the system to understand what is required and they communicate these requirements to systems designers. System requirements documents produced by analysts act as communicative devices between analysts and users, and analysts and software programmers.

Analysts need to develop knowledge of the problem domain to establish valid system requirements. They need to know what tasks are to be supported by a new IS and map workflows or business processes. For applications that span the whole organization, a team of analysts normally work together to develop an understanding of as-is and to-be business processes. They work with business analysts who share information on new business models to be realized in systemic terms.

Analysts undertake systems analysis activities to elicit requirements. The order in which they are done vary according to the elicitation techniques and methodology used. The people in the problem domain have to be consulted to understand what they want the system to do and existing IS analysed to assess deficiencies. Requirements may have to be negotiated because of different data and information stakeholder demands. Once the requirements are established they have to be documented and validated. The validation is used to ensure that the documented requirements are the 'right' ones, and to ensure consistency and completeness to prevent costly mistakes. The resultant requirements form the system specification.

6.3.3 System types and requirements

The process of determining system requirements varies according to the types of IS developed. Though the aim is the same, the process varies if a bespoke system is to be developed or a commercial package is to be installed. In all cases, the aim is to determine what a new IS is required to do and ensure that it satisfies business requirements. The different types of IS for which requirements are established are:

- bespoke IS;
- Commercial-Off-The-Shelf Software (COTS);
- business process re-engineering;
- eCommerce and eBusiness systems;
- knowledge management systems.

Establishing system requirements for COTS is more difficult than for bespoke IS because COTS are pre-designed IS based on business best practices. COTS developers establish what is best practice in a particular business application domain, for example customer relationship management or organizational knowledge management. They then design IS accordingly. Organizations that buy COTS may not have the same practices embedded in the purchased COTS. They will be confronted with the difficult decision of changing existing practices to meet the COTS system or to tailor the COTS to existing practices. Determining system requirements in either case is highly problematical. It needs to be done by skilled and experienced analysts. Most organizations adopting a COTS package change their existing practices and tailor the COTS system.

The span for analysing system requirements will vary according to the type of IS to be developed. The span may be:

- **The whole organization.** For IS to support business processes, eCommerce, eBusiness, or knowledge management the whole organization is the problem domain. Such IS often have a strategic purpose aligned with business strategy.
- **Department, section or group.** Where specific IS for knowledge management or stock control is to be developed the span is limited to departments or sections.
- **Particular business process.** When a specific process like a customer relationship management system is to be supported the span is limited to all the activities that are relevant for customer relationship management, but include the whole organization.

6.3.4 People's knowledge of requirements

Requirements elicitation techniques are designed to gather information from physical documents in the organization and from people. The physical sources are relatively easier to identify and manage. Sourcing information from people though is problematic. There are a variety of reasons why people's knowledge of system requirements may be limited. These are briefly stated here and dealt with critically later in section 6.8.3:

- People may not know what they want – people lack prescience.
- People only know what is required in context – when they actually do a job.
- Peoples' and organization's requirements may change because of internal or external factors.
- People are unable to communicate requirements in technical terms.

Systems ontology and methodologies make assumptions about people. They assume that people are capable of perfectly knowing how IS can support them. This issue poses significant problems for analysts and requirements management. Requirements management is a significant problem in systems project management. The people issues and physical sources of information need to be managed to ensure that the 'right' requirements are elicited.

6.4 Business and system requirements

It is important for a new IS to be aligned with business needs. At a strategy level IT and IS planning needs to be aligned with business strategy. Analysts have to determine system requirements at the business strategy and system levels.

Some methodologies stipulate requirements analysis at the business and system levels. Conducting requirements analysis at both levels ensures that IS are aligned with business needs. IS developed on the basis of system level requirements alone fail to meet business objectives and consequently fail to be used as designed.

In the Yourdon System Method, systems analysts have to conduct an enterprise requirements analysis and a system requirements analysis. In the Information Engineering methodology, there is a specific stage of executive requirements analysis. It enables executives to state their business level requirements. Critical Success Factors are identified at both the business and system levels. SSADM also has business and system levels requirements analysis, though the system level techniques are better developed because of the engineering focus of the methodology. In object-oriented analysis the Kozar's Requirements Model spans from abstract rationale for requirements to detailed requirements. The model relies on an existing business model and business mission statement. Techniques for determining system requirements in structured systems ontology vary. They are simply listed here:

- documentation analysis
- observation
- interviews
- questionnaire.

6.5 Object-oriented requirements techniques and frameworks

Object-oriented system requirements techniques are common to all object-oriented methodologies. Analysts can also make use of the structured techniques listed above. Their specific tasks are to:

- identify core objects;
- define the object structures and associations between objects;
- define objects' attributes;
- define objects' methods and services;
- determine the messages or communication between objects;
- refine the class model.

These tasks are completed by a variety of techniques including those listed for structured analysis. In addition, other techniques include: structured meetings, scenarios, prototyping, group brainstorming, voice and emails. These techniques are used within frameworks for requirements determination in object-oriented analysis, which include:

- requirements determination subactivities;
- PIECES framework;
- Kozar's Requirements Model;
- object-oriented requirements modelling activities.

6.6 Systems modelling tools

System requirements methods and techniques are supported with tools. The tools are largely automated or software packages called CASE tools. They are used to automate diagramming in structured and object-oriented analyses. Entity-relationship diagrams, data flow diagrams, entity-relationship models, entity life histories, use case, and interaction diagrams can all be developed with CASE tools. The aim of CASE tools is to improve software productivity. Other objectives include:

- improve software quality;
- shorten IS development process;
- reduce costs;
- generate program code from design;
- support and automate systems project management.

CASE tools are divided into upper CASE tools and lower CASE tools. Upper CASE tools are used to create integrated systems model diagrams and to store information regarding the system components. Lower CASE tools are design-phase tools. They are used to create systems model diagrams and generate code directly for database system functionality. Integrated-CASE or I-CASE tools contain both upper CASE and lower CASE functionality. I-CASE tools are designed to support the entire SDLC process and RAD. As yet there are few web-based CASE tools available, and their quality is still improving.

Analysts use CASE tools to communicate requirements to users, software programmers and stakeholders. As communicative devices CASE tools provide a standard form and objectified knowledge of requirement. They facilitate communication between IS project team members and between analysts and potential users of IS. Other benefits of CASE tools are speed of development, ease of alteration of diagrams, consistency and validity.

CASE provides a centralized source of information on systems designs called the CASE repository or data dictionary. This contains screen and report designs and any amendments made. CASE tools are available for various approaches to systems analysis and design. The following subsections describe examples for structured, object-oriented and agile programming.

6.6.1 SELECT

SELECT is an I-CASE tool. It can be used with a variety of methodologies and is available for structured and object-oriented analyses, and XP. It is used in the Structured Analysis Design Implementation (SADIE) methodology in the form of SELECT Yourdon CASE tool. SADIE requires much manual diagramming which can become tedious. SELECT is designed to automate the process to achieve cost and time efficiencies.

6.6.2 Database tools

The use of the data dictionary in database management systems has evolved. The extended data dictionary is a systems repository rather than a simple data dictionary. As a systems repository it is considered to be a CASE tool for information on systems processes and programs. Systems repositories are useful for managing IS projects, in particular ensuring system integration.

CASE tools are used for indexes and clustering in physical data models. The E*rwin* CASE tool has an index screen for database tables, which is used to generate code required to construct indexes in a Database Management System (DBMS). A DBMS is software for data storage and its manipulation.

6.6.3 eXtreme programming

The SELECT Scope Manager is used in XP. It is used to minimize time, cost and development by managing the scope of software development. It is used to automate the Story and Task scope management in XP. It creates a structured log of the stories, ideas, documents and analysis generated in XP. It also allows 'what-if' analysis.

6.6.4 Object-oriented case tools

CASE tools for object-oriented analysis cover systems analysis, design and implementation, and they extend to code generation and scriptable HTML reports. They are used to develop models of software and business systems, and are available for specific object-oriented methodologies.

The Bridgepoint tool is designed to support the Shaler-Mellor Method of Object-Oriented Analysis and Recursive Design. CRC (Class, Responsibility, Collaboration) CASE tools are used to create CRC cards and scenarios involving object interactions, to define objects, and run interactional scenarios to check systems designs.

SELECT Enterprise is used to design and develop complex object-oriented applications using UML. The Enabler function frees designers from having to remember to save work as it is done automatically.

6.7 Alternative requirements techniques

Structured and object-oriented system requirements make difficult ontological assumptions about the problem domain. An example is the assumption that people know what they want the system to do. People are expected to be capable and willing to state system requirements. In practice, these assumptions are problematical. Alternative requirements elicitation techniques like prototyping and eXtreme programming have been proposed to address requirements elicitation problems and other problems in requirements gathering.

The alternative techniques compress the SDLC. Quicker IS development is assumed to lead to lower costs. They shorten the time taken to establish requirements or place less emphasis on establishing requirements from people and other physical sources by providing the system product earlier. Involving people and providing the system product earlier reduces reliance on peoples' memory and knowledge of requirements, and enables them to see the development process and product.

6.7.1 Rapid application development

RAD is closely related to systems project management. It shortens the software development process and reduces systems project management activities. It is used to develop small and enterprise-wide systems.

RAD has four phases: requirements planning, user design, construction and cutover. CASE tools and prototyping are used in the user design and construction phases. Unlike the SDLC, or structured and object-oriented analyses, RAD requires the involvement of people working in the problem domain. They become participants in the IS development process. Their participation is essential in the requirements phase, design phase and the system-building phase. In the requirements planning phase both analysts and people conduct Joint Requirements Planning (JRP). During the design phase their participation is limited to non-technical issues through the JAD technique.

RAD is closely linked to prototyping. The purpose of allowing people to participate is to provide them with the required product early. As potential users they can assess the product early and give comments for its further enhancement. Unlike products resulting from the SDLC, the RAD process results in various versions of the product to help potential users see it, make comments and voice requirements.

6.7.2 Prototyping

It is difficult for people to visualize what a system will do from a list of requirements contained in a requirements document. In prototyping an original version of the proposed system is developed to explore and validate system requirements. A prototype of a new IS is built to understand and define the problem itself.

A prototype is used to demonstrate the system to users. People and analysts can learn more about the business problem and what the system should do by using the prototype in the application domain. Most organizations find prototyping is worth the cost in learning about system requirements, especially for IS that are strategically important.

Prototyping is used to understand system requirements iteratively. By using the prototype people can see the system as it is currently understood and use the experience to elaborate what the eventual system should do. Actual work situations can be tested with the prototype to determine whether it satisfies business needs.

In incremental prototyping or evolutionary prototyping system requirements are established iteratively. As each version of the prototype is assessed it provides clearer understanding of what is required. This information is used to develop the next version, and so on. In incremental prototyping the actual prototypes eventually becomes the implemented system.

A throwaway prototype is a single version of the proposed system. It is used for a specific purpose like determining requirements. A throwaway prototype is used to understand deeper system requirements that cannot be easily established by normal techniques. Most prototypes used for eliciting requirements are not used as eventual IS, they are 'thrown away' after requirements have been understood. Though some may result in the eventual system.

6.7.3 Agile Software Development and eXtreme programming

Agile Software Development (ASD) is gaining credibility in some problem domains, where highly structured and formalist methods do not work. It can be traced to adaptive systems development. ASD focuses on people and the social context rather than processes and methods for software development.

It is a response to the problem of 'analysis paralysis' in structured and object-oriented systems ontology. The phrase 'analysis paralysis' captures the problems in the actual application of objective systems ontology. In ASD system requirement is conducted incrementally and iteratively in small work packages with users and stakeholders. It focuses on the process of development rather than the software, and rather than focus on plans and detailed planned action, it plans for change.

XP has a similar focus on people rather than processes. Whereas ASD focuses on people in the problem domain, XP focuses on the systems project team too. It recognizes the social context in which IS is developed, and attempts to keep things simple to ease communication in the project team and with customers, who are involved with the team on a daily basis. They also make system decisions and changes. This close involvement of 'customers' is thought to lead to a shared understanding of the system.

People write stories of their work experiences, which are used instead of requirements documents. These stories are used to determine the scope, content and estimates. The 'stories' are very short, about three sentences, and are used to estimate how long it would take to implement relevant code.

Like ASD, XP focuses on small releases but its practices are based on metaphors that help the team to perceive the system being developed in simple terms. Each story is allocated up to a three week estimate for implementation, where a story is estimated to take longer than three weeks it is simplified into more than one story. In contrast, if a story is estimated to take less than a week to implement, it is combined with other stories.

6.8 System requirements and the Critical Framework

The Critical Framework is useful to bring to surface and assess critically assumptions on system requirements made in structured and object oriented systems ontology. Figure 6.1 is the Critical Framework populated with critical reflection on system requirements. As the bottom layer shows, many questions concerning the four themes of criticality arise from assumption made in structured and object-oriented analyses.

Analysts need to question what is meant by system requirements. The PCF can be enhanced by assessing the value and limitations of the available tools and techniques. Structured and object-oriented requirements analyses is based on objective systems ontology. They assume that requirements *exist* as separate entities in the application domain, whether in the minds of people or in physical documents in the organization. A critical appreciation of system requirements needs to address:

- What are system requirements?
- How do system requirements relate to human and organized activity?
- How does IT and IS relate to humans and organization?

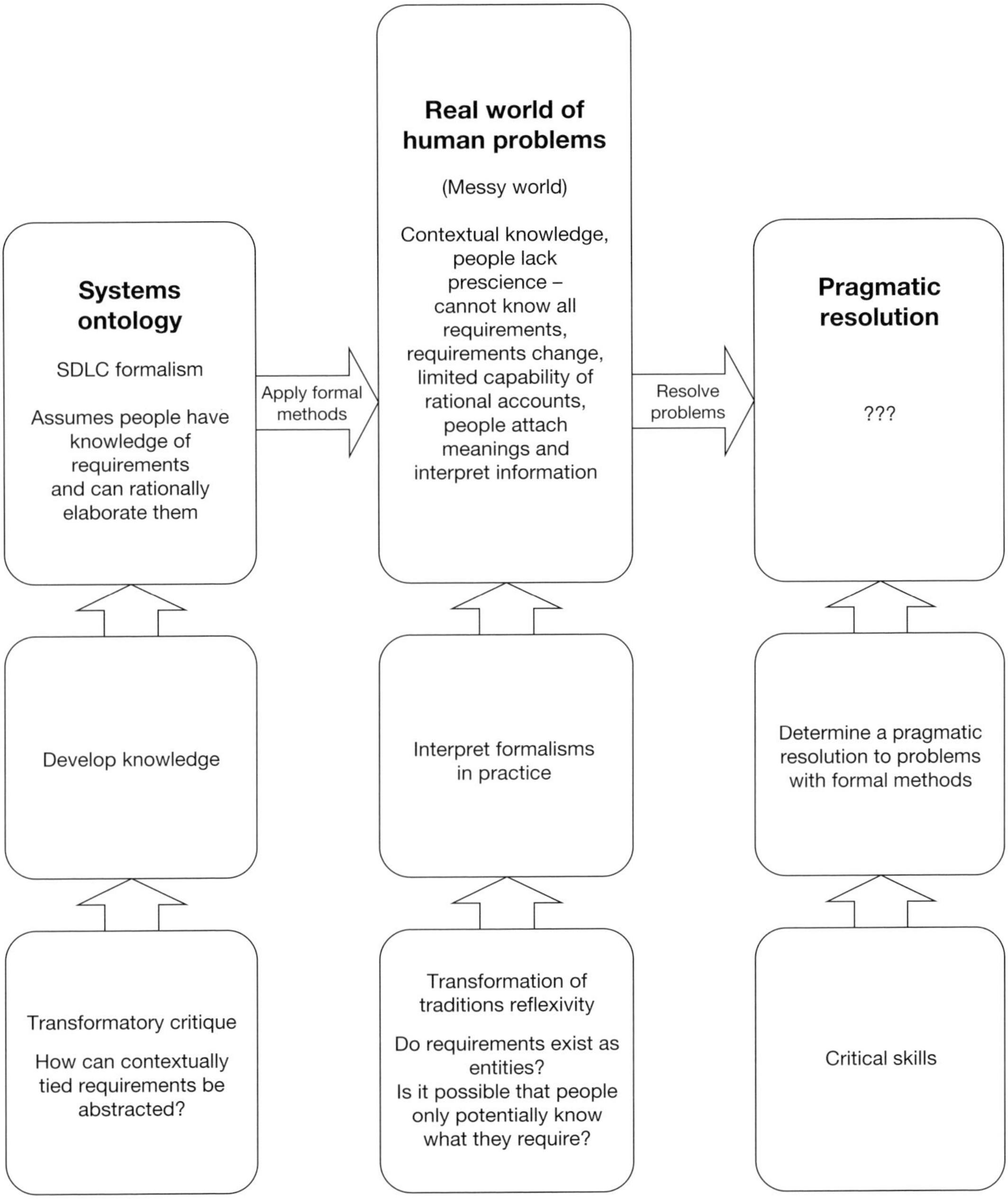

Figure 6.1 Critical framework: people's knowledge of system requirements in the real world

- Is requirements analysis a timebox event or continuous?
- Are requirements independent entities in the problem domain or inherent in humans?

6.8.1 What systems analysis?

Analysis is difficult. It requires deliberate segmentation based on clear purpose of what is observed. Analysing something obvious is even more difficult because the obvious becomes so natural that it is difficult to differentiate it from oneself, or even becomes 'embodied'. For example, analyse how you ride a bicycle or drive a car – they are embodied acts. Analyses leading to innovation, something new, for example a new IS, is the most challenging. It is difficult because it is not possible to draw on relevant previous experience or knowledge.

The complexity of analysing what a new IS should do is caused by various factors. Analysts have to approach people to determine system requirements. The intentions of analysts and the people they depend on to establish requirements are not clear to each other. It is possible that people may withhold important information. Organizational context is another factor impinging on establishing requirements. It becomes more difficult to establish system requirements in changing contexts compared with stable contexts. Stakeholders are a critical factor, as they will be vying for influence and even control over what the system should do. Other factors to consider are actual work, workflows and business processes. All these have to be understood in context. Other issues are summarized in Table 6.1.

6.8.2 Systems ontology

The SDLC, prototyping and XP make different assumptions of the application domain and suppose many characteristics that constitute systems ontology. The SDLC works on the basic premise of objectification of system requirements. In comparing the SDLC and prototyping it can be seen that prototyping allows for requirements to emerge. It does not assume requirements as entities existing in the application domain. In the SDLC once requirements are established as entities a further erroneous assumption is made that they are fixed.

XP is a radically different process for 'learning' about the problem domain and system requirements. Its systems ontology is significantly different. It challenges the SDLC's assumption that analysts are best placed to perform requirements analysis from their positions of technical power. It assumes that requirements need to be learnt rather than assumed to exist independently and elicited by analysts. Unlike the SDLC, XP makes use of human techniques to learn what system requirements are relevant. By using human stories and narratives provided by people it makes IS development more accessible to people and reduces the burden on them to specify requirements.

The term requirements engineering is used in structured systems ontology. There are some fundamental problems with the underlying premise of requirements engineering and the consequent systems ontology. It assumes that users are capable of explicitly and unambiguously stating what they expect a new IS to do. When designing artefacts like bridges it is possible to specify technical aspects and functions. Though even civil engineers sometimes have to learn from their mistakes, as the sway in London's millennium bridge testifies. The design of an IS is different because the computer system has to satisfy human needs, which themselves being based on human intent, purpose and usage are difficult to establish. They raise uncertainty that the engineering metaphor cannot adequately address.

Table 6.1 Issues in requirements analysis

Issues	*Systems ontology*
Abstraction Requirements as abstractions of the application domain. How can contextually tied requirements be abstracted?	The SDLC and structured and object-oriented analyses based on it assume requirements as abstractions. XP and ASD relates directly to the context by relying on stories and narratives.
Objectification Requirements as entities in the application domain. Is it possible to elicit requirements from an application domain? Do they exist as entities? Can requirements be represented and are current notation languages adequate for representing requirements?	The SDLC and structured and object-oriented analyses based on it assume requirements are entities in the application domain, and they can be objectified. They assume it is possible to design notation languages to model objectified requirements. XP and ASD recognize the human basis of requirements.
Specification Requirements can be specified. Do people working in the application domain *know* what is meant by 'requirements' and do they *know* what they require? Can they *agree* requirements among themselves and with other stakeholders? There is a limit to what an individual or group, let alone organization, can specify rationally. Beyond that things become apparent in the *doing* act only, or 'embodied knowledge' or situated action.	The SDLC and structured and object-oriented analyses based on it assume that people do know what they require from the system – reports, frequency, granularity. They assume that agreed specification is possible. Prototyping, XP and ASD all recognize the incremental revelation of requirements. They do not depend on a time-based system specification document.
Engineering Requirements as engineering. Are requirements sufficiently tangible to be engineered? Analysts may fragment requirements because of the methods they use (they should develop integrated requirements). Analysts may not develop a complete set of requirements. Analysts do not ask detailed questions to determine requirements – 'what information do you need from the system?' is too general.	The SDLC and structured and object-oriented analyses based on it assume that principles of artefact engineering can be applied to requirements because requirements lead to a physical system. They assume that a complete set of requirements can be elicited and detailed system functionality specified. XP and ASD abandon the engineering metaphor. They give primacy to people and interaction rather than rigour and method.
Fixed Requirements as reified objects. How can requirements keep pace with business objectives and technological opportunity?	The SDLC and structured and object-oriented analyses based on it assumes that once identified requirements are fixed. Real human problems need to acknowledge emerging and changing requirements. XP and ASD recognize this during development. Evolutionary and adaptive systems enable functional changes post-implementation.

The issues outlined in Table 6.1 are not an exhaustive list nor are the questions they raise limited to those detailed. They need to be further explored in terms of the Critical Framework, and in particular in terms of systems ontology and organization ontology.

6.8.3 Systems ontology and organization ontology

Analysts need to address how to incorporate the phenomenological characteristics of human interaction into the systems development process and the eventual IS developed. Phenomenological aspects of human behaviour are not accounted for by the SDLC, structured and objected-oriented systems ontology. Assumptions of *rational* human behaviour negate the *interpretation* that humans place on information and knowledge.

Notions of system requirements are rooted in systems ontology. The issues in Table 6.1 arise because methodologies assume and define things in the problem domain in particular ways. In the wider sense, the problem domain is the organization for which assumptions and definitions are made. The problem arises when these assumptions and definitions, or organization ontology, is contrary to systems ontology. For example, assumptions about people in system requirements gathering are problematic in practice. The assumption that people have the capability to communicate how IS can support them is one example that poses significant problems.

The main purpose of the SDLC and structured analysis is to provide: (a) a complete set of system requirements; (b) system requirements that are unambiguous; and (c) no redundant data in the system requirements and eventual systems design. These three are the 'elusive trinity' because no structured approach is capable of delivering them in real situations – human problems. They assume not only that system requirements can be identified, but also that once identified system requirements will remain constant until the end of the systems development project.

Important questions on what a new IS is expected to do or how system requirements can be established become operational within certain systems ontology, which may not reflect actuality. Figure 6.1 illustrates the basic contradiction between systems ontology and the real world (organization ontology). It shows that assumptions made in systems ontology are contrary to the actual experience of gathering requirements in the human problems.

This contradiction is reflected in communicative problems between analysts and people and between analysts and other systems project team members. The achievement of system project objectives is based on assuming a common understanding among team members and stakeholders. The initial problem of communicating among professional developers is overcome in the SDLC by developing standards. The SDLC and structured and object-oriented systems ontology provide a consistency of standards that enables professional developers to communicate from a common, objective base. These standards make it easier to monitor and control a system project.

Defining appropriate systems ontology for organizations is reflected in developments in systems modelling for databases. Table 6.2 shows the different database model types that have been used in IS. Their progressive development suggests a significant gap between systems ontology and organization ontology. Practitioners and researchers assume that formal models are necessary. Until formal modelling can reflect real situations adequately the gap between systems ontology and organization ontology will persist.

Analysts need to be aware that there are competing modelling notations and analysis

Table 6.2 Database model types used in Information Systems

Model type	*Comments*
Hierarchical	Hierarchy is an ordered tree and intuitive. As a modelling notation it is rooted in mathematics, which does not reflect the richness of human intent and organization goals and activities.
Network	Network is composed of a record type and a set type. As a modelling language it is rooted in mathematics, which emphasizes the representation of rich human and organizational context as logical relations.
Relational	A relation is a construct to represent the associations among different entities and their attributes. As above, its mathematical roots require human and organizational information to be represented as tables and relations between the tables.
Object-oriented	A class is composed of object with similar attributes. Objects are related through the messages they pass to each other for particular services or operations required. Interacting and collaborating objects form a system. Objects seem capable of representing rich human contexts such as 'aspects' and 'roles'.
Semantic	Semantic databases can be based on the concept of 'semantic net'. They have become popular because of XML, and make use of XML-algebra for formal definitions of data operations and transformations to prove completeness and correctness of design methods.
Post-relational	Post-relational databases enable the convergence of object and SQL technology. An example is Matisse, which combines Object, XML and SQL technologies within a single database.
Deductive	Deductive databases are an extension of relational databases. They support complex data modelling not suitable for relational databases. A deductive database consists of a database (facts), knowledgebase (rules) and an inference engine (the logic), which is used to derive information from the database using the knowledgebase. An example of an inference engine is Datalog, used to develop a relational deductive database system.

techniques for translating business needs into system requirements. They reflect the developing knowledge and practice in systems ontology, and the unsolved problem of understanding the problem domain. It is recognized that systems ontology needs to reflect reality but it is constrained by instrumental, technological and conceptual tools. System requirements gathering techniques assume that the resultant statement of requirements bear synergy with the problem domain. Very large and expensive IS projects have failed at the business level despite this focus on business.

6.8.4 Methods of enquiry

SSM is rooted in academic interpretive action research and it proffers to seek epistemological relevance. It relates the work of systems analysts directly to systems analysis and the problems analysts encounter in investigating 'human activity system'. It bases investigations of human activity systems on a 'declared-in-advance epistemological framework'. This is an 'intellectual framework of ideas'. The framework forms the basis for defining and expressing knowledge. It may be subsequently modified on the basis of research findings.

Analysts similarly need to be aware of any 'intellectual framework of ideas' and the role of methods of inquiry underpinning the systems ontology they adopt. Most commercial methodologies do not venture into questions of epistemology. When they do, the search is superficial to give academic credence. As systems analysis techniques purport to establish knowledge of system requirements it is pertinent to question the validity of these techniques. Analysts need to reflect on epistemic issues relating to determination of system requirements. Knowledge of how to develop IS and knowledge of the IS to be developed are questions of epistemology. They require critical study to assess practical validity.

6.9 Personal Critical Framework development

6.9.1 Personal constructs for system requirements

Activity A

Table 6.3 is a sample repertory grid for system requirements. Reproduce the grid on a spreadsheet and add further columns and their polar opposites that you consider relevant. To objectify system requirements personal constructs, complete the grid by following the details on how to use a repertory grid in section 1.10.1.

Analysts can use knowledge from practical experiences of requirements elicitation to probe the themes in the Critical Framework, which may lead to revising personal constructs and determine an appropriate pragmatic resolution to problems. They should begin the objectification process by asking questions similar to:

- Why did I choose questionnaires to elicit requirements?
- What evidence is there that questionnaires are successful?
- Can people relate to questionnaires or did it distance them?
- How do questionnaires relate to my other personal constructs about peoples' knowledge and their ability to communicate it?

Table 6.3 Personal constructs for system requirements

Pole 1	*Systems analyst*	*Organization*	*Employees*	*Requirements*	*Observation*	*Document analysis*	*Interviewing*	*Pole 2*
Planned								Emergent
Articulated								Hidden
Objective								Subjective
Complete								Incomplete
Ambiguous								Unambiguous
Static								Changeable
Perfect								Imperfect
Abstract								Context

6.9.2 Problem domains and systems ontology

Questions

1 With reference to Activity A, discuss how ontological knowledge of organization affects the design of instruments for systems analysis and design.
2 Critically compare the ontological basis of requirements gathering in SSADM and Multiview methodologies.

Activity A

In pairs, each person:

- Identify a problem domain familiar to you, a university department or company department.
- Record your knowledge of the domain in terms of the assumptions you make of its organization, people, work and information required, and any other aspect you choose.
- Compare your ontological knowledge of the department with a trusted peer.
- Would you revise your ontology? In either case, explain why.

6.9.3 IS types and requirements

Questions

1 How would you proceed to establish system requirements for a large multinational car manufacturing company wanting to buy a COTS supply chain management package?
2 Describe the requirements gathering techniques you would use to establish system requirements for a medical doctor's patients' records management. Justify your choices.
3 Comment on how you would overcome the problem of changes made to the system after freezing the design. How would you ensure that such changes do not remain undocumented features of an IS?

Activity A

- Describe the major functions for a student record system and locate your description in appropriate systems ontology.
- Surf the net for a COTS student record system.
- Assess whether the COTS system meets the requirements you identified.
- Critically evaluate the COTS systems ontology with your description.

6.9.4 Aligning IS and business objectives

IT/IS planning is intricately related to corporate strategic planning. A company needs to plan its IT/IS development to meet business objectives.

Questions

1 How are IS projects selected in your organization? In considering this question, think about the following:
Does the organization have a mission statement and a corporate strategic business plan (three years or one year)? How is the mission statement and business plan used in IS planning? Can you identify specific projects to exemplify?
2 What IT/IS planning activity are you aware of in your organization?
What role do potential users of IS play in selecting IS projects?

3 Identify a formal IT/IS planning process that includes users and critically discuss its relevance.
Is IS project selection a 'top-down' (set by senior management) or 'bottom-up' (determined by operational employees) approach in your organization, or a combination of the two?
What problems can you identify in the IS/IT planning process? Are important users or stakeholders facilitated to take part in the planning process, if so how?

6.9.5 Requirements techniques

Questions

1 Discuss the comparative merits of observation, interviews and questionnaires for requirements elicitation.
2 Describe in detail the situations in which you would use each type.
3 Structured analysis aims to use a minimum number of tools to specify a system. Discuss the veracity of this position.

Activity A

Design interviews to establish requirements for an admission system or human resource management system from:

- An admissions tutor in a university department or head of a human resource department.
- The admissions principle in the university's central department for admissions or the director of human resources in a company.

Activity B

- Surf the internet to find sites on RAD and XP.
- Critically compare RAD and XP to identify differences in systems ontology between the two for requirements gathering.
- What assumptions of the problem domain does RAD make that XP does not?

6.9.6 CASE tools

Activity A

Deciding which CASE tools to use poses problems. Some advice is to be clear of the purpose and to keep the criteria simple.

- Develop criteria for selecting CASE tools for requirements analysis.
- Search the internet to identify available CASE tools for structured and object-oriented requirements analyses.
- Justify your choices for each approach.

6.9.7 Representations of reality

Questions

1 Discuss whether it is better to write software programs directly from the problem domain as in ASD practice, or from systems models produced by systems analysts as structured and object-oriented analyses do.
2 Discuss how you think your PCF affects your ways of seeing a particular problem domain. Assess how the component and personal construct relationships affect requirements elicitation.

6.9.8 Internet sources

See www.agilelalliance.org and www.agilemanifesto.org/history.html for ASD perspective and philosophy (accessed 25 March 2004).

See www.dsdm.org for details on adaptive systems development (accessed 25 March 2004).

See http://www.xprogramming.com/xpmag/whatisxp.htm for information on the philosophy and principles of XP. Also see http://www.jera.com/techinfo/xpfaq.html for a simple introduction to XP (accessed 25 March 2004).

6.9.9 Further reading

Checkland, P. and Holwell, S. (1998) *Information, Systems and Information Systems*, Chichester: Wiley.

Chapter 7

Structured data modelling

7.1 Learning outcomes

After completing this chapter you should be able to:

- Interpret the problem domain in terms of logical data models, entity types, attributes and relationships.
- Evaluate critically the entity-relationship modelling technique.
- Assess the importance of logical data modelling for database design.
- Describe the normalization process.
- Develop simple E-R models.

Structured analysis focuses on transaction data in the problem domain. The structured modelling process allows for the redesign and integration of complex IS using data and process models. The term 'logical' means removed of physical characteristics. Logical data modelling is the process of logically describing the problem domain. Logical data modelling separates design from systems implementation. They are used to develop logical data models. This separation of the logical modelling of the system from its implementation will be discussed critically.

Structured systems analysis and design introduced the discipline of planning in data and process models to IS development. The focus on data in the problem domain makes structured analysis 'data-centric'. Techniques and tools have been devised to interpret data as pre-existing entities in the problem domain.

7.2 Introduction

A logical data model depicts data in the problem domain that is considered of interest or important to business people. It provides analysts and designers with knowledge and understanding of relevant data to be processed and the processes for processing it in a new IS. Business *data* and *processes* in the operation and management of the organization are used to develop the data and process systems models. It is also called an 'infological' data model because it is a data or information perspective separated from physical implementation.

Structured analysis assumes that it is possible to plan and develop complete logical data models for a new IS before it is implemented. It requires systems analysts to:

- Identify and define the problem domain – people, processes, documents.
- Develop 'infological' or logical data and process models.
- Convert the logical models into the physical design for computer implementation.

Analysts' task is to identify 'things of interest' in the problem domain and design data structures suitable for computer processing. Logical data models are an initial diagrammatic picture of a new IS. The purpose of logical data models is to determine what data is to be processed and decide how it is organized.

The process of data modelling makes use of the decomposition problem-solving strategy or the top-down problem-solving. If the problem were to be solved as a whole it would prove to be too complex. In the decomposition strategy, a complex IS problem is taken and it is decomposed into smaller parts. Breaking the problem down into simpler, smaller problems enables the sub-problems to be solved independently of each other. The independent solutions are then combined to provide an overall solution to the problem. The modelling process is not expected to produce the final system data models the first time. It relies on an iterative model-building process.

7.3 Transaction data

Organizations generate two types of transactions data. One, it results from transaction within an organization, for example hiring an employee. Two, it also results from transactions between an organization and its customers, business partners or government agency, for example making a sale or buying raw materials from a supplier. Transactions data covers products or services, employees, customers, finance and the government. An organization will typically generate hundreds of thousands or millions of such transactions depending on its size. Not all transactions data is recorded.

Transaction data that is legally required or considered relevant for managing is recorded and processed to produce information and knowledge. The executive and management in an organization will be interested in analyses and summaries of the transaction data or other special information to support decision-making and management. Managers will need it to support business operations management. For example, credit controllers need an analysis of customers by their payment record to check the credit worthiness of customers. The analysis of unpaid invoices can be structured in descending order of overdue date. Identifying, collecting and processing such data for management information is done in logical data modelling.

In structured systems ontology transactions data are regarded as independent items. They are part and parcel of a conceptual entity derived from data or processes in the actual situation – problem domain. An entity is any 'thing of interest' in the problem domain that needs to be represented in an IS. For example, in a customer relationship management system, 'customer' is an entity and the 'product' or 'service' bought by the customer are examples of other entities. The selling of a product is data that is related to the customer entity.

Understanding the business rules in an organization identifies things or entities of interest. Operational issues in turn determine business rules. For example, in an internet book club, decisions have to be made on whether a member can place more than one order. The policy decision becomes a business rule that can be used to identify entity types, relationships, and cardinality.

Logical data modelling is used to develop an integrated picture of an organization's data for corporate databases and various IS that draw on the databases. Most organizations store transactions data on computer databases to provide a central store which various IS access to produce operational and management information. Such database design requires an integrated approach to enable the development of integrated IS that provide consistent information to managers. Failures in the integrity and consistency of information can result in poor management decisions, which affect a company's performance.

7.3.1 Structured data and process systems models

A model is an analytical conceptual tool. It does not capture the problem domain *as it is* but as an abstract representation of it. A systems model is made using a formal notation and is normally diagrammatic. The modelling notation used is not the actual computer programming language used to write the eventual computer programs. In this respect structured analysis is computer-independent. It is separated from the computer implementation of the system. The logical system data and process models have to be subsequently converted into computer design and code.

From analysts' perspective, an important feature of systems models is that they are objectified representations of the problem domain. (Objectification is not the same as objectivity.) Objectification is the process of externalizing mental ideas or constructs for others to share. An early justification for the development of structured analysis was that objectified data and process models enable communication among systems project professionals and between systems analysts and people.

In structured analysis logical data models are designed first and then the logical process models. They are developed to facilitate communication, check assumptions and understanding, and test proposals. As the actual problem domain cannot be comprehended and understood in its entirety, aspects of it are taken and represented in a model. Only details of the problem domain that are of interest are modelled, other details are not modelled. The purpose of a model is to:

- develop knowledge and understanding;
- enable communication;
- analytically generate alternatives;
- determine relationships between entities;
- design implementable systems models.

Data and process models are developed to gather knowledge and understanding of the transaction data in a problem domain. The models are not to be understood as the *actual* application domain. It is because the actual application domain is complex and difficult to comprehend that models are developed. The models resulting from structured analysis are used to inform systems design.

The various data and process models developed provide three different perspectives on a system. The perspectives are data, process and entity life. Together they provide an overview of a new IS. They are related and enable checks to confirm consistency. For example, process and data stores in dataflow diagrams in process modelling are related to entities in an E-R diagram in data modelling. An entity in data modelling is related to an entity life history model if its life history is modelled. A process is related to an entity life history as events in it.

7.4 Logical data modelling

Techniques and tools for structured logical data modelling have been developed to represent data in the problem domain as logical data models. The techniques are supported with CASE tools. They prescribe systems analysis activities for analysts and consist of the following:

- entity relationship diagrams;
- logical database descriptions;
- data dictionary for the project.

The term 'logical' refers to data and processes in the problem domain from which all implementation dependent – physical – characteristics are removed. The term 'structure' emphasizes the need to provide frameworks that analysts can use. The frameworks describe activities, detail steps and stages, and define inputs and outputs required for systems modelling.

An analyst's task is to develop entity-relationship diagrams (data models) using the entity-relationship modelling technique. Two systems analysts may develop two completely different entity-relationship models of the same problem domain. The entity-relationship modelling technique is individually (subjectively) applied leading to individual (subjective) logical data models. Subjectivity is not considered to be a problem in structured logical data modelling. Experienced data modellers are regarded highly for their ability to determine relevant business rules that lead to appropriate logical data models.

7.4.1 Entity-relationship modelling

Analysts develop entity-relationship diagrams or E-R models of transactions data in the problem domain. E-R modelling serves to represent things of interest to the organization (entities) and determines their relationships with each other. A result of the E-R model is data tables that enable analysts to define the necessary set of operations on the data to produce operations and management information. In a typical E-R modelling process the analyst will:

- identify the problem domain, or relevant sections of the domain;
- identify things of interest in it;
- identify and name entity types;
- determine each entity type's attributes;
- determine the relationship types between entities;
- draw an entity diagram model;
- normalize the entities and relationships.

E-R modelling is an iterative process. A final E-R model will be the result of several iterations, especially when developing the model into third normal form or higher.

Entity

An entity is anything in the application domain that can be represented by name, attributes and relationship. It is an abstraction of a thing in the problem domain that generates data that is stored. An entity is important because it provides information on actual data to be processed in a new IS. Entities are classes of concrete or abstract things in the problem domain. They can be physical things in the problem domain or ideas or concepts relevant to the operation and management of an organization.

Both technical modelling skill and knowledge of the problem domain are required to identify entity types. Data is generated and stored for each occurrence of entity types in the problem domain. Analysts examine transactions data to identify relevant entities, which are shown in capital letters in descriptions of a system. Analysts decide whether a thing of interest in the problem domain is an entity or an attribute, and they give a name to the identified entity type.

Predicates are used to determine entity types because they describe a problem domain. For example, as thousands of sales occur in a company, an abstraction capable of recording all the sales is required. Consider a sales transaction description that an analyst might use to identify relevant entity types:

> A customer makes a purchase of a DVD unit. It costs £99.00. The sales assistant records the sale on the till, which generates a sales receipt for the customer.

In this description of a sales operation, transactions data on the sale of goods can be analysed to identify various physical and abstract entity types. Entity types can be identified from a text document in the problem domain by underlining the nouns, as shown in the box above. The sale or purchase is an entity. The sales assistant is an entity. The customer who makes the purchase is an entity. The item purchased and sales receipt are entities. The sale price is an attribute of the sale entity with a value (cost) of £99.00.

The underlined entities can be named as entity types in capital letters: SALE, SALES_ASSISTANT, CUSTOMER, ITEM and SALES_RECEIPT. An entity type is a group of similar objects in the problem domain, for example CUSTOMER entity type.

The identification of entities and entity types is non-trivial. Consider the following description of an employee:

> An employee Freda Ericsson is an experienced financial accountant. She works in the finance department. She is one of the senior managers of the finance department.

The same person in the organization may have different roles. Freda is both an experienced financial accountant and a senior manager. Analysts need to determine how best to organize useful information by skilfully identifying entity types for database design. They have to decide whether to create separate entity types for these two roles of the individual in the organization, say FINANCIAL_ACCOUNTANT and SENIOR_MANAGER, or to create one entity type, say FINANCIAL_ACCOUNTANT and add an attribute for management skills – senior manager. The latter is an efficient solution in terms of normalization.

Relationship

Relationships between entities establish the meaning or semantics of the logical data model. Analysts identify the relationships between the entities. As entity types have no meaning on their own, they need to be expressed as relations for business operations and management information. Defining relationships is an important feature of logical data modelling because it determines the eventual usefulness of the IS to people.

Relationships can be identified from facts in the problem domain. For example, it is a fact that the SALE and ITEM entity types are *related* to the CUSTOMER entity type, because a sale occurs when a customer makes a purchase of an item. This fact is an example of a business rule. Business rules form a significant basis for establishing relationships between entity types.

Managers rely on relevant relationships definitions for operational and management information. For example, a sales executive interested in generating more sales to make the organization profitable will seek answers to certain questions. They may ask questions such as: 'Which item is selling more?' 'Which customers are buying the most?' Information on the former question will help to produce more of the popular item. Information on the latter question will help to target customers who generate the most income. Such information can only be generated if appropriate relationships between entity types are determined.

The value of an IS occurs when it is able to provide valid, accurate and timely information to decision-makers. So the 'right' relationships need to be modelled. Since logical data modelling is a subjective process, analysts require knowledge of the problem domain and its detailed operations to ensure that entity types are appropriately identified and related.

Attribute and keys

Analysts identify the properties or attributes of entity types. Attributes are emboldened in a description of the problem domain. The attributes of the SALE entity type are: DATE, <u>RECEIPT_NUMBER</u>, ITEMS, CREDIT_CARD_NUMBER, AMOUNT, VAT. Attributes have values that provide details of particular sales. Each sale made will have values recorded for each attribute, even if the value is zero. For example, a customer may make a cash purchase so there will be no value recorded in the CREDIT_CARD_NUMBER attribute. Attributes appear in the data dictionary as primitives.

As many sales will be made each sale needs to be identified by an attribute uniquely, this is called the RELATION KEY or KEY ATTRIBUTE, shown as underlined. In the SALE entity type, the <u>RECEIPT_NUMBER</u> is the key attribute. Its value or receipt number can be used to uniquely identify any particular sale, no other sale will have the same value for receipt number.

An attribute key can consist of more than one attribute of an entity type. The attributes of the CUSTOMER entity type may be: <u>NAME</u>, <u>CUSTOMER_NUMBER</u>, ADDRESS, TELEPHONE_NUMBER. The key attribute for the CUSTOMER entity type is the group of attributes CUSTOMER_NUMBER and NAME. The attribute NAME cannot be the key attribute because there may be two customers with the same name John Smith or Risha Patel. CUSTOMER_NUMBER and NAME can be used to identify uniquely a customer.

> A hotel receives a guest from a booking agent. The guest registers with the hotel clerk who takes the guest's name and address, and records the length of stay and allocates a room.

Consider the following:

In the above description of the hotel problem domain, an analyst may identify GUEST, EMPLOYEE, ROOM and AGENT as entity types. The attributes of the GUEST entity type may include: FIRST_NAME, FAMILY_NAME, ADDRESS, <u>GUEST_NUMBER</u>, REGISTRATION_DATE, and ROOM_NUMBER. The attributes identified of EMPLOYEE may include: NAME, <u>NATIONAL_INSURANCE_NUMBER</u>, ADDRESS, SKILLS, SALARY, LINE_MANAGER. The key attribute for the GUEST is GUEST_NUMBER and for the EMPLOYEE it is the NATIONAL_INSURANCE_NUMBER.

Attributes have a *value*. Table 7.1 shows the attributes and values of the entity type GUEST. The entity and its attributes are a record of the facts about the entity.

Entity types are defined in table types similar to Table 7.1, but without the value column. The key attribute normally appears first and the other attributes below it. If a key is composed

Table 7.1 Entity type, attributes, values

Entity type	*Attributes*	*Value*
GUEST	GUEST_number	007007
	GUEST_first_name	Risha
	GUEST_family_name	Patel
	GUEST_address	1 Kensington Place
	GUEST_registration_date	20/12/03
	GUEST_room_number	234

of multiple attributes they are listed above the other attributes.

7.4.2 Types of relationships

The relationship in an E-R model is used to represent an association between entity types. A relationship between entity types is named. In the hotel example, the GUEST entity type is related to the ROOM entity type by the relationship OCCUPIES, as shown in the E-R diagram in Figure 7.1. An E-R diagram is interpreted by reading every relationship in both directions. A GUEST occupies a ROOM or a ROOM is occupied by a GUEST. A GUEST may occupy one ROOM.

The relationships between entity types vary in accordance with logical, social, organizational and physical aspects of the problem domain. Relationships are defined on the basis of the business rules established by analysts. Business rules are how the organization operates based on its objectives, policies and procedures. In a hotel, rooms are for occupation by guests and earn a rent for the company.

The relationship types are called the cardinality of the relationship because a relationship between entity types has a specific number. Cardinality defines the numeric relationship between occurrences of the entities on either side of the relationship line, they are defined in Table 7.2. For example, in the E-R diagram in Figure 7.1, the relationship between the GUEST and ROOM entity types is one-to-one (1:1) because a guest normally occupies one room. (It is possible for an individual guest to occupy more than one room, say for example when someone may book several rooms for a wedding function.) Another example of a one-to-one relationship is that one person is responsible for a department's budget.

Analysts have to determine the cardinality of the relationship between entity types. They need to examine each entity type in turn to check how many entities occur. The information to determine the cardinality of the relationship between entity types is found in the logic of the problem domain or the business rules established in conceptual data models. In the example from section 7.4.2, hotel occupancy determines how the identified entity types are related, as shown in Figure 7.1.

7.4.3 Normalization

Normalization is used to develop relational data structures for relational database design. Normalization is a series of steps followed to obtain a database design that allows efficient storage and access of data in a relational database. It is based on relational calculus and it is theoretically derived. Analysts need to be aware of normalization but do not necessarily need to explain how it works.

Figure 7.1 Naming entity type relationships

Table 7.2 Entity relationship types

Cardinality of relationship	*Description*
One-to-one (1:1)	The one-to-one cardinality is defined for entity types when one occurrence of the entity type is related to one other occurrence of a different entity type.
	CUSTOMER transacts one SALE (1:1)
	CUSTOMER has one INSURANCE POLICY (1:1)
	EMPLOYEE has one NI_Number (1:1)
	NI_Number has one EMPLOYEE (1:1)
One-to-many (1:m)	The one-to-many cardinality is defined for entity types when one occurrence of the entity type is related to many occurrences of a different entity type.
	SALE contains many ITEM (1:m)
	CUSTOMER can buy many ITEM (1:m) (each item is bought by one customer).
	DEPARTMENT has many EMPLOYEES (1:m)
	DEPARTMENT has many EMPLOYEES (1:m) but each EMPLOYEE belongs to one DEPARTMENT.
Many-to-many (m:n)	The many-to-many cardinality is defined for entity types when occurrences of multiple entities in a entity type are related to multiple occurrences of another entity type.
	STUDENT can enrol on many MODULES (m:n)
	PART can have many COLOURS (m:n)
	EMPLOYEE can be on many PROJECTS (m:n)

Normalization is the process by which entity tables are simplified so that each item of data stored for an entity is in its simplest form. The purpose of normalization is to remove redundant data items from a set of entity type tables to make database storage efficient. It is used to eliminate data redundancy by removing entity attributes that may be stored more than once in relational tables. The normalization process is based on functional dependencies and relation keys or key attributes.

There are various forms of normalization. An entity type table in a relational database system is in normal form if it satisfies certain constraints as defined in Table 7.3.

Normalization is extended to 4NF and 5NF. In 4NF the check is for multivalued dependency. In 5NF the check is for multivalued

Table 7.3 Relational data structure normalization

Normal form	*Comments*
First normal form (1NF)	No repeating groups of attributes and all attributes entered. Eliminate repeating groups by putting each repeating group in a separate table and connect it with a one-to-many relationship. Every piece of information stored in each of the rows of the table should be **atomic** or indivisible.
Second normal form (2NF)	If it is in 1NF and every entity attribute that is not part of the key depends on the whole key. Second normal form eliminates functional dependencies on a partial key by putting the attributes in a separate table form those that are dependent on the whole key.
Third normal form (3NF)	If it is in 2NF and each entity attribute not part of the key attribute depends on the key attribute. Third normal form eliminates functional dependencies on non-key attributes by putting them in a separate table. At this stage, all non-key fields are dependent on the key, the whole key and nothing but the key.

redundancy. These higher orders of normalization ensure that relational data structures contain the simplest and non-redundant data structures.

In the CUSTOMER entity type, the attributes name and address are functionally dependent on the key attribute customer number. There are no redundant data items for the CUSTOMER entity type, as for each CUSTOMER there is only one record of name and address.

The five rules of normalization stated in simple terms are:

1 *Eliminate repeating groups*. Make a separate table for each set of related attributes, and give each table a primary key.
2 *Eliminate redundant data*. If an attribute depends on only part of a multi-valued key, remove it to a separate table.
3 *Eliminate columns not dependent on key*. If attributes do not contribute to a description of the key, remove them to a separate table.
4 *Isolate independent multiple relationships*. No table may contain two or more l:m or m:n relationships that are not directly related.
5 *Isolate semantically related multiple relationships*. There may be practical constraints on information that justify separating logically related many-to-many relationships.

7.4.4 Logical database design

The logical database model is a layer of design between people who use the data in a database and the physical database. People who use the data are employees, managers, and executives. They each have different needs from the data – this is called their 'logical view' of the data. The physical database is the way that data is actually stored in a database on a computer. The term 'physical' refers to implementation dependent characteristics of data and processing. It describes how data is stored and processed on computer disks, tapes and paper.

The logical database model allows different logical views of the stored data in the database. For example, a credit controller needs information on payments made or outstanding and a production manager needs information on quantities and types of products ordered. The logical database contains the overall structure of the database not the physical storage.

In the logical database, similar objects from the problem domain are collected into entity types and represented by their cardinality. E-R models contribute to the system architecture design of a logical database. They are used to develop the logical views. The E-R models depict the database structure. They need to be converted to logical record structures, or the 'logical database' for the detailed physical database design.

The CUSTOMER entity type from section 7.4.1 with its attributes is translated into a logical record structure, shown in Table 7.4. The logical record will store data or values on the customer's NAME, FIRST_NAME, CUSTOMER_NUMBER, ADDRESS, AND TELEPHONE_NUMBER. The same is done for the SALE and ITEM entity types.

7.5 Documentation

The deliverables from system requirements analysis, data and process modelling activities are tangible products. Documentation is a deliverable from data and process modelling. The different kinds of documents produced in SSADM as an example are:

- diagrammatic reports
- forms
- matrices
- narrative reports.

Documentation is easier with CASE tools, which produce higher quality documents that can be rapidly amended while preserving integrity and consistency across diagrams. The deliverable from data modelling is a set of E-R diagrams and from process modelling a set of coherent, inter-related data flow diagrams.

7.5.1 Data dictionary

The data dictionary is a key element of a system project. It is a data resource for systems analysts and systems designers, and other team members. It contains information on the IS to be developed and other sources of information on it, and on existing systems. It contains all the required information to develop an IS, including data definitions and systems programs, and databases information. As it contains information on the data used in the system or other IS, its content is also called 'meta-data'. Meta-data is data that describes or explains other data.

A relational database design results in many entity type tables that become difficult to manage and access. The data dictionary stores information on these tables. It contains information on the logical data and process models, and collation of other system information. In terms of entity types it would contain information on the CUSTOMER, SALE, ITEM entities from section 7.4.1. It contains information on the logical process models and data definitions.

Table 7.4 Logical database records based on entity definitions

CUSTOMER_ SURNAME	*FIRST_NAME*	*CUSTOMER_ NUMBER*	*ADDRESS*	*TELEPHONE_ NUMBER*
Romonavich	George	001007	1 Chelsea Place	0208 5758 4890
Patel	Risha	007007	1 Kensington Place	0208 3892 9240
Brown	Andrew	006911	2 Tedlow Gardens	02789 881 998

7.6 Data modelling and the Critical Framework

7.6.1 Systems ontology

The systems ontology component of the Critical Framework is the basis for uncovering and examining premises in logical data modelling. It supposes certain systems ontology that analysts need to appreciate and evaluate. Figure 7.2 is the Critical Framework populated with critical reflection on logical data modelling and the E-R notation. As the bottom layer shows, many questions concerning the four themes of criticality arise from the assumption of logical data modelling.

To develop knowledge and practice, analysts using structured logical data modelling require particular personal constructs. They need to understand the objectivist systems ontology of structured logical data modelling. Analysts need to consider 'objectivity' as a particular personal construct and its development in practice. Objectivity is the ability of a systems analyst to be detached and impartial of the subject being investigated, and to analyse the problem domain with detachment and impartiality, free of any bias.

Development of E-R diagrams is central in structured logical data modelling. E-R modelling assumes an objective problem domain where data exist independently of analysts and people. Whether data entity types or data-centric view is a sufficient characterization of the problem domain is an issue for analysts to consider in terms of critical skills for applying structured instruments.

Analysts job is to determine business rules, identify entity types, their attributes and relationships. A critical issue is how this can be achieved. It is possible that two analysts may develop entirely different E-R diagrams containing different entity types, attributes and relationships for the same problem domain. Two different E-R diagrams on the same problem domain raise questions about the independent existence of entities and objective modelling. A team of analysts working together would need to reach consensus. This means that in practice – the real human problems – subjective definitions have a significant role in structured analysis.

7.6.2 Real problem domains

Representations of reality are only as sophisticated as the notation language used. The E-R notation is based on a mathematical representation of reality – relational calculus. It relies on a representative sample, the business rules extracted from the facts in the problem domain, to define data structures. It is arguable whether human activity in organizations can be represented in mathematical terms and whether a representative sample is sufficient for IS in a dynamic business environment. The UML object-oriented notation leads to an arguably better reflection of reality and human activity. Its development constitutes a transformatory critique of ontological knowledge and practice.

Structured logical data analysis and systems ontology define human problems – organization and its human activity – in terms of entity types. It further assumes that three types of objects can represent the problem domain: (a) entity type, (b) relationship, and (c) attribute. As analysts have to learn to interpret the problem domain in terms of these three objects, they need to consider whether they are sufficient to represent the complex social reality that is human organization.

Structured data analysis assumes that systems analysts can create logical data models that can be translated into systems implementation models. Analysis in actual situations poses barriers for analysts. A significant barrier is communication with people. E-R diagrams are intended to facilitate communication but people do not generally

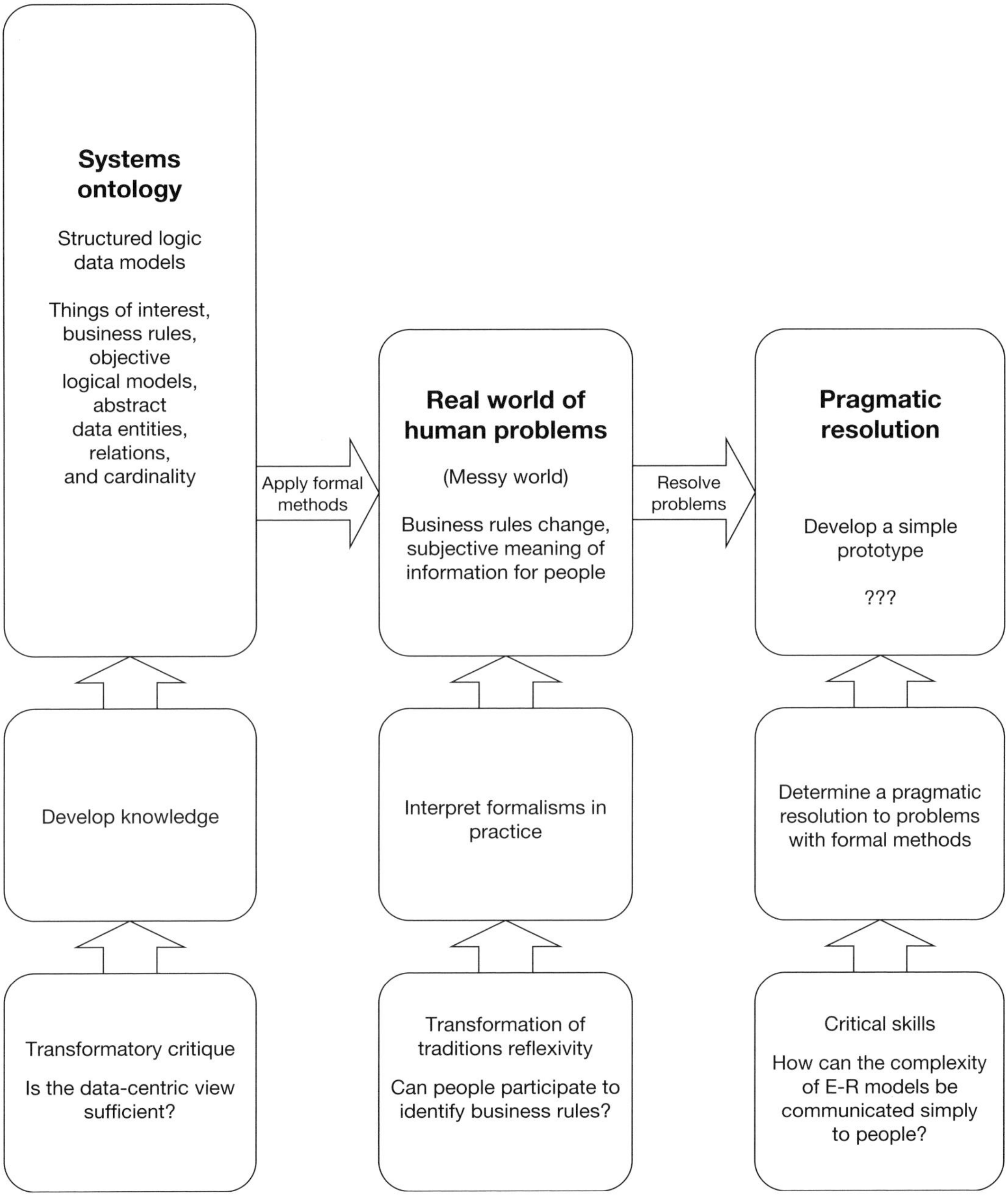

Figure 7.2 Critical framework: systems ontology based on logical data modelling

understand them, and even analysts find non-trivial E-R diagrams difficult to interpret.

Analysts have to question business operations and processes to seek out inefficiencies and improve effectiveness. Representing the problem domain as E-R models poses difficulties for improving organizational efficiency and effectiveness. E-R models are not rich enough to enable analysis of organization. Structured logical data modelling itself does not facilitate this search for business performance improvements. Analysis should produce alternative business process models that take advantage of IT. Radically alternate uses of IT based on business process notations, rather than structured notations, have been spurred by business research, for example BPR, and business practice, for example eCommerce. An example is the Role Activity Diagram (RAD) notation which depicts roles, activities and relationhips.

7.6.3 The representative power of notations

The representative and communicative power of notation languages is significant for systems ontology. Notation languages provide a common language with which analysts can communicate with other systems project team members, users and stakeholders. Analysts' use of E-R diagrams as communicative tools need purposeful thought. It is easier for other analysts, programmers and database administrators to share the mutual and common understanding of E-R diagrams. It is not so transparent for people. They are not technically capable of understanding the E-R notation and so cannot make intelligible observations or comments.

There are limitations of notation languages to represent the problem domain and function as communicative tools. The E-R notation is not sufficiently rich to represent a problem domain. Its focus on data and data effectiveness is at the expense of reflecting social and organizational elements in a problem domain. The social and individual dimension is now accepted knowledge and practice of IS. These limitations have led to other forms of IS development like XP. XP is an example of transformatory critique and refashioning of traditions originating within the practitioner community. It makes use of 'user stories', which are used to establish system requirements and inform the IS development.

7.6.4 Analysts' objectivity

During the 1990s the new IS models completely transformed organization on the basis of BPR knowledge resulting in human resource cost savings. The systems models analysts produce aim to achieve such business efficiency effectiveness, and improved productivity. Detachment and impartiality is sometimes onerous on analysts who have to point out the problems with current IS and recommend cost savings or job redundancies. Objective practice is questionable when analysts work closely with people.

To be objective analysts need to predetermine analysis activities. They need to make plans of how a particular systems analysis will be undertaken. Despite the objectivity criteria, two systems analysts may develop two completely different E-R diagrams for the same problem domain. Rather than being objective, the E-R notation is subjectively applied. Subjectivity is recognized and permitted in alternative systems ontology, for example ASD or prototyping to some extent. In prototyping systems models are developed iteratively and in consultation with people. The emphasis on individuals rather than formalism in ASD recognizes subjective views.

7.6.5 Corporate databases

The challenge in data modelling is to develop relevant data models and databases such that they cater for past and future data requirements. Companies have tried to develop 'enterprise-wide information architecture', but mostly failed because of system integration problems, and incurred high costs. One persistent problem is the localized or fragmented data stores needed for various IS and the lack of integration.

In actual situations there is a constant need to update databases to reflect localized business change. Also, new data models are emerging that lead to a transformatory critique of E-R modelling. For example, image and voice data are now a significant element of organizational data too, requiring new data models combining traditional transaction data, image and sound.

The development of corporate databases through data analysis is problematic and less plausible with constant business change. One solution has been to consistently redefine data, but this is costly to do. Another is to optimize data structures to deliver speed and flexibility, but this results in more time spent doing data modelling, whereas the trend is to shorten IS development time.

7.7 Personal Critical Framework development

7.7.1 Personal constructs for data, information and knowledge

Table 7.5 is a sample repertory grid for data, information, and knowledge. Reproduce the grid on a spreadsheet and add further columns and their polar opposites that you consider relevant. To objectify personal constructs in data, information and knowledge, complete the grid by following the details on how to use a repertory grid in section 1.10.1.

7.7.2 Systems ontology and the problem domain

Questions

1 Evaluate whether E-R modelling is a sufficient notation language to represent a problem domain.
2 As an analyst, discuss whether you would want to develop objective data models.

Table 7.5 Personal constructs for data, information and knowledge

Pole 1	*Data*	*Information*	*Knowledge*	*Business transaction*	*Notation language*	*Model*	*Actual*	*Meaning*	*Pole 2*
Stable									Fluid
Can capture									Cannot capture
Can model									Cannot model
Apply									Cannot apply
Process									Cannot process
Logical									Physical
Fact									Interpretation
Abstract									Real

Provide analytical examples to illustrate your position.

3 Structured logical data modelling is based on the premise that an optimal and efficient solution can be obtained for a problem. Critically discuss.

Activity A

Analysts have to decide what 'things of interest' to include in an E-R model. The choices made affect how the problem domain is conceived, or how they define the real world. E-R models can distort the issues in the problem domain. For a problem domain of your choice:

- Identify and justify entity types, attributes and relationships.
- In what ways do your choices limit representation of real human problems?
- What is the significance of the limitation on decision-makers, who will eventually use the information produced, and the organization?

Activity B

Structured logical data modelling requires analysts to:

1 Identify the relevant physical organization – people, transactions, documents – or the problem domain.
2 Develop 'infological' or logical data models.
3 Convert the logical models into the physical design for computer implementation.

In a group:

- Identify a problem domain, say university student records or supply chain management in a company.
- Make a list of the physical things of interest in the problem domain.
- In 'infological' terms, explain why the things you identified are of interest.
- Make a list of the entity types of interest, for each entity type determine its attributes, and state the cardinality of the relationships between entities.
- With a peer discuss whether 'infological' is a relevant personal construct for you.

7.7.3 Conceptual and logical data models

Questions

1 The notion of 'model' is a significant intellectual tool for framing, analysing and understanding actual human problems. Critically evaluate the role of modelling in IS development.
2 Evaluate the importance of abstract models in the process of IS development. (Investigate ASD as an alternative.)

Activity A

Before E-R diagrams can be developed, analysts need to develop conceptual data models from the problem domain based on the 'business rules'. Consider the following facts or things of interest in the Human Resources department problem domain: department, divisions, employees, function, roles and skills.

Some gathered information on the things of interest is:

An employee always has a job title.
An employee always has zero or more skills.
An employee always has one and only one role.

Sample Facts: Risha Patel is a software programmer currently working in the IS department as a project manager. Brenda Jones is a database specialist currently working in the IS department as a database administrator.

- Define the area of interest.
- Define the things of interest.
- Analyse the things of interest and identify the corresponding tables.
- Establish the relationships between the tables.

7.7.4 E-R models

Questions

1 In real human problems practice and the application of formalism like E-R notation to develop IS is important. Discuss the value formalism has for you as a systems analyst.
2 Explain why you would decide to represent an object in the problem domain as one or more entity types rather than as attributes of an entity type.

Activity A

In pairs each person:

- Independently look at Figure 7.3 and write a description of the problem domain you think the E-R model represents.
- What assumptions have you made in your description?
- Compare your description and assumptions with your peer.
- Discuss the problems subjectivity raises for interpreting E-R data models.

Activity B

- Select an organization familiar to you.
- What are the entity types of interest in it?
- List the attributes of each entity type.
- Underline the attribute in each entity type that is its unique identifier.

7.7.5 Separation of systems design from implementation

Questions

1 Evaluate the benefits to structured systems ontology of computer-independent design.
2 What kind of problems might arise in translating or transforming analysis models into implementations (design models)?

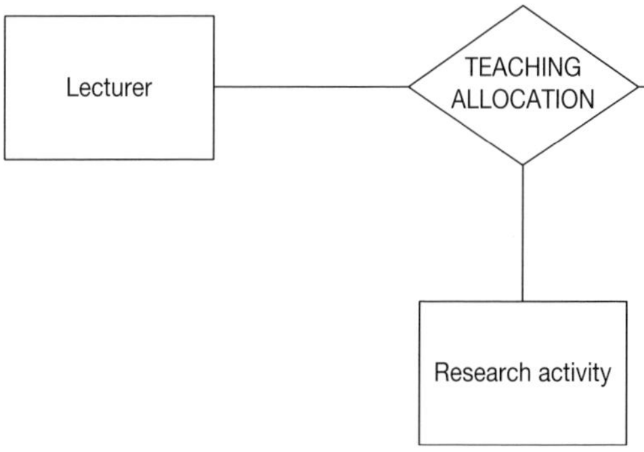

Figure 7.3 Teaching allocation relationship

7.7.6 Internet sources

An excellent source for definitions relevant to data modelling is the following website: http://www.datamodel.org/DataModelDictionary.html. Also look at www.datamodel.org as it offers a community of data modellers.

Surf http://www.databaseanswers.com/data_model. It provides free data models. You can examine a variety of structured and object data models and the associated information on database design.

7.7.7 Further reading

Schmidt, B. (1998) *Data Modelling for Information Professionals*, New York: Prentice Hall.

Reingruber, M. and Gregory, W. (1994) *The Data Modelling Handbook: a Best-Practice Approach to Building Quality Data Models*, Winchester: Wiley.

Chapter 8

Structured process modelling

8.1 Learning outcomes

After completing this chapter you should be able to:

- Interpret the problem domain in terms of processes and data flow diagrams.
- Develop simple DFD diagrams with DFD notation and logic modelling techniques.
- Analytically evaluate the role of process modelling in IS development.
- Critically assess how well data flow diagrams describe a problem domain.

Process modelling is important in structured analysis. It determines how a new IS will function to produce required information. It draws on the E-R models from logical data modelling to define system functions and data processing.

8.2 Introduction

A logical process model depicts processes that transform data into other data or information. Like logical data modelling, structured logical process modelling is the development of non-implementation specific logical process models. The models are of the logical process required to complete business tasks or processes. Process modelling is a descriptive act in which the organization is interpreted in terms of processes. Descriptive process models 'describe' how processes are enacted in the problem domain.

Logical process modelling removes physical references from the process description such as who does what, using what medium, and where in the process. If physical aspects were to be modelled it would restrict systems design decisions later. The purpose of logical process modelling is to objectify or determine all the functions or data transformation required in the system. Process models are used as a communicative device among the systems project team and with people.

Processes are dynamic and reflect the nature of business organizations. A typical business process in a manufacturing company is a sales order process, for which transaction data is generated. It involves a customer contacting the company to place an order. Sales make a record of the details of the order, accounts generate an invoice, production produce the goods, and transportation ship the finished goods to the customer.

Logical, also called 'infological', process models provide a process perspective of a new IS. Processes can be defined in terms of what they produce. A sales order process results in a sales order being generated, or a market research process results in a marketing plan. Analysts refer to the series of actions required to generate a sales order or a market plan as 'logical business processes' or processes. The term 'process' describes the series of actions that people take to complete a task, like taking a customer order or producing a marketing plan. Taking an order for a sale requires a clerk to take details of the customer, number and type of units required, and to pass them onto production. In efficiency terms, process modelling covers actions taken to achieve an objective such as customer relationship management, the sales process or a market research process.

8.3 Process modelling techniques

Process modelling techniques are formalism for representing processes as systems process models. The developed process systems models demarcate the things to be performed by the computer system from things that people do manually. They set the boundary between human activity and computer processing. Such models describe what data transformation the system will do and how the data will be transformed.

The process modelling technique used in structured systems analysis is Data Flow Diagrams (DFDs). Another structured technique is State Transition Diagrams. The DFD provides a graphical representation of processes or functions in the problem domain that will compose a new IS. It shows data transformation through processes that capture, manipulate, store and distribute data between components within a system and between a system and its environment.

8.4 Data flow diagrams

The DFD is used to depict the scope of a new IS, show the interaction of humans with the system, and the system with its environment. Analysts' task is to identify the processes, develop process models and write documentation to record the findings. DFD depict how data is processed in system terms with initial inputs, and outputs from processes used as inputs for other processes. The DFD depicts:

- context
- system boundary
- processes
- inputs
- outputs
- subsystems
- interface.

A DFD depicts how logical data move around in a system. It describes logical data at rest, moving, and being transformed or processed. There are four types of DFD:

- Current physical: description of the existing physical system, if one exists.
- Current logical: description of the 'infological' existing system, if one exists.
- New logical: description of the new 'infological' system required.
- New physical: description of the new physical system.

Usually all four types are developed by analysts. The current physical process model is developed to gather information about the problem domain. The current logical process model is developed to separate the logical processes from the physical things they depend on to be accomplished. These models can be analysed to examine current processes and workflows. The new logical process models are developed to depict system requirements, and they remove any inefficiency observed in current physical and logical process models. The new physical process models are developed to describe implementation features.

8.4.1 DFD notation

There are many notations for drawing DFD. The one adopted here is shown in Figure 8.1. Definitions of the symbols are given in Table 8.1.

Analysts make some typical errors in drawing DFD. These are summarized as:

- connecting an external source directly to a data store;
- showing a process with no outputs;
- showing a process with no inputs;
- connecting a data store directly to another data store.

Following the rules listed in Table 8.1 can prevent such modelling errors.

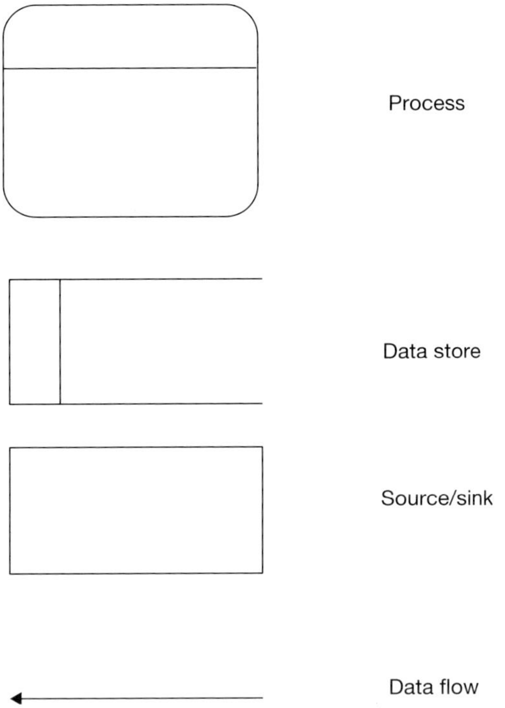

Figure 8.1 DFD notation

8.4.2 Levels, decomposition and balanced DFD

Detailed DFDs are drawn at different levels of granularity using the decomposition or the divide and conquer problem-solving strategy. It enables a complete set of requirements to be captured. Decomposition results in levels of DFD that contain further detailed description of the process and sub-processes. This method is called functional decomposition or levelling. It is an iterative process consisting of breaking the description or perspective of a system down into finer and finer detail. Decomposition of processes is performed until no further benefit is gained by further decomposing a DFD. The result is a set of hierarchically related DFD in which one process of a given DFD is explained in greater detail in another DFD.

The extent of levelling depends on the level of abstraction required. Skilled analysts are able to decide which processes to decompose further and when to stop decomposing a particular process. The general rule is to stop when the most primitive activities of the process is reached. This is the lowest logical or elementary level of the process, which cannot be further decomposed.

Levelling begins with a Level 0 diagram. The Level 0 diagram, also called a context diagram, is a top level DFD. It shows the process node – process 0 – as a general representation of the entire system in terms of external entities. The process node at Level 0 is a black box with inputs and outputs. The details of how it converts inputs into outputs are not known but become clearer when the process is decomposed further. It depicts time, volume and frequency issues in the system, and shows coupling

Table 8.1 Definitions of DFD symbol notations and DFD drawing rules

Symbol label	*Definition and drawing rules*
Process	A process is the work or actions performed on data so that they are transformed, stored or distributed within or outside of the system. Every process is triggered by an event. Every process uses or transforms data. *Rules* A process cannot only have outputs. A process cannot only have inputs. A process has a verb phrase label.
Data store	A data store is data at rest. *Rules* Data cannot move directly from one data store to another data store. A process can only move it. Data cannot move directly from an outside source to a data store or move directly from a data store to an outside source. It must be received by a process and placed into a data store and placed into an outside source by a process. Data cannot move directly into an outside sink from a data store. A process must put it there. A data store has a noun phrase label.
Data flow	A data flow is data moving from one process in a system to another. Also referred to as data that move together. *Rules* Data flow can only flow in one direction. A fork indicates a copy of the same data flowing. The same data coming together in a coming location is a join. Data flow must be processed before returning to the same process it came from. Data flow to a store means updating either through deletion or change. Data flow from a store means retrieve or use. A data flow has a noun phrase label.
Source/sink	A source is the origin of data and a sink is the destination of data. *Rules* Data cannot move directly from a source to a sink. A process must put it there. An object with only outputs is a source. An object with only inputs is a sink. A source or sink has a noun phrase label.

Source: Adapted from Celko, J. (1987) 'I Data Flow Diagrams', *Computer Language* 4 (January): 41–43.

and decoupling of entities. Entities are coupled when one entity is bonded with another. They are decoupled when the bond is weaker. Coupling is important during systems design when programs have to be designed and coded. Decoupled programs are easier to maintain when changes in the problem domain need to be reflected in them.

When the Level 0 DFD is decomposed it is necessary to conserve the inputs and outputs in subsequent sub-processes. Data flows present at one level should also be present in a subsequent decomposed level. This is called balanced DFD. An unbalanced set of DFD result when inputs or outputs at one level are not reflected at a further decomposed level. An example is when a process in the context diagram has one external source but at level 1 the same process has two external sources.

An example context diagram for a university PC inventory system is shown in Figure 8.2. It depicts a business problem domain. It shows the overall 'system', the system boundaries, external entities and inputs and outputs. Everything within the boundary, Process 0, will be performed by a new IS and everything outside the boundary is the system's environment – supplier and maintenance contractor. Demarcation of the boundary is the first decision for physical design, because it shows what the computer will do. The boundary may cut across processes indicating that some processes will be computerized in part only. This is more explicit in the SSADM methodology.

Points at which humans interact with the computer system over the boundary become the human–computer interaction interfaces (user interface), or system interfaces with other IS, or sources of data. Analysts establish a list of events in the system's environment to which the system needs to respond. These events may number into the thousands for complex systems.

At Level 1 shown in Figure 8.3, the node process in Level 0 is decomposed into other processes and each sub-process has its business objectives. Two processes are modelled called Clerk Logs Invoice and Accounts Pay Invoice. A data store is created called Invoice File. The arrows show the flow of data between the processes and between the processes and the data store. Subsequent decomposition leads to functional primitives or the elementary level, which through process logic modelling, are converted, through systems design, into programme code.

8.4.3 Entity life history

As a DFD is weak at identifying events or data, entity life history (ELH) diagrams are used to ensure that the system has a response for all the internal and external events. An event is any

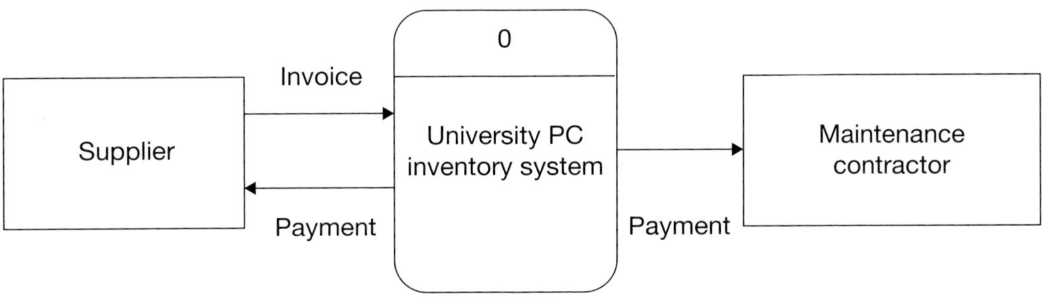

Figure 8.2 DFD context diagram for a PC inventory system

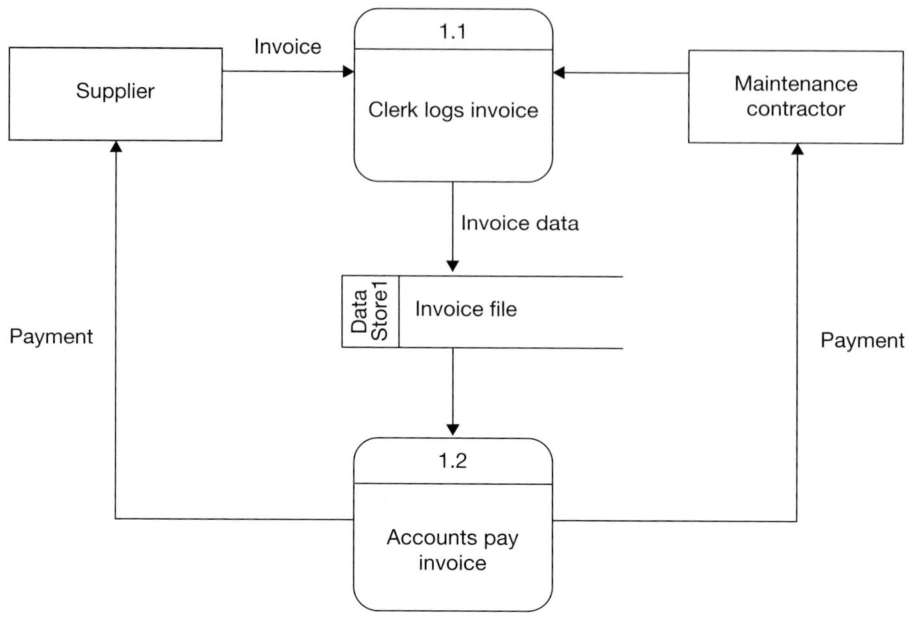

Figure 8.3 Level 1 DFD for PC inventory system

occurrence in the entity's life. For example, for the entity type INVOICE, logging an invoice and settling the invoice are events. It is the 'life history' of the entity in the system, and can be drawn for logical or physical entities. In structured systems analysis, the life of an entity consists of three types of events sufficient to represent the complexity of any life:

- sequence
- selection
- iteration.

Figure 8.4 shows the basic life of the invoice entity with two nodes' details shown to third level. Using SSADM notation it shows sequence, selections and iteration. It must have a 'birth', events that happen during its 'life', and a time of its 'death'. Selection is shown with a '0' in the box and iteration is shown with a '*'. For example, an invoice is received, processed and paid. The received invoice is logged as either for cash payment on receipt or three month credit, so it is shown as a selection with a '0'. Processing an invoice is iterative, so it is shown with a '*'.

ELH diagrams ensure that errors in the DFD are corrected before the system boundary is fixed. Such diagrams can be complex with several levels of detail. They also show the time dimension of events. Analysts normally only develop ELH for entities that are important or complicated.

The E-R model, DFD and ELH diagrams must be balanced. They must reflect the same content and be logically consistent. Such abstract data and process models enable analysts to 'walkthrough' the systems design with potential users to verify system requirements.

8.5 Process logic modelling techniques

When a DFD is decomposed to its elementary or primitive level, analysts have to describe its

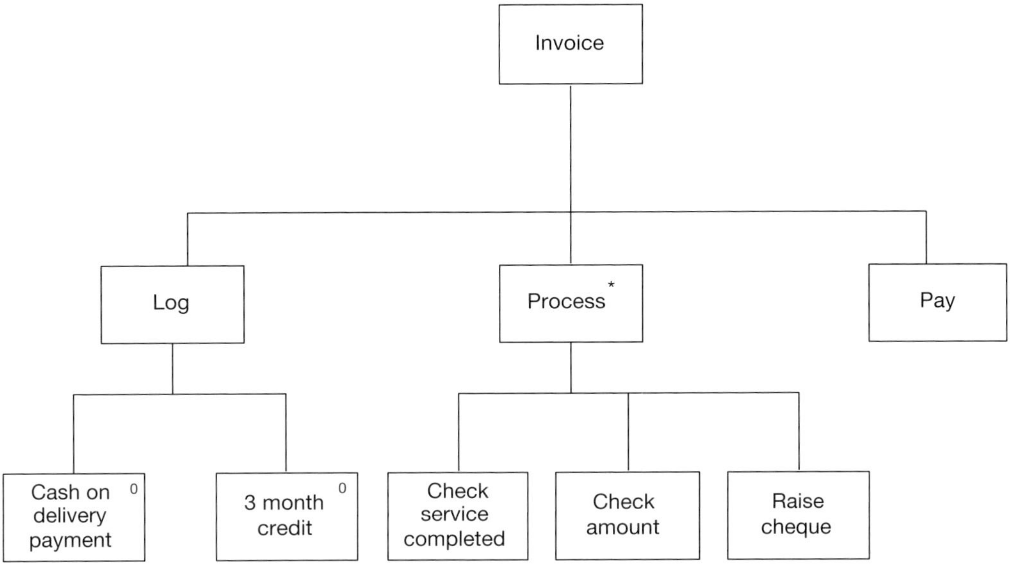

Figure 8.4 Entity life history form

process logic. The action taken to transform transaction data into business information is termed business logic. Process logic may be described as the data-to-information transformation and decisions on data. It is not quite pseudocode or programming language specific but sufficiently detailed to describe business logic. The term process logic describes the business actions taken on data, or what a process does and how it does it. Determining process logic results in a model descriptive of how the process is controlled and the detailed action taken on data in it.

For example, in a customer order process, decomposition may result in a sub-process for calculating discounts given. The discount policy may consist of giving percentage discounts depending on the size of the order and distance of delivery. So there will be various discount decisions. Process logic modelling is used to provide detailed descriptions of these discount decision actions.

There are many notations for depicting process logic. They enable clear and unambiguous descriptions of activities on data in a process. As the techniques are not complementary, analysts need to select an appropriate one for describing the logic of elementary processes. The choice will depend on the complexity of the business logic in the process. The logical models can be developed using a variety of notations:

- structured English
- decision tables
- decision trees
- action diagrams
- Nassi-Schneiderman diagrams
- entity life cycles
- flow charts.

Logic modelling involves representing the internal structure and functionality of the processes depicted in the DFD. Like the DFD, logic diagrams require iterative refinement. Analysts share their work with people to seek feedback for correctness checking. Process logic models serve to provide an unambiguous and thorough explanation of the system's specification. The deliverable from logic modelling are structured descriptions and diagrams of the logic of each DFD process. Logic models show the temporal dimension of system – when something happens and how it happens.

8.5.1 Structured English

Structured English (SE) is a modified form of English used to specify process logic. SE is program-like structured English descriptions of processes that use simple sentence constructions and logical constructs. It is used to model the business policies or rules in the problem domain and to refine models of business processes. Process logic can be refined with SE to a level where all parties can understand unambiguously the process being described. This is regarded as a benefit for communication of process logic modelling with SE.

SE uses action verbs like read, write and move to describe action on data, and noun-phrases like stock-item or high-value-customer to describe entities. It uses four types of logical constructs: sequence, conditional statements, repetition, and case. Examples are shown for business policies in Table 8.2.

The logical constructs used in SE are also used in structured programming, which makes SE suitable for process implementation, or program code. SE is not suitable for describing process logic with large number of variables in the process. In such situations decision tables are more appropriate.

8.5.2 Decision tables

A decision table is used where the business process logic is more complex. A decision table is based on a declarative format that is clear and suitable for communication and program specification. Decision tables are used for displaying large amounts of complex data. A decision table has a condition stub, action stub and condition rules, as shown in Figure 8.5 (a). A limited-entry decision table with process logic is shown in Figure 8.5 (b).

In Figure 8.5 (b), the decision table shows a PC manufacturing company's discount policy. The number of condition entries depends on the number of conditions and the binary (Y or N) entries in decision tables. Since there are three conditions and two possibilities for each, the number of columns required is $2^3 = 8$.

Reading down the first condition entry column, a customer with a sale of more than £20,000, who has been with the company for more than two years, and has a repeat sale is given the maximum discount of 10 per cent. In the eighth column a new customer with a sale of less than £20,000, who has not been with the company for two years, and makes their first sale is given no discount. Other columns show the other permutations. During the iterative development of the table it is possible to reduce the number of columns similar to the functional decomposition in DFD.

Extended entry decision tables are used when there is not a simple binary yes or no condition entry. Some process logic requires further differentiation. For example, the sale could be differentiated in three ways: < £5,000, ≥ £5,000 but < £10,000, and ≥ 10,000. These conditions would be placed in the condition entry stub and the condition would be the size of sale.

8.5.3 Decision trees

A decision tree graphically depicts a decision or choice situation as a connected series of nodes and branches. It is good for displaying process logic graphically but becomes cumbersome to follow when the decisions are more complex. Analysts have to think of the process logic as

Table 8.2 Structured English

```
Sequence
DO
        READ next stock-item
        ADD new order from order-file to stock-file
UNTIL End-of-file
General Form
DO A
DO B
Conditional statement
BEGIN IF
        IF stock-level is less then minimum-order-level
        THEN GENERATE new order
        ELSE DO nothing
END IF
General Form
If X is true
        Then Do Y
        ELSE Do Z
Repetition
READ Stock-File
WHILE NOT End-of-File DO
        BEGIN IF
                IF stock-level is less then minimum-order-level
                THEN GENERATE new order
                ELSE DO nothing
        END IF
END WHILE
General Form
Do A While B is true
Case
READ Stock-level
CASE 1 (stock-level greater than minimum-order-level): Do nothing
CASE 2 (stock-level equals minimum-order-level): Do nothing
CASE 3 (stock-level is less then minimum-order-level): GENERATE new order
CASE 4 (no stock): ENTER emergency order
General Form
Case 1 A > 1,000 Do B
Case 2 A < 1,000 Do C
Case 3 A = 1,000 Do D
```

(a)

Condition stub	Condition entries
Action stub	Action entries

(b)

Condition stub	**Condition entries**							
Sale > £20,000	Y	Y	Y	Y	N	N	N	N
Customer for > 2 years	Y	Y	N	N	Y	Y	N	N
Repeat sale	Y	N	Y	N	Y	N	Y	N
Action stub								
Discount 0%				X			X	X
Discount 3%					X	X		
Discount 5%			X					
Discount 7%		X						
Discount 10%	X							

Figure 8.5 Decision tables

conditional alternatives with resultant actions. Figure 8.6 shows a schematic example decision tree. A node is numbered and has two or more alternative paths. The action is the final policy decision reached along a particular branch. Probabilities can be attached along the paths to show the likelihood of a particular outcome.

Decision trees are better communicative tools because people in the problem domain can easily follow them. As they graphically show the basic business logic, people can follow them and verify them with analysts. An illustrated example of a university examination board's condonement policy with conditions and actions completed is shown in Figure 8.7.

8.5.4 Action diagrams

Action diagrams use a limited set of English to depict process logic. This set includes the structured programming constructs sequence, condition, repetition, case and repeat . . . until,

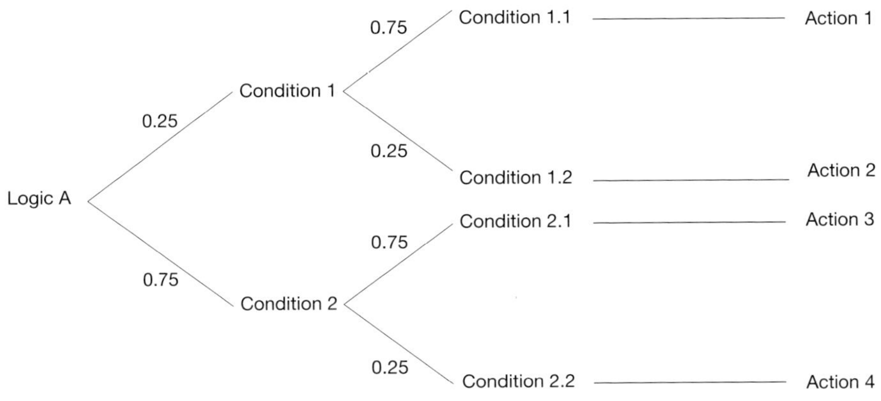

Figure 8.6 Schematic decision tree

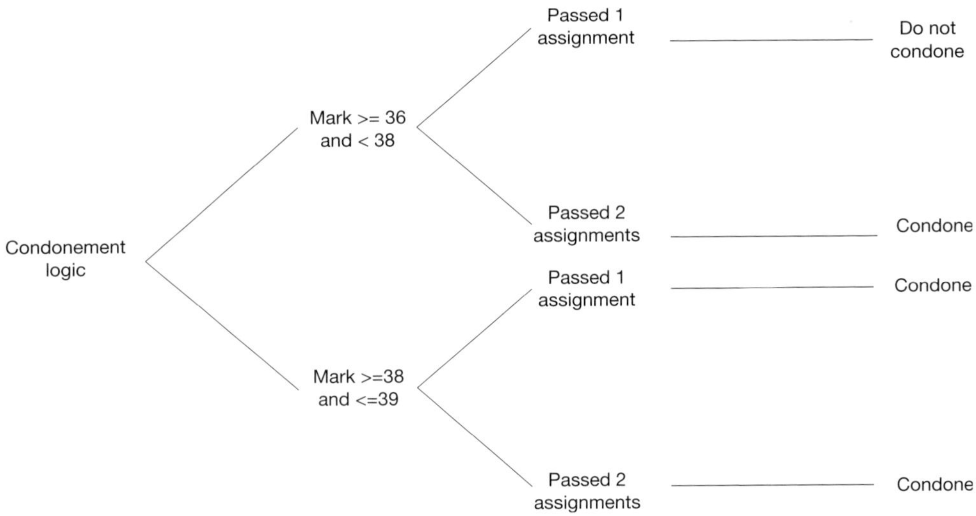

Figure 8.7 Student module condonement logic

similar to SE. An action diagram is good for showing simultaneously the detail and overview of process logic. Figure 8.8 shows an insurance risk calculation example using sequence only. It is drawn using a bracket indicating control, which can be nested to show a hierarchical structure. The bracket envelops a set of actions that are performed in sequence.

Action diagrams can be complex, depending on the business logic to be modelled and whether database functions are modelled. When

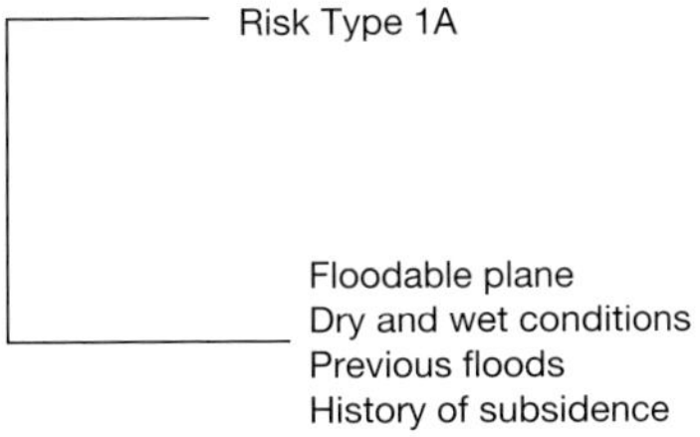

Figure 8.8 Action diagram

conditions, repetition, case and repeat are part of the decision logic, the diagrams can be long and nested and difficult to read and interpret.

8.6 SSADM technical products

Knowledge of SSADM technical products is useful for an overview of IS development, covering data, process and logic techniques. They are listed in Table 8.3 in sequence to reflect the phases of IS development. SSADM contains project management, technical and quality products. Among the technical products are application products. The application products contain the systems analysis and design products, or the system data and process models. E-R data modelling and DFD process modelling are SSADM systems analysis techniques.

SSADM complementary techniques include Quality Assurance Review, Formal Documentation, Project Control Methods, and CASE tools. The DFD notation used in SSADM is shown in Figure 8.9.

Table 8.3 SSADM technical products

SSADM techniques and models
Logical data models
Data flow diagrams
Requirements definition
Function definition
Specification prototyping
Relational data analysis
Entity/event modelling (entity life histories, effect correspondence diagrams)
Business and technical options
Dialogue design
Update and enquiry process models
Physical data design
Physical process specification
Physical design control

8.7 Process modelling and the Critical Framework

Structured systems ontology aims to provide 'a common value system' for IS developers. It assumes that knowledge and information is perfectible. Analysts need to analytically evaluate and critically examine process modelling and logic modelling and its structured systems ontology to develop personal constructs for PCF. The level of ontological questioning depends on how nearer to the 'truth' analysts want to get. Some assumptions made by structured process modelling and issues are examined here.

Figure 8.10 is the Critical Framework populated with critical reflection on structured and process modelling. It shows basic ontological assumptions in process modelling. Many of these are questionable because they are based on rationality, plans and plan-based human action. As the bottom layer shows, many questions concerning the four themes of criticality arise from the assumption of objective processes

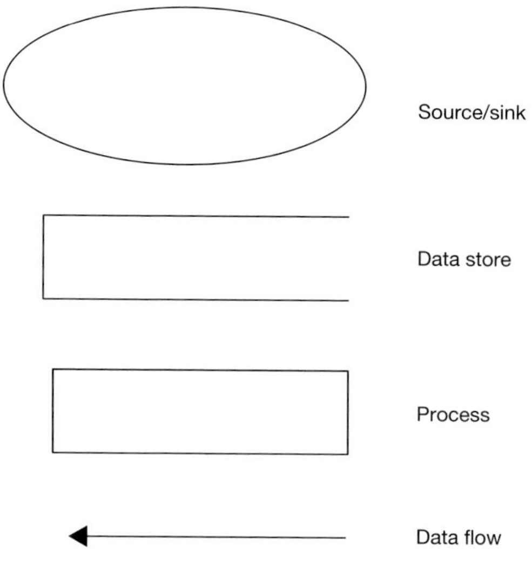

Figure 8.9 SSADM data flow diagram notation

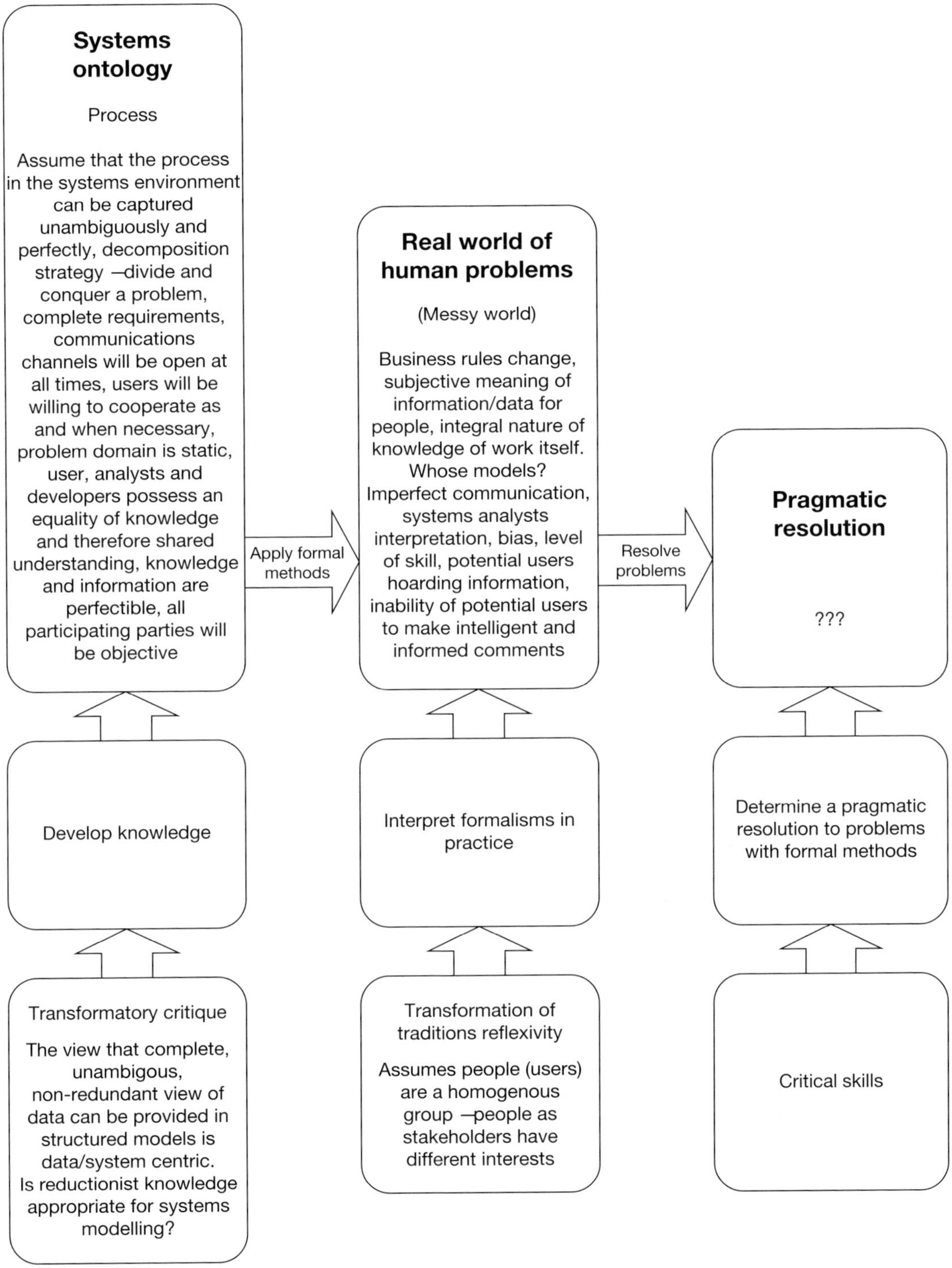

Figure 8.10 Critical framework: systems ontology and process capture

and business logic. The critical observation arises from the basic assumption of rational human action in structured systems ontology.

The 'process' concept has wider application. It is not restricted to structured systems ontology. Analysts should consider alternative process modelling techniques including Oracle's CASE designer tool, RAD and JAD. The notion of process extends beyond structured systems ontology to cover a wider scope in systems analysis. Some prominent process description techniques include: Petri-nets, Role Activity Diagrams (RADs), and IDEF0. There are four types of process modelling techniques, though a particular technique may reflect more than one category:

- functional
- behavioural
- organizational
- informational.

Process description techniques have improved the scope for innovation using IT. Earlier applications of computers were to automate existing tasks and procedures. IT combined with process knowledge now enables radically innovative applications, especially in BPR, eCommerce and knowledge management systems. These applications have transformed organizations with significant electronic infrastructure and informational and knowledge content.

8.7.1 Reductionism

Descriptive structured data and process models are derived using reductionism as a problem-solving strategy, or functional decomposition. They simply focus on the required IS functions modelled on business data because they are thought to provide more stable design and reduced data redundancy. Reductionist models are abstractions that do not reflect complex social, political and organizational reality. They marginalize critically important human factors in IS.

Structured systems ontology is broken down into three sequentially derived components. Structured systems analysis is done first and Transform Analysis is applied to it to derive the structured systems design. Structured systems ontology assumes that systems design can be derived from systems analysis. To progress to the design stage the Transform Analysis needs to be completed. This step is used to convert systems analysis diagrams into systems design diagrams. The conversion assumes that a series of rules can be applied to transform, for example, DFD, into systems design diagrams. This is questionable since structured systems analysis is mostly logical with no implications for physical software design.

Reductionism is the opposite of holism. Holism is the view that the whole exhibits properties that are more than just the individual parts put together, allowing for emergent properties of human action. Structured techniques do not adequately reflect the holism of the problem domain because they decompose it into parts such as data and processes. They do not reflect the holistic nature of interconnected organizational behaviour and its emergent properties. SSM is designed to model such emergent properties and is used in the Multiview methodology.

8.7.2 Representation

Structured systems ontology is composed of data and processes. It seeks to represent the problem domain as objectified data and process models, but the actual situation is composed of peoples' meanings, which are not represented in structured models. The relationship element of E-R diagrams aims to add semantics to the data, but it does so only in the context of abstracted data entities.

Structured systems ontology requires analysts' to be objective and detached to represent the problem domain. Information arising from systems analysis, though, needs to be transformed into systems design decisions. Analysts may emphasize some features of the problem domain more than others, and even downplay others to facilitate communication with people. New IS are constrained by subjective strategic business input.

Structured systems ontology does not acknowledge that logical modelling requires a 'creative leap' where analysts have to exercise imagination. They have a 'blank canvas' which they have to fill with systems models. The creative leap goes beyond simple objective and detached modelling. Analysts' intuition and judgement, based on experience, are part of the creative leap. They create something that does not yet exist.

A central issue is 'freezing' the representation as systems models. Once the models are developed and the system requirements 'agreed', the modelling stops. In actuality there is always some change required or emergent issues to be addressed in a system project. Stopping modelling in structured systems ontology has consequences because the actual situation does not 'stop'. Entities and relationships may change and processes may become superseded by managers' decisions or competitors' actions. Business logic may change to maintain competitiveness or react to market conditions.

Critical observations of the logical data and process modelling techniques can be made too. The Critical Framework in Figure 8.11 depicts a sample. Structured analysis was developed to provide a common value system, promoting a uniform view of what is or is not desirable in a system using structured methods and techniques. Ironically, logical models can be unclear, fuzzy and confusing because, through abstraction, they become too detached from the problem domain. Advocates have argued that the term 'essence' is a better description than models. Such semantic differentiation though is superficial and it fails to address the limited capability of structured systems ontology to represent real situations.

There are limits to descriptive modelling techniques. The representational capacity of structured English is limited when a company has several actions related to a particular set of conditions. The resultant logic model is a complex and cumbersome set of statements that can confuse even systems analysts, leaving potential users with no scope to make intelligent and informed comment. The complexity of most business logic makes SE unsuitable for modelling process logic.

The DFD lacks information on how long data takes to move from one process to another and it is unable to explain the direction of flow. The notation does not represent interrupts or signals. It does not capture when IS need to be synchronized and coordinated to achieve tasks. Consequently, modelling techniques have been devised to overcome some of the temporal problems in the DFD. State-transition diagrams, control process and control flows all enable the capture of interrupts and signals in a system and the depiction of synchronization and coordinated behaviour to achieve goals.

A DFD can be unreasonably continually decomposed. The decision to stop decomposing DFD is taken by the analyst and this may lead to less overt requirements being overlooked. Too much time can be spent modelling the current physical and current logical system, especially if a complex system with more than 100 events is to be modelled. It can make the consequent models cumbersome to decipher for both analysts and people and it can be politically dangerous and may cause the system project to be cancelled.

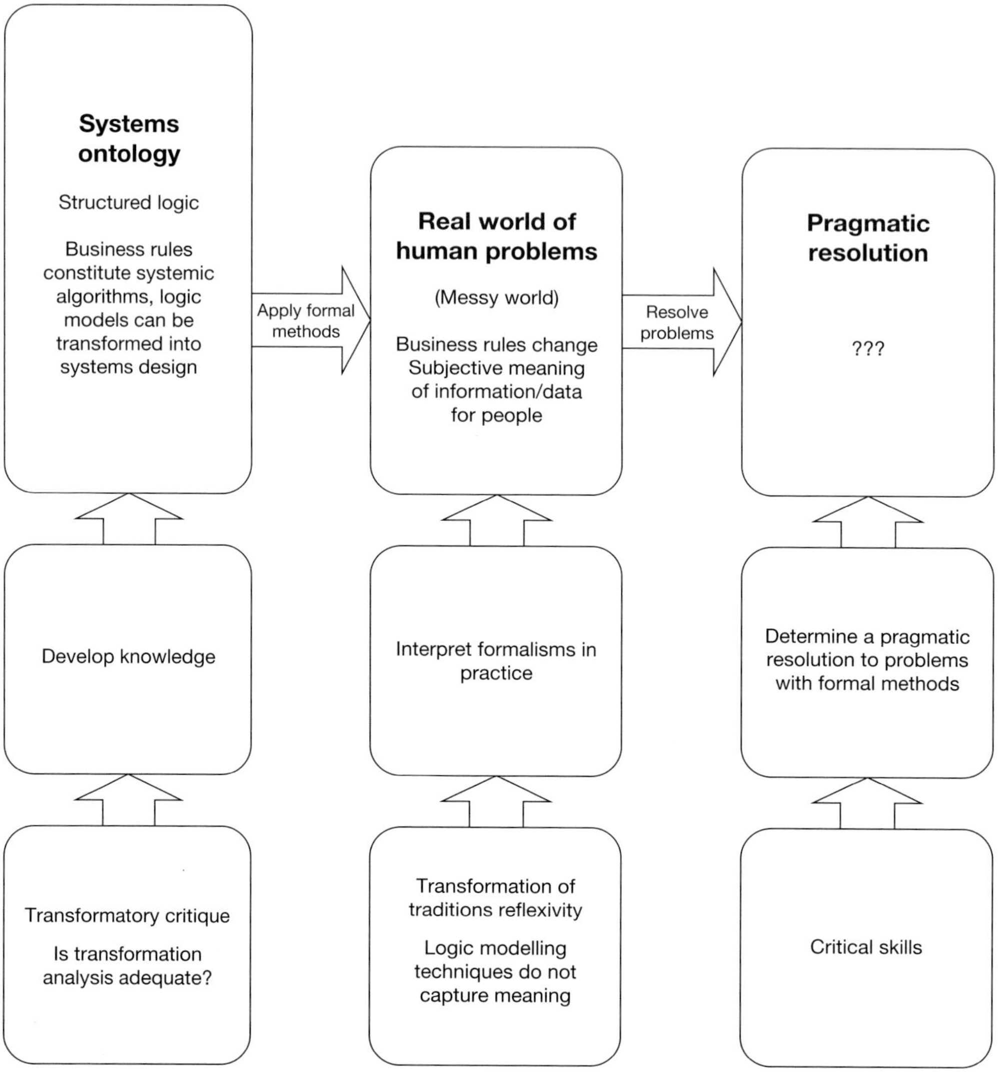

Figure 8.11 Critical framework: perfect and unambiguous logical modelling

Advocates of the structured approach regard the data dictionary as critical, without which the process models are a 'rough sketch'. It is used to show what data is processed and how it is processed. In practice the use of a data dictionary among potential users and practitioners is limited and it is more likely to be used in conjunction with developed systems models to verify requirements.

Other issues in representation include:

- The DFD has no explicit mechanism for identifying reusable components to facilitate design by extension or reuse as in object-oriented IS development.
- Though Level 0 and Level 1 DFD identify human–computer interaction interfaces, they provide no guidance on developing the user-interface.

- The setting of a system boundary is arbitrary, and may overlook real business performance improvement opportunities.

8.7.3 Repeatability

Structured, object-oriented and even ASD systems ontology assumes the problem domain will repeat itself once modelled and implemented in a new IS. The repeatability flaw arises because of reliance on planned action and passive models based on an assumed static real world of human problems.

Structured methods assume that the problem domain is repeatable. This assumption is possible within structured systems ontology because of the allied assumption that the problem domain is static. Structured systems ontology assumes that logical data and process models, once implemented, will repeat in the problem domain in the form of a new IS.

The assumption that the problem domain is predictable or repeatable has posed difficulties for researchers and practitioners. It is highly questionable in the context of modern changing organization. The social and economic context of organization mean that modelled entities, relationships, and processes may require changing. New entities and relationships may need to be added and others amended or removed. Such change causes problems during the modelling process, and is certainly problematical once an IS is implemented.

8.7.4 Representative sample

Structured modelling assumes that a representative sample can be produced and relied on to make data and process models to represent the problem domain. The E-R and DFD techniques assume that data and processes need to be modelled once – the representative sample. How this sample is determined and evaluated depends on the decisions made by steering committees, project managers, senior analysts and senior designers. At local level the autonomy of analysts influences it too, because they decide what features of the problem domain to emphasize or avoid in systems models.

The representative sample results in static or passive models. A passive model once developed is detached or independent of the subject – the problem domain. Changes that happen in the problem domain or emergent features are not reflected in the developed models subsequently. So the models become outdated, resulting in the problem of legacy systems. The representative sample assumption is related to the repeatability assumption. It further assumes that it is possible to develop knowledge of information needs in advance of human activity that has yet to happen.

The E-R and DFD diagrams separate systems modelling activity from the actual physical implementation of required functions. Separating analysis and design from its physical implementation is questionable because it leads to passive models. Emerging approaches seek to develop active models in which the systems design and implementation are intertwined, for example in ASD.

To overcome the problem of passive models resulting from the repeatability and representative sample problems, other process modelling approaches develop active models. An active model maintains the relationship with the subject – the problem domain – after it has been created. This allows systems developers to keep developed IS in tune with the needs of the problem domain. Structured systems ontology does not include active systems models. Object oriented systems ontology is capable of developing active models through deferred classes.

The power structures within an organization, and the problem domain, affect people's influence on the models developed. Power and politics is not recognized in the structured systems ontology. The power structures are prominent during requirements determination and modelling. Stakeholders and people with the most power exert the most influence on the models developed. Such power relations do not lead to a representative sample.

8.7.5 Shared understanding

Structured systems ontology assumes people and practitioners have shared understanding of required new IS. A system project and its management relies on the same assumption. This assumption lacks validity from several perspectives and raises the pertinent issue of ownership of models depicted in Figure 8.12. Structured systems ontology gives ownership of systems models to developers. The technical nature of the structured IS development process inherently means that only developers can understand the models.

Structured systems ontology assumption of homogeneity of people is erroneous. It is assumed people have a shared understanding of the problem domain, which analysts only need to elicit as data and process models. Research reveals that system projects have stakeholders with varying interest and power to influence a new IS development. Rather than a shared understanding the problem domain is composed of social conflict and power relations between stakeholders, users and developers, which some researchers argue results in a 'winner and loser'.

The necessary shared understanding breaks down when E-R and DFD techniques and logic modelling are applied. Their focus is on functions and algorithms. The DFD does not consider relationships between data that is stored in many places in an actual organization. This relationship between data sources and data sinks exists in one form physically and is interpreted in another form by analysts in logical models, creating differences in understanding between people and practitioners. Similarly, the DFD does not capture the temporal aspects of physical data. The sequence in which data is captured, stored and processed in logical models varies with actual situations, creating further differences in understanding.

Logical data and process models are analysts' interpretation of the problem domain, raising an inherent problem with assumed shared understanding. Structured systems ontology assumes that systems analysts have no preconceptions or pre-resolved solutions to the current problem. This is questionable. Analysts' individuality, and even ego, plays a role in systems modelling, especially during the 'creative leap' that leads to new designs. This may cause analysts to misrepresent a required IS. It may lead to biased requirements, biased towards technical needs, or the actual requirements not being captured because they are not technically elegant or feasible. This in turn can raise questions of ownership of IS. Related to this is that structured systems ontology assumes that systems analysts are best placed to develop systems models.

Other real world issues are depicted in Figure 8.12. One example is that structured systems ontology assumes people have the capability to know what they require and are capable of objectifying system requirements. Some researchers argue that potential users cannot know what they want and that, if they do, then there is limit on communication with analysts and objectification. The philosopher Polayni is known for formulating our limits on communication as the adage: 'We know more than we can tell.' The ontological truth of this adage has fundamental consequences for structured methods that rely on explicit information and knowledge for IS development.

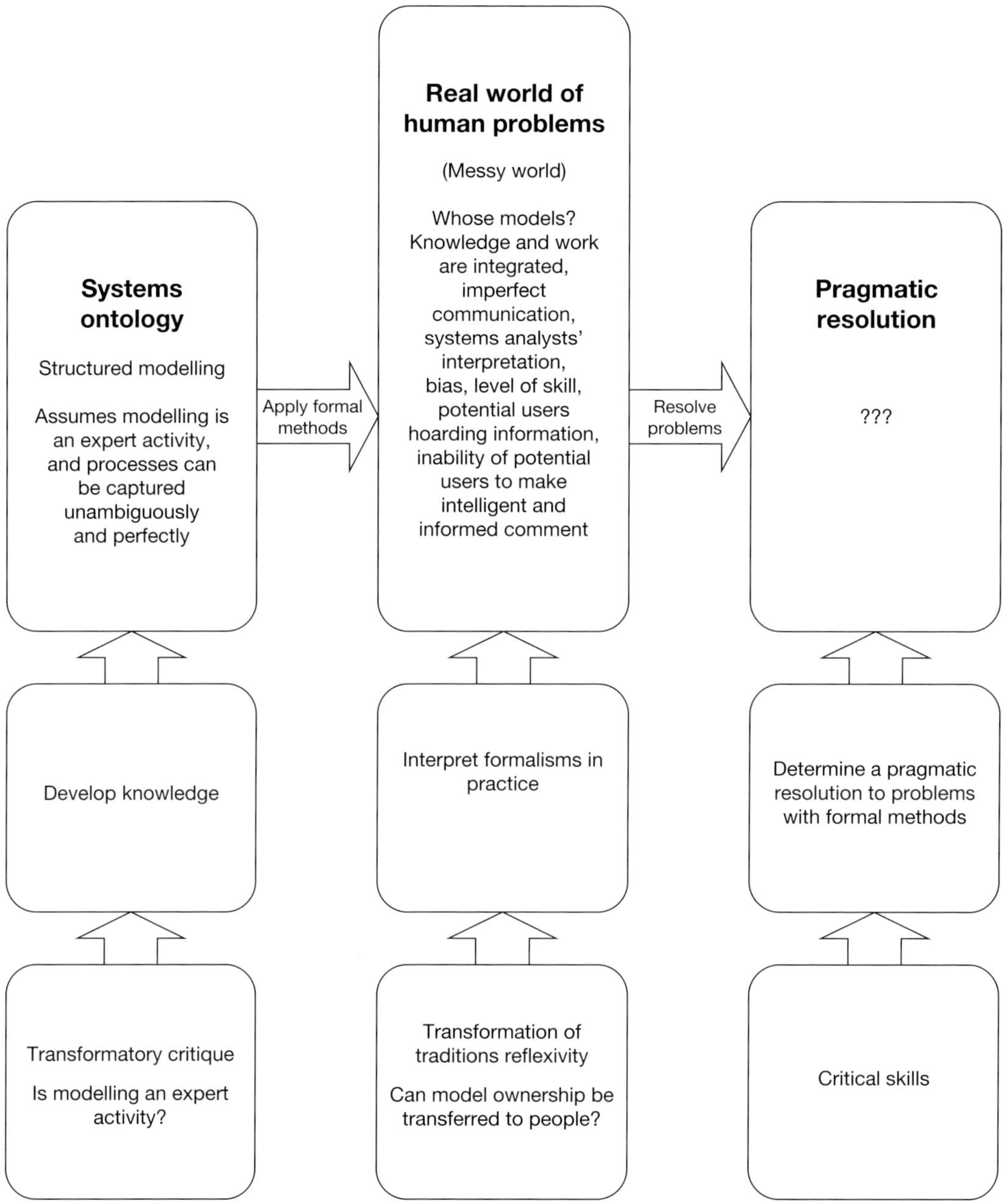

Figure 8.12 Critical framework: whose process models?

8.7.6 Communicative device

The shared understanding and diagrams are communicative devices based on engendering a 'common value system'. Structured systems ontology assumes that modelling techniques resolve communicative difficulties between people and analysts and among developers. This assumption lacks practice evidence. Advocates assume an equality of knowledge among developers and people, enabling the latter to make informed comments on developed systems models presented to them.

The effectiveness of these techniques as communicative devices is not formally tested. People lack technical knowledge to understand the E-R and DFD logical models. They do not have equal technical expertise in interpreting complex logical models. Even analysts have to learn how to develop the DFD and process logic. An inappropriate learning curve within system projects constraints will impede meaningful communication.

Shortcomings in knowledge may compromise completeness, unambiguity and non-redundancy of developed models. Structured methods assume that people have disclosed all the information required to develop a new IS. In actual situations, job security and resistance to change are only two issues that question the assumption. Lack of understanding among people means they feel they are acting on trust, which has consequences for confidence and future use of developed IS.

The usefulness of instruments is determined by the context of the problem, purpose of use and objectives to be achieved. Analysts need to determine the modelling objectives and choose an appropriate technique, because each technique produces different outcomes. The outcomes depend on the constructs each technique provides for representing and reasoning about the problem domain, and whether analysts understand its features and can apply them.

8.7.7 Human intention

Structured systems ontology does not acknowledge diverse human intention. It assumes all the parties concerned are interested in achieving system project objectives. The real situation is composed of human purpose and intention, often unclear to even organizational decision-makers themselves. As Figure 8.13 shows, when complex business decisions need to be made purpose and intention are not always clear in real human problems.

Structured systems ontology does not recognize developers' intention. Analysts have preconceived ideas, either implicit or explicit, and make assumptions about the problem domain. These cover organization, the type of IS required, the design problem, and what is expected of them. They also behave politically to achieve their aims.

Structured systems ontology assumes people have a clear understanding of system requirements. The scope of large IS projects is difficult to set because of lack of clear understanding of the organizational problem and of what is required to resolve it. 'Creeping requirements', the realization of new requirements after requirements analysis is completed, has contributed to the failure of complex system projects. The problem is compounded by the assumption of mutual intelligibility of the problem and its resolution among people and developers.

Acquiring explicit knowledge of system requirements is complicated by social conflict. Structured systems ontology assumes people will participate as and when necessary. Individuals, groups or stakeholders may deliberately withhold information because of internal conflict or differences with developers. Structured systems ontology does not acknowledge social factors like human intention and conflict in IS development.

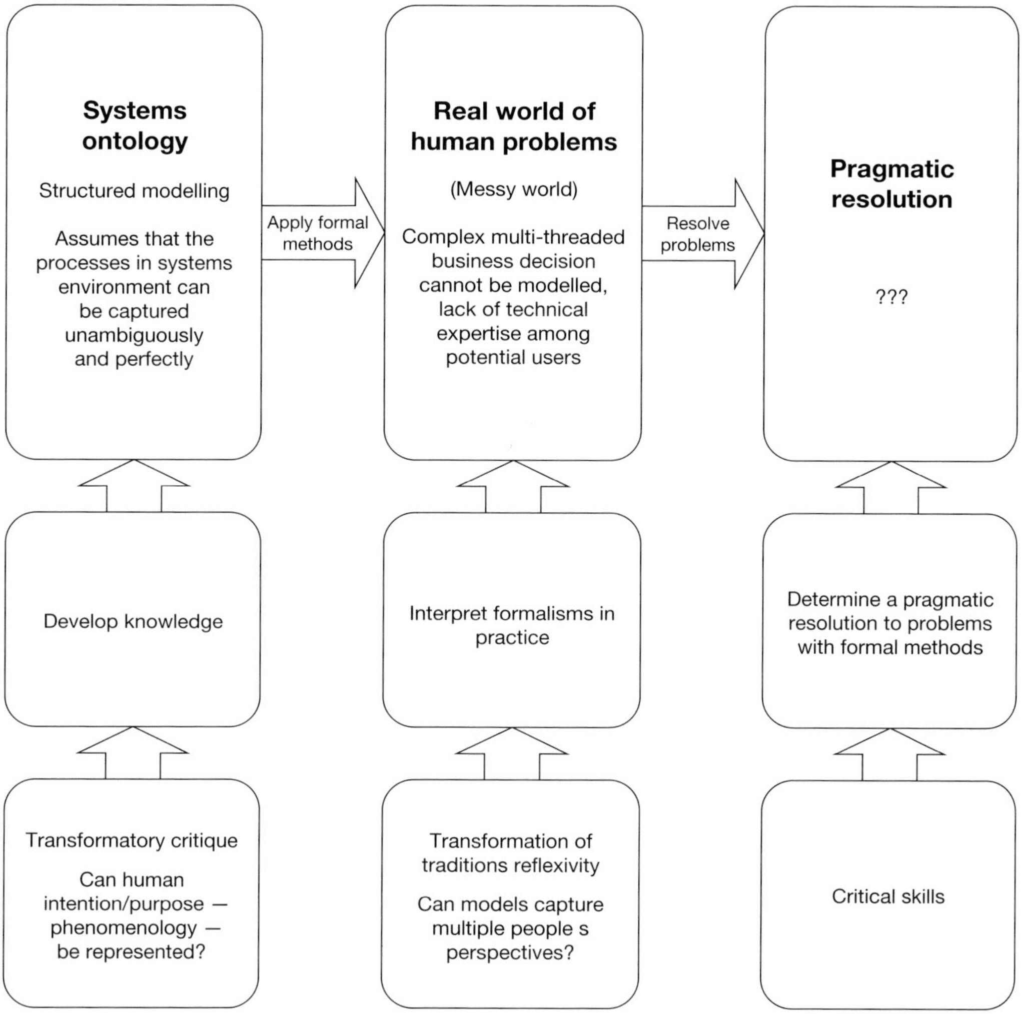

Figure 8.13 Critical framework: ambiguity and imperfection in the real world of human problems

Developing representative models of human intention is problematic but interpretive approaches are emerging. Agile or eXtreme Programming makes use of stories from people to design algorithms. SSM seeks mutual understanding through its root definition and rich picture techniques. The root definition is a concise description of the system, which is then modelled to reflect multiple views and seek consensus.

8.8 Personal Critical Framework development

8.8.1 Personal constructs for process modelling

Table 8.4 is a sample repertory grid for process modelling. Reproduce the grid on a spreadsheet and add further columns and their polar opposites that you consider relevant. To objectify personal constructs in process modelling,

Table 8.4 Personal constructs for process modelling

Pole 1	*Activity*	*Process*	*Flow*	*Business logic*	*Entity*	*System boundary*	*Representative sample*	*Data store*	*Pole 2*
Stable									Fluid
Can capture									Cannot capture
Can model									Cannot model
Logical									Physical
Mutually intelligible									Individual or group intelligibility
Decomposition									Whole

complete the grid by following the details on how to use a repertory grid in section 1.10.1.

8.8.2 Defining IS

Questions

1 Critically evaluate how DFD process models are defined and represent functionality.
2 Assess the validity of the claim that the DFD and process logic diagrams make communication between systems analysts and people easier.
3 What ethical conduct would you use when communicating with people in the problem domain?
4 What would you do to improve the quality of DFD diagrams produced to facilitate communication?
5 What innovative ways could you devise to interact with people to elicit process requirements?
6 What value does separation of systems analysis and implementation add? What is your personal construct on the separation of development activities?
7 Why spend time modelling a way of working, as required in current physical and current logical modelling, which will be replaced anyway?

Activity A

Think of an organization or section within it. Find examples of: processes, dataflow between processes, and data sources and sinks. Investigate other aspects or roles that should be reflected in a systems model. Discuss consequences of leaving them out of process models.

8.8.3 Systems and real world ontology

Question

1 Analyse whether reductionism is consistent with the notion of systems.
What assumptions of the problem domain does reductionism make? Are they appropriate?
What part could reductionism play in your personal construct for establishing knowledge of a problem domain?

2 Critically appraise the premise that process models and process logic can be rationally objectified.
To what extent is human rationality capable of capturing system requirements completely, unambiguously and with non-redundant data?
3 Discuss the ownership of logical data and process models. Though developed by systems analysts, what can be done for people to perceive ownership?
4 To what extent does people's lack of intelligibility of logical data and process models affect the relevance and quality of a new IS?

Activity A

GobiDesert is a medium sized internet bookseller. It sells books over the internet and has its UK book distribution centre in Slough near Heathrow Airport. It wants to develop an IS to provide information on creditors. It buys its stock of books, stationery, computer equipment and other operational needs from various suppliers. The accounts department responsible for paying suppliers receives invoices and checks balances owing. It then sends a cheque by post to the supplier or authorizes an electronic transfer of monies for the amount owning.

Using the above description and making your own assumptions, in two or more separate groups:

- Draw a Level 0 DFD for a new accountants payable system.
- Mark the system boundary and identify where humans interact with the system. Make a separate record of these points as human–computer interfaces.
- Decompose the context DFD to Level 1, ensuring balanced DFD.
- Critically compare your solution with a peer group and discuss your group's reasoning. What is the role of interpretation in process modelling and what consequences does it have for objective systems ontology?

8.8.4 Logical modelling

Questions

1 Evaluate the value of distinguishing between logical and physical DFD.
2 Compare critically decisions tables and decision trees for process logic modelling.
3 What criteria would you use to select logic modelling techniques?
4 How can a visual tool improve structured English to reduce the complexity of the resultant models?

Activity A

A bank has to decide what level of reminder to send to a customer whose payment is overdue – a legally worded letter, a reminder or no letter. Its business rules are: If the customer's monthly payment is overdue by two months then send a reminder. If the monthly payment is overdue by three months then send a legally worded letter and reduce the credit worthiness rating of the customer. If the customer has made a payment recently then raise their credit rating. Introduce other business rules you require. Use a version of the DFD notation and structured English accessible to you.

- Draw the Level 0 and 1 DFDs.
- Write the process logic for one or more processes in structured English. Use control, selection and iteration structures in the logic model.
- Discuss how you would communicate practically your process logic diagrams to people in the problem domain.

8.8.5 Relating structured data, process modelling and databases

Questions

Logical modelling removes all the physical details from the problem domain.

1 Describe how E-R diagrams relate to DFD process modelling.
2 What information do E-R diagrams and the data dictionary provide to develop a database?

8.8.6 Further reading

Warboys, B., Kawelek, P., Robertson, I. and Greenwood, R. (1999) *Business Information System, a process approach*, Maidenhead: McGrawHill.

Chapter 9

Object modelling

9.1 Learning outcomes

After completing this chapter you should be able to:

- Analytically evaluate descriptions of problem domains in object-oriented terms.
- Develop a simple class model using UML.
- Explain how object-oriented modelling enables reuse of software.
- Apply critical skills to object modelling.
- Compare object-oriented analysis with structured analysis in terms of transformatory critique.

9.2 Introduction

Object-oriented systems ontology originates in object-oriented programming. Software developers' drive to improve programming led to the notion of object-orientation as a basis for writing programs. It enabled programming efficiencies and intuitive programming.

Object-oriented systems ontology assumes that reality, the problem domain, is composed of classes or objects and relationships between objects. The collaborating relationships between objects define an object-oriented system. Philosophically, conceptually and practically object-oriented systems analysis and design differs from structured systems ontology.

The purpose of object-oriented systems analysis and design is to find objects and relationships in the problem domain and develop an object model or class model representative of it. The class model is meant to define *what* the system is required to do in terms of fulfilling system requirements. The actual practice of object-oriented systems analysis and design is not strictly stipulated in terms of sequential analysis and design steps as in structured systems ontology.

Object orientation stems form Classification Theory. The theory explains how people classify their everyday experiences. It asserts that people use three methods for organizing experiences:

- Experience is differentiated in terms of objects and their characteristics.
- An object is demarcated in terms of the whole and its parts.

- Groups of objects form a class and a class is distinguished from other classes of objects.

People's intuitive use of objects and classes to organize personal experiences is considered an advantage in object-orientation. It facilitates understanding of object-orientation based on intuition and personal experiences. Object-oriented analysis and design draws on this underpinning and familiarity of cognitive process that both analysts and people have according to Classification Theory.

Object-oriented systems analysis and design supports the development of object-oriented IS. The term object-oriented systems ontology describes IS development, project management, programming and systems. In systems analysis and design it encompasses requirements analysis, interrelationships between objects, operation specification, design and implementation.

9.3 An object-oriented system

An object-oriented system is composed of independent, interrelated, collaborative objects. It has classes, objects, data and operations. An object-oriented system is described in terms of behavioural language. An object is said to behave in a certain way and the whole system behaves as intended. A system behaves as interrelated objects collaborating to provide services to other objects in the system and, if required, to other systems.

Analysts need to understand the notion of object behaviour because it is the basis for identifying services provided by objects. A service is some information processing provided by an object when requested by another object. An object is responsible for providing services. It processes temporary data and permanent data.

In object-oriented systems ontology, analysts investigate the problem domain to find objects, patterns, responsibilities and scenarios. These are then defined in terms of classes, objects, data and operations in a class model. A class model is an integrated model of data and operations in which process flows, data and process are combined into one model to represent the problem domain.

The physical implementation of an object-oriented system contains the data and **methods** required to perform the system specification. Drawing on the integrated class model, the system is designed so that objects contain the data required to complete the required information processing.

9.4 Patterns and the generalization–specialization pattern

The notion of patterns is central in object-oriented systems ontology. A pattern is a template that can be reused. Patterns are used to organize and relate classes. A pattern is an abstraction of classes from the problem domain depicting stereotypical class responsibilities and interactions. It is repeatable within the same problem domain or across problem domains. The generalization–specialization (gen–spec) is one of several patterns in object orientation, others include:

- whole–part
- participant–transaction
- place–transaction
- transaction–transaction line item
- peer–peer.

The gen–spec template is shown in Figure 9.1. It shows the inheritance relationship in terms of classes. The figure shows a class (the parent) and class with objects (the child), the branches, using Coad's notation. A class is shown as a rectangle and a class with objects is shown as concentric rounded rectangles – with the inner rectangle being the object.

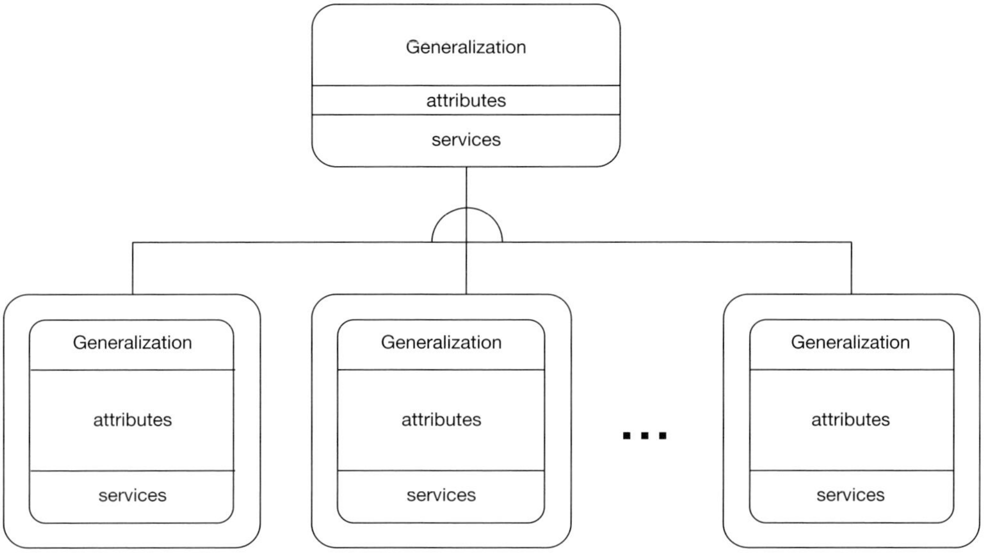

Figure 9.1 Generalization–specialization pattern

The gen–spec is a hierarchical parent–child pattern used in object-oriented systems analysis. The purpose of creating a gen–spec, and other patterns, is to characterize or develop ontological knowledge of the problem domain. This ontological knowledge is assumed to be valid across problem domains and over time. Consequently, patterns are reusable. Reusability is a significant characteristic of object-oriented systems ontology. Programming code developed on the basis of patterns can be reused in other systems and as components in component-based systems development.

Inheritance is used in object orientation to manage complexity. In hierarchical terms inheritance is the principal that all lower-level classes inherit the attributes and services of the top class. A lower class in a hierarchy can inherit attributes and services from a higher class. Inheritance is based on the parent–child pattern that is one way, from the parent to the child. The gen–spec pattern is an example.

A problem domain illustration of a gen–spec pattern is shown in Figure 9.2. It shows how the gen–spec template is used in the insurance problem domain to identify the parent (insurance policy) and the child (car insurance, house insurance and content insurance), and the inheritance relationship. The parent–child pattern is a common everyday relationship, but it is powerful in object-oriented systems ontology because it affords inheritance.

9.5 Classes and objects

An object-oriented system consists of classes and objects. The problem domain is analysed to determine relevant classes and objects capable of storing data and processing it. Modelling is critical and strategic, especially in database modelling. Object and class modelling is relatively less well-developed though the UML is a significant development in object-oriented modelling. Classes and objects in this section are drawn using the UML notation (Coad's notation was used in the previous section to provide comparative contrast).

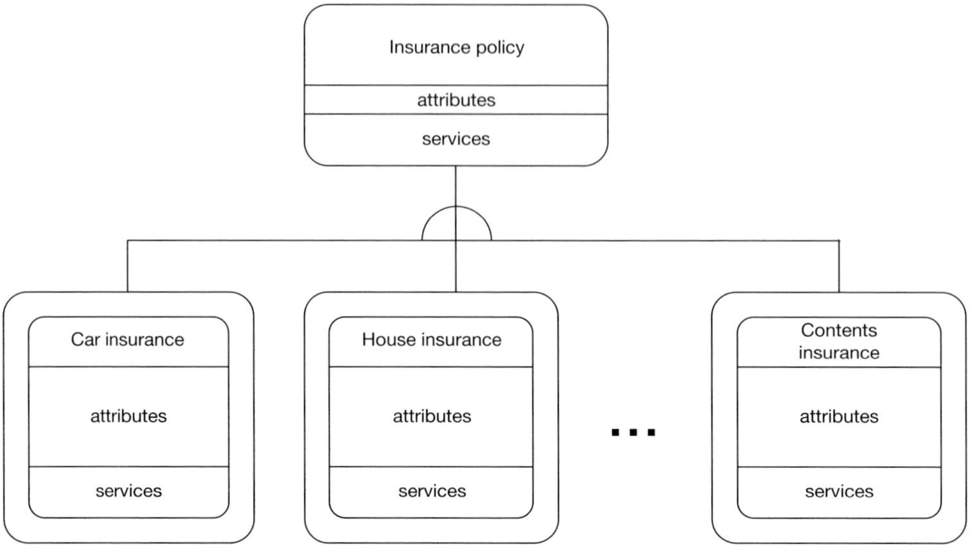

Figure 9.2 Gen–spec example for insurance policy

9.5.1 Objects

An object-oriented system is a set of interacting and collaborating objects. An object is an abstraction of some thing of interest in the problem domain for which information is kept and which interacts with other objects. An object can be anything tangible or conceptual that is relevant to the problem domain. Example physical abstract objects are an insurance policy and customer. They are related because a customer draws out an insurance policy. Risk and premium are examples of related conceptual abstract objects. The premium is high when the risk is high. An object contains the data and the operations.

An object is drawn as a rectangle as shown in Figure 9.3. In programming terms, an object is referred to as an instance of a class. An object has three sections: the name of the object and its class underlined, its attributes, and its operations (services). In naming an object, the object name is first with capital letter, followed by an colon and then the class name. So an object is distinguished from a class by the *objectName:className* underlined label. It is not necessary to show all three components in a model. The rectangle notation could cause confusion between a class, which is similarly depicted, and an object, but this is now the accepted UML convention.

An example of the car policy object is shown in Figure 9.4. The attributes are listed but the operations are not specified. Example operations could be making a claim, changing address details or recording an accident and updating accident history.

Attributes are qualities that describe an object. For an object called insurancePolicy the attributes may be customerSurname, customerFirstName, policyNumber, risk and premium. The attributes have data types. For example, customerFirstName is a string type and policyNumber is an integer type. Each of these attributes has a value or state, which is its data. customerFirstName will have a value that is the name of a customer, for example Risha or Jackie. PolicyNumber will have a numeric

Figure 9.3 A diagrammatic representation of an object

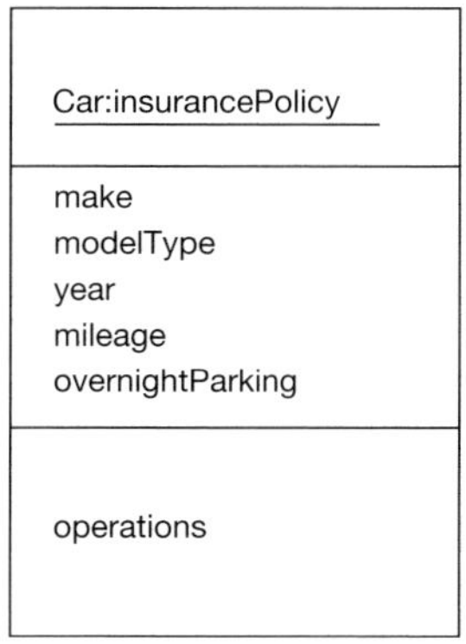

Figure 9.4 An example object for car insurance

number like 457987. The actual values or data are hidden in the object. Storing data and operations in an object is called encapsulation.

An example object from the insurance problem domain is an insurancePolicy shown in Figure 9.5. Its attributes are customerSurname, customerFirstName, policyNumber, premium and maturityDate. Example operations are calculatePremium or changeAddress. Analysts determine what data and operations to encapsulate in an object during systems analysis. A class model is progressively developed in object-oriented systems analysis and design, changes to object definitions can be made.

An object is responsible for services. A service is a set of instructions for performing action on data. Services are also called operations for logical specification of an object's behaviour during systems analysis and methods when the operation is implemented. An example operation for an insurance policy is the procedure to calculate premium. It contains information necessary to carry out the procedure or operation. The operation is invoked by a request from the object itself or from other objects with a message.

During systems analysis analysts need to determine which attributes of an object generate

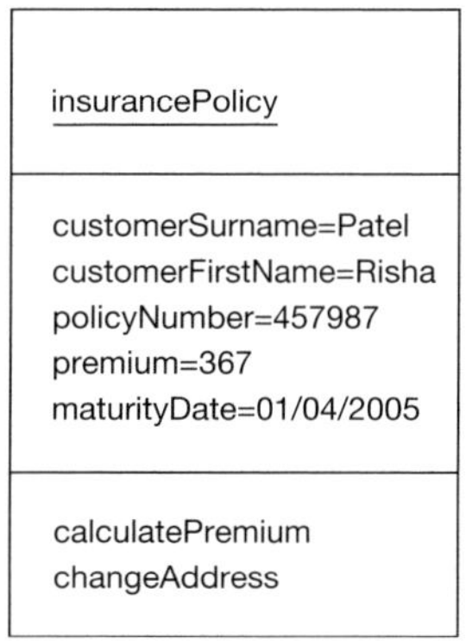

Figure 9.5 Insurance policy object

persistent data. Persistent data is data that is initially input into a system or generated by it that needs to be stored. Information on persistent data is used to create a database containing such data so that it can be used by objects or analytically queried in the database for information. In the insurancePolicy object all the attributes would probably be persistent data. Some applications or attributes may contain transient data. Such data is not required after the program has been executed and so does not need to be stored.

9.5.2 Classes

An object in a system is an instance of a class. A class is a classification of a set of abstracted objects with common attributes. A particular class is a specification of objects with the same or common attributes. It is also called an object type, but object types do not contain any methods. The term class is normally used to define a group of similar objects. In object-oriented systems analysis analysts' task is to draw an abstract class diagram of the problem domain.

The template for a class is shown in Figure 9.6. It shows a solid rectangle with three sections. The class name is in bold type in the top part, the list of attributes is in the middle and the list of operations is in the bottom section.

The set of home insurance policies offered by a company can be categorized as insurance. Those with the same characteristics can be labelled a certain class, for example marine insurance, house insurance or contents insurance. An example class for contents insurance is shown in Figure 9.7.

Some example attributes of the Content Insurance class are customer's first name, surname, policy number, address and number of the house insured. To preserve the description found in the problem domain, attributes and services can be listed in the class as compressed words with a lower case for the first word, no spaces, and a capital letter for each subsequent word. So, customerFirstName, customerSurname or policyNumber are acceptable. Two or more nouns and adjectives and nouns can be compressed.

Classes can be either static or dynamic. Static classification does not allow an object to change type. Dynamic classification allows objects to change type. Dynamic classification is useful for problem domains where an object type needs to be changed. For example, in the case of a class Professor it is possible that a professor in the problem domain becomes the head of department.

9.5.3 Polymorphic classes and objects

Polymorphism refers to objects behaving in varying ways in response to different messages received. It is used in object systems ontology to make subsystems independent and enables the de-coupling of subsystems to facilitate system extensibility.

A polymorphic class is designed to receive different type messages from objects and produce the required result. So a polymorphic object can change its behaviour to respond to the message

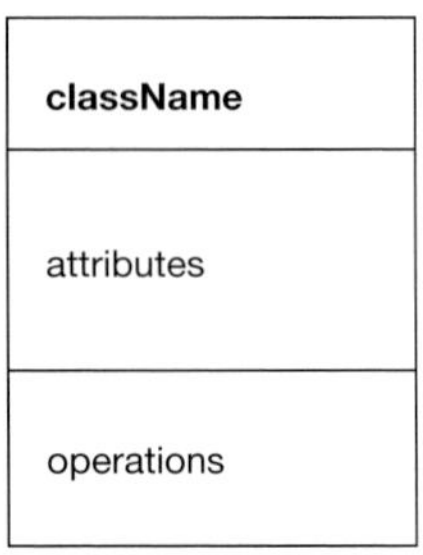

Figure 9.6 Class template

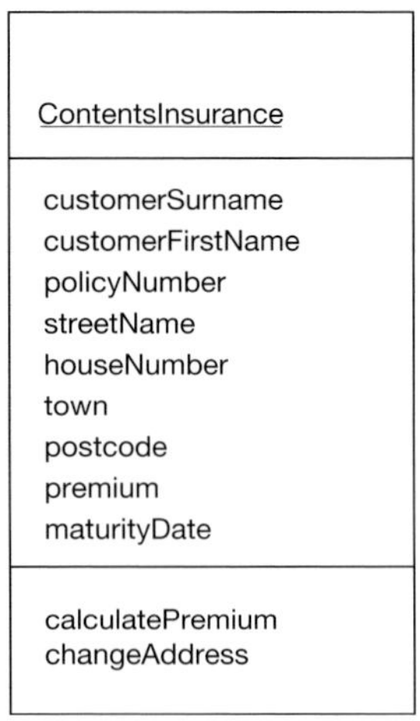

Figure 9.7 Home contents insurance policy class

it receives. The object sending the message does not have to concern itself with the internal structure, its operations, of the polymorphic object. It can simply send its service request and not need to know how the receiving object will respond.

In polymorphic object behaviour there is only one instance of a class and one or more attributes or operations are moved to subclasses so that they can be redefined differently. In Figure 9.8 a class with subclasses is shown. The operation calculateRiskType() is modelled in the subclasses, rather than the main class, because in each subclass it can be polymorphically redefined. Similarly, the attribute insuranceType is modelled in the subclasses because it can be variously defined as required to describe different types of insurance policy. The effect of polymorphic objects is to receive a message and produce the required output.

Analysts have to investigate the problem domain to find polymorphic objects. Polymorphism is a challenge for practising analysts. They have to define classes and a class model capable of polymorphic behaviour, usually this is based on inheritance of behaviour from superclasses to subclasses. Its appropriate use during systems analysis should lead to an extensible IS, a much demanded feature in business IS.

9.5.4 Modelling operations logic

The operations in an object may be abstract or concrete. Abstract operations do not have implemented methods. Operations in an object-oriented system are simple structures because collaborating classes enable simplification. The total operations in the system can be analysed in terms of specific object responsibilities rather than as some main system functionality. Operations can be specified or modelled using:

- decisions tables
- pre- and post-conditions
- structured English
- pseudo-code
- activity diagrams.

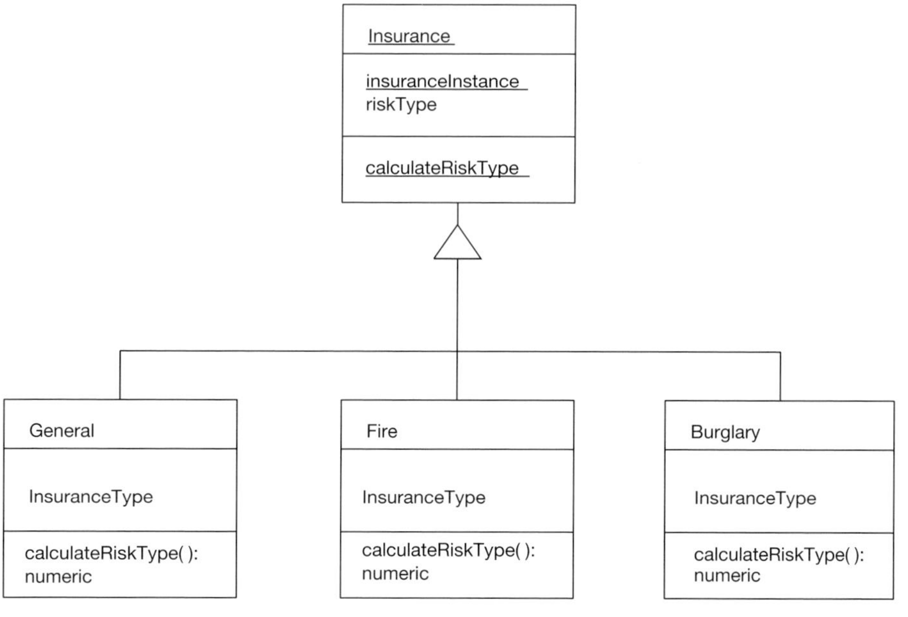

Figure 9.8 Polymorphic class and subclasses

The first two techniques are non-algorithmic. They cannot be easily translated into program code. The second three techniques are algorithmic – they provide structures that can be used to design program and systems implementation. The decision tables and structured English techniques are used structured systems analysis (see section 6.5).

9.5.5 Links and associations

When things in the problem domain are related, they can be modelled as an association between classes. An association between classes is a generalization of a relationship. For example, an insurance policy is linked to a client. It generalizes the specific links between instances of objects in the real situation.

9.5.6 Multiplicity

Association **multiplicity** is used to describe cardinal relationships between objects. Links between objects need to reflect the problem domain constraints of how objects can be related. Multiplicity describes the number of instances of objects that can participate in an association. One client can have one or more insurance policies, for example. Association multiplicity contains information on such structures in the actual situation – problem domain. Association multiplicity is determined by the business rules in the problem domain.

In object-oriented systems analysis the data dictionary is a store for information on classes, objects, attributes, operations, links and associations. The class model would normally contain much of this information, but analysts gather much other information that may not be depicted in the class model. Such information is stored in the data dictionary. For example, the detail of how an operation would be carried out is usually not shown in class diagram, but would be stored in the data dictionary.

9.6 Responsibilities and scenarios

An object is attributed with responsibility in an object-oriented system. Responsibility has three aspects: attributes, relationships and services. What the object knows about itself is referred to as its attributes (see section 9.5.1). Who the object knows in a problem domain is its relationship. A problem domain will consist of many objects and a particular object will be related to one or more of them. This is the relationship the object has with other objects. A relationship is a connection between an object and other objects. For example, the customer object is related to the insurancePolicy object because a customer has an insurance policy. What the object does is its services. The services are provided to other objects.

A scenario is a time ordered set of interactions between objects designed to fulfil specific responsibilities. For example, there are procedural steps in drawing a new insurance policy for a customer that involves interaction between the customer, insurancePolicy, risk and premium objects at certain times. Modelling this interaction is called a scenario.

9.6.1 Techniques for identifying services

Identifying the details of the services in a class is done with specialized techniques and some borrowed from structured systems analysis. Systems analysts make use of these techniques during systems analysis. (Decision trees and decision tables are discussed in section 8.5.) They include:

- scenario
- use case and use case scripts or descriptions
- structured English
- decision tables

- decision trees
- state-transition diagrams.

Scenario is a specialized object-oriented technique for documenting and identifying a class's service details. A scenario depicts an interaction between the system and a user as an ordered sequence of actions. It is used in Coad's methodology but use case and use case scripts are a similar technique used in UML. A scenario describes a specific time-ordered sequence of object interactions that fulfils a specific processing requirement in an IS. A scenario view is quite complex for any particular object interactions.

Service scenarios are developed to describe how objects interact to complete a task. For example, the insurancePolicy object will send a message to the Risk object to invoke the calculateRisk service. The Risk object will process the result, calculated risk. The notation for a UML use case diagram is given in Figure 9.9 because it is more widely used than Coad's services scenario and an example is given in Figure 9.10.

The use case in Figure 9.10 shows a user goal of processing an insurance policy. It is a set of scenarios tied together by the common user goal. It shows a scenario consisting of: raise new policy, update policy, calculate risk and make a claim. Each of these can be described further with use case scripts. A use case diagram is a snapshot of one aspect of information requirements. Many such use case diagrams are produced to describe a new IS.

A use case script or description can be used to elaborate a set of interactions. An example use case script for online purchase in an eCommerce system is shown in Figure 9.11. It shows the primary scenario in steps 1 to 10. These are the necessary steps to complete a process. A use case may have one or more paths. In the example, there are two alternatives. One for credit card authorization failure and the other for a previous customer who is offered discounts and whose details are automatically displayed to facilitate the purchase.

A use case script is able to capture more detail than a use case diagram. The use case script in Figure 9.11 shows alternatives for example. It could also show other information. The UML does not specify a standard script format. So it is possible to check for any

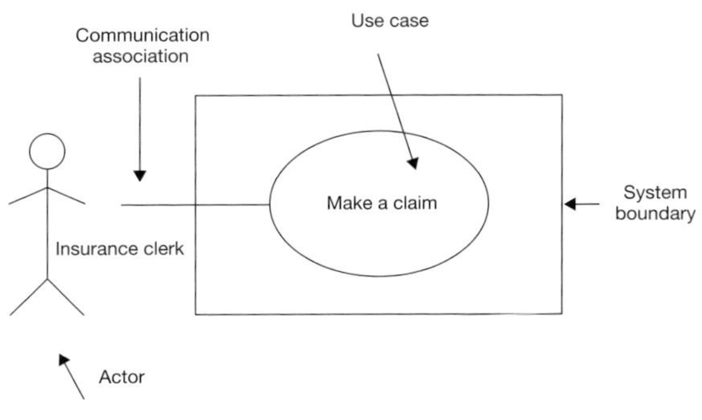

Figure 9.9 UML notation for use case diagram

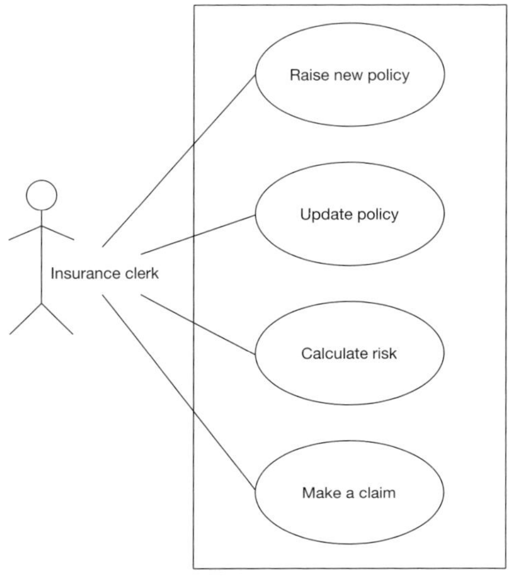

Figure 9.10 Use case example for processing an insurance policy

Purchase online

Customer browses the online book catalogue and places one or more items in the shopping trolley
Customer selects purchase icon
System prompts for further purchases
System presents pricing information, including customer details information and shipping
Customer enters credit card details
System checks authorization
System authorizes purchase
System confirms sale
System prompts for further purchases or other service
System completes sale and sends confirming email

Alternative: Authorization failure
At step 5, system fails to authorize credit card.
Allow customer to re-enter credit card details three times

Alternative: Previous customer
2a System presents discount policy decision for customer with six months purchase history
4a System displays current customer information
4b Customer may accept or override displayed defaults
Return to primary scenario at step 5

Alternative: Other service
9a System presents CD department
9b Customer enters CD department

Figure 9.11 Use case script for online purchase

preconditions before the primary use case starts. For example, only members of a book club can make purchases, so the system would check for status before allowing purchase to be made. In this regard use case scripts enable detailed descriptions or modelling of actual situations.

Structured English or pseudocode is used in structured systems analysis. It is an action-oriented form of the English language. It is also used in object-oriented class modelling because it describes 'what' the object needs to do to complete a service request. Figure 9.12 shows part of the pseudocode for processing an insurance claim for a car accident.

```
For Each newclaim
        Record accident details
        Assess liability
If liability=other then contact other company
Else commence proceedings
Endif
issue claim form
End For Each
```

Figure 9.12 Description of a service in structured English

9.7 Unified Modelling Language

UML is a notation language for developing robust industry standard software systems. Ironically, it is capable of modelling such complex system even though it is based on the 'less is more' idea of coping with complexity. It does not contain a vast range of notation symbols but is capable of developing rich semantic models of the problem domain. It is used to develop class models for small, medium and large industry standard software. With it analysts can specify, visualize and document models of systems. UML can be used to explore system ideas or to specify a detailed design. It is a third generation object-oriented modelling language based on the MOF(tm) metamodel and has become the industry standard. It extends notations used in Booch, OMT and Objectory methods. An upgrade to UML 2.0 is in progress.

UML is intended to provide a single and common object-oriented modelling notation language. It has 12 standard diagram types that can be applied to a wide range of application domains and database design. It is used to specify, construct and document classes and objects in the problem domain. Its standard diagram types lead to well-designed software architecture. It can be used to design class models independent of specific hardware platform or specifically for a particular proprietary platform. The diagram types are categorized into static, dynamic, and organization and management, as shown in Figure 9.13.

UML is still evolving. Many of the techniques do not have specific standards, for example use case scripts, which can be interpreted and used variously by analysts. The value of UML for analysts in system projects is that it is 'methodology-independent'. It can be used to develop systems models regardless of the methodology being used. It can be used with a methodology but it is not a methodology itself. Structure is important in UML. It is used to deal with the complexity of designing large systems and it enables code reuse.

Static application structure (four standard diagrams)
Class diagram
Object diagram
Component diagram
Deployment diagram
Dynamic object behaviour (five standard diagrams)
Use case diagram
Sequence diagram
Activity diagram
Collaboration diagram
Statechart diagram
Organization and management of application modules (three standard diagrams)
Package diagram
Subis
Models

Figure 9.13 UML diagram types and diagrams

Formalism is based on the proposition that natural language is unsuitable for communicating system requirements or modelling. Natural language is imprecise and ambiguous. Proponents of notation languages claim that a class model developed with UML diagrams eases communication between analysts and people because it is specific and formal. Classes can be named in the language of users, rather than as abstract names. Some UML tools make it possible to visualize defined classes, objects and relationships. Visualization further makes it easier to communicate with diagrams.

9.7.1 Components and reusability

Object-orientation facilitates the design and use of components. In some object-oriented IS development software components are reused. A component is a self-contained module of software code designed for reuse. Software components are assets and are normally developed for enterprise applications, business process redesign and web services.

The practical advantage of developing reusable components is that though they are developed for a particular application or problem domain, they can be stored in a library and reused in other applications that require the same or similar functionality. The Object Modelling Group (OMG) provides a central source for developing a marketplace for re-usable components.

The concept of reusability is extended to reusable requirements and patterns like gen–spec. Analysts can make use of reusable requirements and patterns, and contribute patterns to a system library. This results in reusable components that can be used in problem domains for which they were not intended, and makes it possible to cut costs of a system project.

Methodologies exist in the commercial marketplace to enable object-oriented component analysis and design. Use cases are used in a top-down approach, but bottom-up approaches also are incorporated to allow existing IS and databases to be reflected in the class model. The aim is to create flexible component architectures that integrate with existing IS and draw on valuable data in legacy systems.

Components are designed in terms of:

- the problem domain component;
- the human interaction component;
- the data management component.

Classes identified by analysts during systems analysis and systems design, and subsequently coded as Java classes, can be componentized using JavaBeans. JavaBeans is an architecture that supports reusable components. JavaBeans are normally used for GUI interface classes.

Analysts' can take advantage of reusability because of object and encapsulation. When including components not designed by the analyst, analysts do not have to worry about how the service request will be met. This is termed decoupling, where the object is designed to work in context but is not intertwined in it. Gen–spec and other patterns can be reused because they are decoupled.

Polymorphism is a critical feature of object-oriented systems ontology that enables reusability. An object is polymorphic if it can vary its behaviour according to the messages it receives. Systems analysts use polymorphism in object-oriented analysis by identifying and defining a class and its subclasses that behave differently depending on the message received.

9.8 Object-oriented analysis and design

Object orientation has influenced systems design, particularly database systems. Object-oriented analysis is based on the view that the problem domain has objects that perform tasks and have attributes, and that the objects are related to other objects in terms of operations to enable tasks to be completed. An object-oriented system is a set of such objects that are interrelated. For example, an insurance policy is taken out by a customer, brokered by an insurance broker, underwritten by an underwriter and administered by an administrator. The parties are related and each party performs operations for other parties.

The aim of Object-Oriented Analysis (OOA) is to identify detailed system requirements by describing the existing physical system and determine how a new IS will become part of it. Data definition involves identifying object, class, relationships and abstraction. Data manipulation involves determining methods and messages. Data definition and data manipulation result in the object or class model.

The problem domain is analysed in terms of objects, patterns, responsibilities and scenarios. Once the analyst identifies objects they are likely to persist through to the design and implementation or programming stage. The analyst's critical role is to identify the objects and classes in the problem domain. Analysts undertake five activities in Coad and Yourdon's (1990) OOA:

- Identify classes and objects in the problem domain.
- Identify structures or relationships between classes and objects.
- Identify subjects or related objects.
- Define attributes or the data elements of objects.
- Define services or how the object behaves.

OOA results in systems models of the structure of objects and their behaviour. Objects can be identified from transcribed records of interviews or text documents by underlining nouns. For example, insurance policy, client or premium are acceptable nouns or noun phrases in an insurance company. The label given to an object depends on the analyst's preference and style, but it is important to preserve the terminology of the problem domain. Naming objects and classes with terminology from the problem domain makes it easier for people to understand a class model.

OOA is not strict planned action like structured analysis. Analysts do not have to perform the five activities in sequence. As eliciting requirements depends on people OOA enables analysts to work with them. They are permitted to explore the problem domain and undertake each activity as and when appropriate information is found or revealed by people. The effort analysts spend on each activity will depend on

Problem domain
The objects here will be those required to address the business problem.

Human interaction
The objects here will be those required to design the human–computer interaction, such as the interface, windows and reports.

Data management
The objects here will be those required for the persistent storage of data, such as databases.

System interaction
The objects here will be those required to enable the present system to interact with other systems.

Figure 9.14 Components of a class model

what information they discover, with some aspects of a class model more deeply explored than others. The iterative OOA process enables analysts to validate and continuously improve a class model with information from people.

A class model is composed of major model components consisting of the problem domain, human interaction, data management and system interaction, shown in Figure 9.14. Analysts identify objects and their responsibilities for the components of a class model.

9.9 Java and object orientation

The Java programming language has become closely associated with object-oriented systems. The origins of object orientation are in Scandinavia and SMALLTALK is a popular object-oriented language. Java though is considered suitable for implementing distributed object models. Object models can be implemented using any programming language, though there will be efficiency gains in some and costs in others. Java is designed for the object-oriented systems environment. It is widely used largely because of the success of the internet and the world wide web.

Unlike other programming languages Java can be many things to many people. It is an object-oriented programming language, containing the syntax and semantics of a programming language. It uses the notions of a virtual machine and byte code to enable it to be executed on different machines. Java is a programming environment too. It contains classes and libraries provided by the language.

Programmers use these classes by 'extending' them to suit their processing requirements. Java is the de facto programming language of the web. Other object-oriented languages can be used, but Java is the preferred language of most commercial software developers for the web. Java treats strings, arrays, windows and integers as objects. For example, the string 'Risha Patel' is an object of the class String.

9.10 Object modelling and the Critical Framework

Object-oriented systems ontology addresses two problems in structured systems ontology. The creation of separate data and process systems models and multiple models in structured systems ontology causes validation problems. This is overcome with the one class model created in object-oriented systems ontology. The encapsulation of data and operations within an object removes validation issues.

The other problem is determining the transition from structured analysis to structured design. This arises because of the Transform Analysis stage, which assumes that rules can be applied to the systems analysis E-R and DFD diagrams to convert them into systems design diagrams. This is overcome in object-oriented systems ontology with the one class model developed during analysis, and progressively developed to include design and programming tasks. In OOA the integration between systems analysis and systems design is now complete, as it is possible to generate implementation program code from analysis diagrams.

9.10.1 One class model

Figure 9.15 is the Critical Framework populated with critical reflection on object-oriented systems ontology. It questions the assumption that real situations can be accounted for by classes, objects, relationships and collaboration between objects. One issue is the temporality of analysis, design and implementation. These activities are concurrent in object systems ontology. The analysis class model is elaborated to include design and implementation features. The notion of analysis in object systems ontology is different

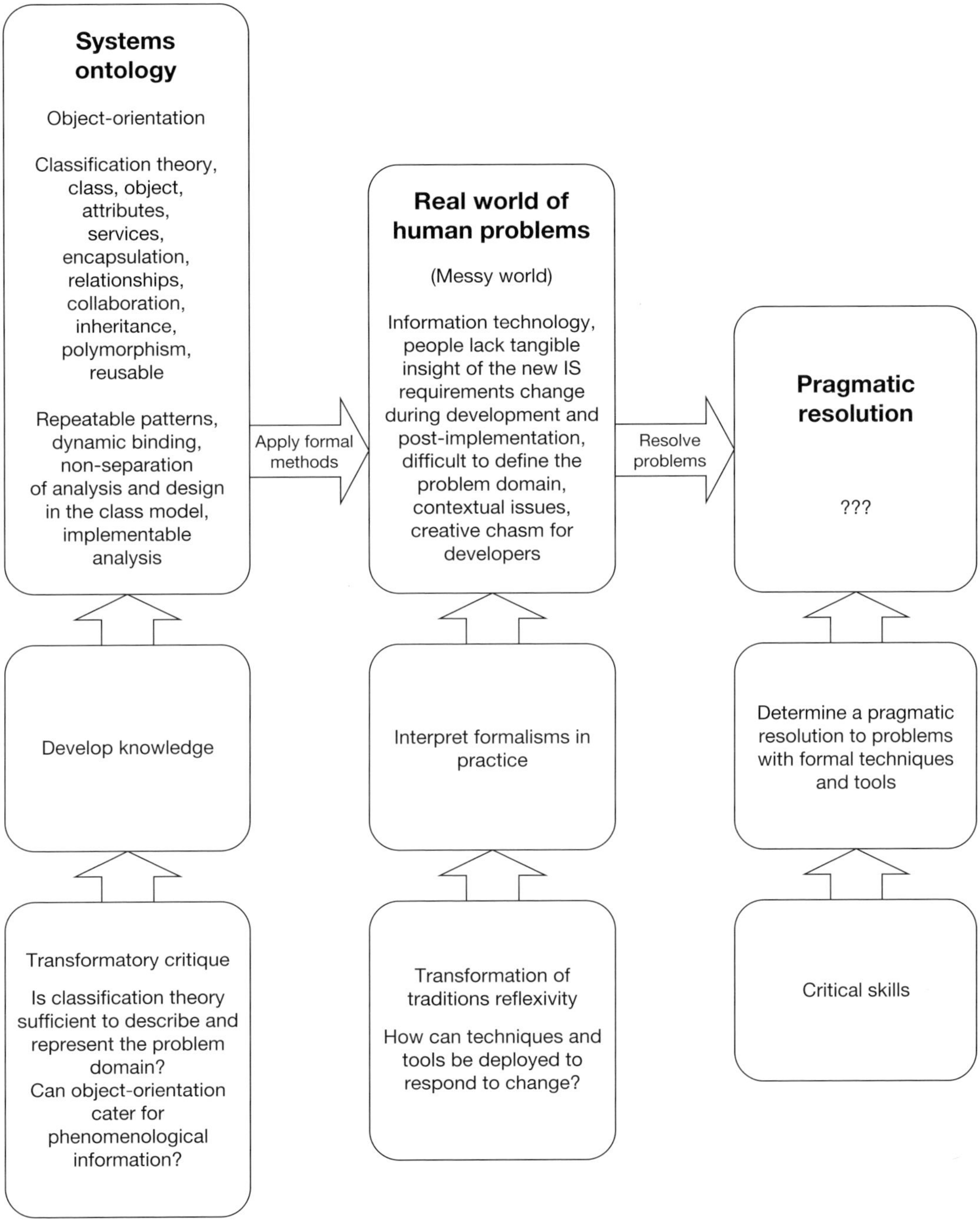

Figure 9.15 Critical framework: object-orientation

from structured systems ontology because analysis contains design and implementation features.

Actual practice in industry though is different. Practitioners maintain a separation between analysis and design despite the concurrency facility in object-oriented analysis and design. In traditional IS development (see section 3.4) such concurrency was the norm, but the force of the SDLC and engineering metaphor has set itself in practitioners' minds. An additional factor reinforcing this separation is the systems project mentality and work breakdown structures.

Compared to structured systems ontology, in object systems ontology analysts interpret and represent the problem domain as classes and objects. A transformatory critique is to question the validity of classification theory to describe and represent the problem domain. For example, is it possible to model human meaning attached to information? There are aspects of object technology such as deferred classes and dynamic binding that can be modelled to reflect change in the problem domain. They cannot reflect the interpretive aspect of IS.

Reusability and patterns lead analysts to think of problem domains as consisting of stereotypical and repeatable object behaviours. Repeatability in object-oriented systems ontology is problematical because it could lead to applying unsuitable patterns in particular problem domains. The problem being that the 'best available' pattern may be applied in situations where it is inappropriate. In problem domains where requirements are ambiguous, a reusable requirement pattern may be applied for convenience.

9.10.2 Representation and problem-solving

Object modelling is considered to be essential, especially for large enterprise-wide applications. Proponents argue that modelling can contribute to the success of a system project, and improve the quality of software produced. It supports scalability, robustness, security and extendibility. Like structured systems ontology, object-oriented systems ontology proposes building a class model before coding.

The basis of this modelling is the meta-model from which UML is defined. A meta-model is a model of the constructs in UML.The meta-model is defined using a metacircular interpreter, which is a language, defined in terms of itself. The meta-model is significant because it contains ontological assumptions about reality. Its power to address problems and develop representative models of the problem domain depend on the validity of these assumptions. For example, the assumption that experience is classified in terms of classes and how one class differs from another is significant because it, in turn, assumes a certain level of ontological objectivity.

Techniques for gathering information to develop class models need to be analytically evaluated to assess whether they are sufficient to identify objects, attributes, operations, associations and multiplicity. Sampling, questionnaires, interviewing, reading or research, and observation seem sufficient, but the quality of information gathered with these techniques is sometimes poor, resulting in class models that are a poor representation of the real human problem situation. Poor quality information contributes to lack of analysts' creativity.

The association definition and association multiplicity are generalizations. Analysts' need to consider how they would determine such generalizations with information obtained from requirements analysis. The available information is essentially limited to the time when requirements analysis is undertaken and to the people, and other sources, available at the time. In real business situations changes occur to how things are related and the multiplicity of their relationship. An issue with structured and object-oriented systems ontology is how such changes can be easily reflected in developed IS.

Problem-solving in object-oriented systems ontology is relatively easier compared to structured analysis. Once classes are identified they tend to remain stable during analysis and design, though attributes, operations and relations may change. This may be interpreted as validating classification theory's power to represent the problem domain. Similarly, the dynamic classification of an object is useful in changing problem domains. Object systems ontology in this respect reflects real human problems. Though it introduces uncertainty in object-oriented systems ontology because an object can be anything and its features are not syntactically defined in an available notation language.

Structured methods and methods of other systems ontology are derived from predicate calculus, for example normalization in E-R models. Object-oriented instruments are not as mathematically rigorous. They are intuitive rather than mathematically rigorous and their popularity attests to their relevance to practice. For example, unlike structured systems ontology where an entity is an abstraction of something of interest, an object does not have to be an abstraction it can be concrete too. The problem exists because an object can have any number of features beyond the basic three features used in the UML notations: name, attributes and operations. The move from mathematical rigour in object-oriented systems ontology is significant because it makes software intuitive rather than formulaic. Its logical progression is evident in storycarding in ASD.

Analysts' task is made challenging because of encapsulation and polymorphism in object-oriented systems ontology. They not only have to identify objects, attributes and operations, but also ensure that polymorphic behaviour is facilitated in a class model. Often, analysis of polymorphic object behaviour is overlooked because the problem domain does not afford itself so transparently to identifying polymorphic things of interest.

Structured and object-oriented process modelling techniques make an erroneous assumption of real situations. They assume that the real situation requires to be modelled once in the form of systems models. In structured systems analysis data and process models are created separately but become frozen. In class models, data and operations are encapsulated within objects. The real situation though is dynamic and 'frozen' models can become obsolete when the system is delivered.

9.10.3 Incremental and iterative development

Pragmatically, object-oriented IS development is not planned but is architecture-driven, incremental and iterative. It can be described as situated action rather than planned action. The analysis, design and implementation activities are not distinct phases but iterative activities. In some cases, this leads to an evolutionary IS development approach, where a system is delivered in phases.

UML may be described as the situated use of modelling rather then planned. It is not a methodology, so it does not prescribe planned action, and it can be used with any object-oriented methodology. Its various diagrams can be used in context, decided by analysts. Unlike the systematic phases of data and process modelling in structured systems analysis, a class model is developed progressively in context. Object encapsulation ensures that data and operations are contained and validated each time a process is added or amended.

This means that systems project management is not planned in the traditional sense used in structured systems ontology, but rather project constraints are readjusted each time a version of a system is delivered. A problem arises when a project manager wants to apply rigorously the work breakdown structure and enforce the PERT activities on a project. In such cases the inherent flexibility of object systems ontology is lost. Practice reflects such inflexibility because IS are developed as projects.

Analysts should allow flexibility in their PCF. Systems ontology, whether structured or object-oriented, is not mutually exclusive. New ideas can be used with existing methods and techniques if it is appropriate. For instance, relational databases support object-orientation. The SQL3 database includes some object-orientation features. This suggests analysts and designers should not think only in ideal types – relational database or object-oriented database – but recognize that knowledge of systems ontology is progressive and can be inclusive.

9.11 Personal Critical Framework development

9.11.1 Personal constructs for object modelling

Table 9.1 is a sample repertory grid for object modelling. Reproduce the grid on a spreadsheet and add further columns and their polar opposites that you consider relevant. To objectify personal constructs in object modelling, complete the grid by following the details on how to use a repertory grid in section 1.10.1.

9.11.2 Systems ontology

Questions

1 Critically evaluate how appropriate object-oriented systems ontology is for representing a business organization's activity and goals.
2 Comparatively evaluate logical data and process modelling in structured systems analysis with class modelling in object-oriented systems analysis.
3 What is the systemic value of encapsulating data and operations in objects?
4 'The purpose of creating a gen–spec is to meet the needs of the problem domain, not neatness or elaborateness.' Evaluate the effectiveness of patterns in object-oriented systems analysis for representing real human problem situations.

Table 9.1 Personal constructs for object modelling

Pole 1	*Object*	*Attributes*	*Class*	*Interaction*	*Pattern*	*Community*	*Hierarchy*	*Operations*	*Inheritance*	*Pole 2*
Pattern										No pattern
Method										No method
Encapsulate										Separate
Reusable										Not reusable
Dynamic										Stable
Grouped										Ungrouped
Service										Independent

Activity A

Comparing structured systems ontology with object orientation, the latter recognizes the creative role of systems analysts. In determining detailed system requirements, analysts identify the objects and classes in the problem domain.

- Make a list of your creative attributes and discuss them with a trusted peer.
- Select an IS familiar to you and comment on its creative aspects.
- What does it do anew that is not in the previous system or manual work?
- Comment on what difference the creative element of the system makes to the organization's performance?

Activity B

Investigate a situation in a problem domain where polymorphic behaviour would be required. Think of a class to represent the situation that should only have one instance, where one or more operations and attributes are modelled to subclasses to produce polymorphic behaviour. You will need:

1. a class
2. subclasses
3. generalization
4. one or more attributes or operations that need to be polymorphically redefined in subclasses.

Once you have the required elements:

- Draw a class diagram.
- Discuss critically whether polymorphic systems ontology reflects real human problems.
- Identify requirements analysis techniques you would use to find polymorphic classes and objects and analytically evaluate their effectiveness to do so.

9.11.3 Planned action versus situated action

Questions

1. UML is not a methodology. How would you use it in object-oriented systems analysis and design?
2. Evaluate the practicality of the iterative development of a class model in an IS project. Base your evaluation on the themes of situated action and planned action.

Activity A

Object-oriented analysis may be described as situated action. Though sequence is important, unlike early versions of the SDLC, iteration is accepted and essential in object-orientation and it is often linked with prototyping. A class model is validated and continuously improved with clients. In two groups:

- Identify a problem domain for which an IS is required or one in which there is an existing IS.
- Make a list of the advantages of object-oriented analysis applicable to the problem domain.
- In a plenary compare your lists and discuss the similarities and differences.

9.11.4 Communicating with diagrams

Object-orientation diagrams are used as a tool to ease communication between systems analysts and people in the problem domain.

Questions

1. Discuss the reasons why natural language is not suitable for communicating system requirements?

2 Assess the effectiveness of the various UML diagrams to communicate system requirements and systems models to people.

3 Critically evaluate the claim that object systems ontology is intuitive and therefore enables easier understanding of the class model by people.

9.11.5 Internet sources

The Object Management Group was set up 'to create a component-based software marketplace'. It is a rich resource for object-oriented analysis, design and reusability: www.omg.org

A site for object-oriented tools is www.rational.com.

9.11.6 Further reading

For a practical text on object-oriented analysis see: Bennett, S., McRobb, S. and Farmer, R. (1999) *Object-Oriented System Analysis and Design – Using UML*, London: McGraw Hill.

Coad, P. and Yourdon, E. (1990) *Object-Oriented Analysis* (2nd edn), Englewood Cliffs, NJ: Yourdon Press, Prentice Hall.

Part IV

Systems design

Part IV covers systems design. Whereas systems analysis focuses on what the system will do, systems design focuses on *how* the system will do it. A design is a solution for a problem. As alternative solutions can be generated, designers aim to generate a solution that is efficient, effective and achievable within monetary and time constraints. Quality standards are used to ensure that the system design is robust and capable of fulfilling system requirements specification.

In structured systems ontology design is concerned with converting the logical designs from systems analysis into logical design and physical design for implementation. It normally begins after decisions on hardware and software platform have been made. Systems design involves making decisions on system performance, computer files, data structures, processes, databases and user interfaces. E-R models, DFD and data dictionary are used to inform these decisions. In object-oriented systems ontology, systems design is the refinement of class diagrams, or class model, created during analysis, but does not necessarily cover actual implementation or physical design. In both, the models resulting from the logical design describe how the parts of the system will work.

Chapter 11 covers user system interfaces design. In structured systems design this requires considering what data input and output is required. In object-orientation, systems design leads to the creation of the human interaction class model and, if required, the system interaction class model to enable communication between systems.

Chapter 12 is on system and data design. In structured systems design, overall system level performance design decisions are made, and file or database design decided. In object-oriented systems design, system level and detailed design decisions are made. Systems design is concerned with overall systems issues like performance. Detailed design is concerned with class design.

Both chapters cover critical skills in praxis. In considering how requirements will be met and how the system will be implemented, designers need to reflect on technical issues like improving system performance, reducing processing times and making efficient use of data storage. The kinds of praxis questions concern, why is a particular hardware and software platform better than another and how will the choice affect the business, is the data organized efficiently to make use of limited storage, and what is the benefit of using platform-independent software.

Chapter 10

Interface, input and output design

10.1 Learning outcomes

After completing this chapter you should be able to:

- Analyse and evaluate the process of refining the analysis class model for design purpose.
- Evaluate the human interaction and system interaction components in object-oriented systems design.
- Critically compare structured and object-oriented systems design.
- Apply critical skills to practical interface, input and output design.

10.2 Introduction

The focus of design is on how to achieve a goal, with the emphasis on the means or mechanisms. In structured systems ontology systems design succeeds systems analysis. Systems design is concerned with *how* system requirements established during systems analysis will be transformed into an implemented system. Systems design also extends to determining what computer hardware will be required to implement the system specification. The transformation of the requirements specification focuses on determining specific hardware and software platforms required to operationalize the new IS.

The system design involves the design of interfaces, programs, data storage (databases), construction, testing and implementation activities. Detailed design documents are created during systems design. Programmers' work from the detailed design documents produced during systems design to write and test software to realize the new IS. This chapter will cover only the design activities.

10.3 Structured design

Interface design consists of designing interfaces between the system and people who will use it and where inter-system communication is needed designing interfaces between the system and other systems. The user interface is the medium through which people interact with a new IS. Programs that require data to complete processes need a user interface for the data to be inputted and outputted. The main principle guiding user interface design is the ease with which people can interact with the new IS.

10.3.1 User interface design

User interface design is based on knowledge of Human–Computer Interaction (HCI). Modern user interfaces now use Graphical User Interfaces (GUI) rather than text commands. Often a user model forms the basis of designing user interfaces. A user model describes and predicts how people interact with a computer system. User models or aspects of it are used to decide what interaction methods, presentation forms and colours to use. They help designers to determine appropriate screen layout, menus, form filling and interactivity.

The purpose of user interface design is to specify how users will interact with the system and how database actions will be triggered by user actions. In SSADM such a model is called the User Object Actions. It is a logical model descriptive of user action. A user action is something in the minds of users that can be carried out on a user object. The GUI actions are the user interface mechanisms used to implement the User Object Actions. The User Object Actions are logical and relate to the GUI actions that are physical.

Usability is a major design issue in user interface design. It addresses the ease with which people can use the system through user interfaces and how to improve interactivity, efficiency, effectiveness and reliability. Usability is concerned with enabling specific users to achieve specified goals with a system. Significantly, usability is concerned with peoples' satisfaction with achieving goals.

Usability provides knowledge on how to design the layout of interfaces, for example, use of colours, grouping related elements and system response times. For interactive systems, usability knowledge is used to decide the kinds and types of interface objects to use, for example, function keys, radio buttons, form-filling. **e-Tailing** for example requires highly interactive systems, in which direct manipulation devices, such as menu selection, are used. Usability issues for online user interfaces include response times and the depth of navigation users have to go through. Response times and depth of navigation both need to be kept to a minimum in e-Tailing.

Designers consider the details of input data and what user interface mechanism will be appropriate to enter it into the new IS. The choice of data entry methods will depend on the timeliness and speed of data entry required. Data entry methods include manual entry, entry via user interfaces, Electronic Data Interchange (EDI), or networked direct entry. The latter two are used in system where a large volume of data needs to be entered. User interface methods are used where a user can enter manageable volumes of data manually.

User interface interactivity is designed at two levels, the primary window and secondary window. A primary window has a border or frame. It contains a title bar, toolbars, a status bar, a menu bar and the content. Where the content extends beyond a screen, horizontal and vertical scroll bars are used. An example primary window is a word processor's window. A secondary window has no bars or frame functions. It can be a dialog box, a message box, a tab folder or a drop-down list. An example secondary window is a logon window. Some examples of interactivity methods or activities are: toolbar button, double click and scrolling button.

In structured systems design, for instance in SSADM, the user interface design is part of the requirements phase, rather than the physical design. User interface design during requirements analysis helps to understand and gather information for user requirements. The physical user interface design in SSADM consists of a window navigation model, window specification and help system specification. The design of windows can be based on window navigation models, which are themselves sometimes based on a function navigation model.

Interactivity is important for business IS. Interaction models of different interaction styles are used to improve interaction between people and a computer system. They are rooted in early studies of ergonomics and information science and technology. Interaction models focus on tasks that people want to achieve on a system and the model user and system, both of which are complex. They seek to understand the interaction in terms of what a user wants and what the system does, and help to identify interaction problems that can be avoided through appropriate design decisions.

A popular interaction model is Norman's rational model, which is intuitive. In the model a person is thought to formulate a plan of action which is executed at the computer system through the user interface. During the execution and at its completion the person assesses the obtained results, and based on the results determine further action. In detail the model is:

- establishing the goal;
- forming the intention;
- specifying the action sequence;
- executing the action;
- perceiving the system state;
- interpreting the system state;
- evaluating the system state with respect to the goals and intentions.

10.3.2 Data output or report design

An IS processes data, supports work and generates management reports. The information generated by the system is termed data output in structured systems ontology. Data output that is useful for operational and management use for an organization is converted into management reports. Managers and executives use management reports to support decision-making, and to control and monitor workflows and business processes.

Designers in consultation with managers determine the format and presentation of such reports. Reports can be daily, weekly, monthly or even real-time on screen. The frequency and granularity of the reports will depend on managers' needs and are initially determined during requirements analysis.

10.4 Object-oriented design

The distinction between the analysis and design phases in structured systems ontology is not made in object-oriented systems ontology. In object-oriented systems design, designers rely on the problem domain class systems model created during systems analysis. Designers create additional classes for data inputs, outputs and associated processes. They make decisions on how data required for classes will be entered, how it will be displayed on screen for people or presented as management reports. Standards are used to produce basic user interface designs like screen or report layout.

Object-oriented user interfaces are normally developed first as **prototypes**. The aim in using prototypes is to gather requirements of what user interfaces need to be capable of doing and then designing and evaluating it through the prototype. Many prototype user interfaces are developed using visual programming languages. A visual programming language environment enables user interfaces to be designed first to understand system requirements. Often this is the practice adopted by systems designers.

10.4.1 Human interaction component

In an object-oriented system, user interfaces, input and output design usually consist of windows and reports objects. An interface class

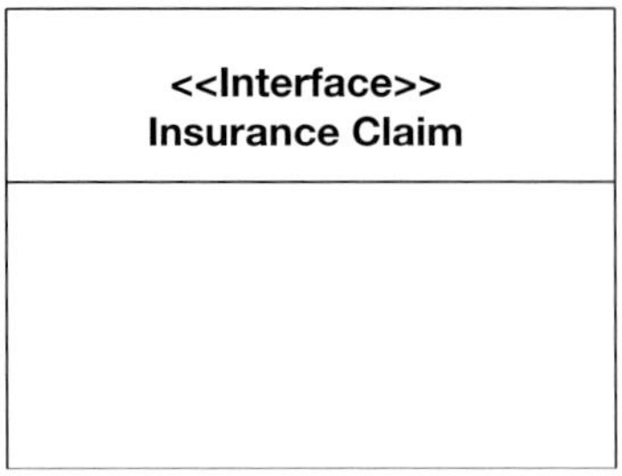

Figure 10.1 An interface class

is shown in Figure 10.1. It is drawn with the word 'Interface' in chevrons. Systems designers identify and design these additional classes and objects. Three types of windows are designed:

- logon and security
- system set-up
- business events.

Security design features are important. The log-on window is used to enable authorized people to access a new IS. Its design is critical to safeguard sensitive data and information. The set-up windows enable systems administrators to:

- Install and operationalize the system either in standalone mode or, more likely, as a networked system.
- Add or remove peripherals such as printer drivers.
- Add or remove people authorized to use the system.
- Manage the system privileges given to particular people.
- Create, maintain and remove persistent data, for example client records.

The business events windows are the main user interfaces to actual systems processing. For an insurance example, a window is needed to enable a new policy insurance to be drafted or a management report to display all the policies with a high or low risk. The business events windows provide the main interaction with the system for business people. Consequently, they need to be designed with usability features.

Though management reports can be displayed on screen, they are normally designed as objects. There may be many kinds of report objects depending on managers' need for information, for example decision-making, monitoring and control needs. Systems designers identify the report objects and attributes, and design the responsibilities (operations) and scenarios. The report objects collaborate with the problem domain objects identified during systems analysis to produce the required information. These interactions are the connections between the problem domain and human interaction components of a class model.

10.5 Interface design and the Critical Framework

10.5.1 Interactivity

Business people interact with an IS through user interfaces. Knowledge of interactivity is therefore significant in systems ontology. Theoretical knowledge on user interface design is based on the broader research area of human–computer interaction. The kind of questions addressed in human–computer interaction concern appropriate ergonomic design and interaction

The social context is significant for IS. Interaction models like Norman's model are based on the assumption that people behave rationally and logically. Interaction with an IS, though, happens in a social and organizational context, which affects both users and the system. Interaction models do not adequately consider this context because they focus on tasks and assume rational systems ontology.

User interface design needs careful consideration to enable human–computer interaction. Figure 10.2 is the Critical Framework populated with critical reflection on human–computer interaction. As the bottom layer shows, many questions arise concerning the four themes of criticality. The systems ontology component contains some examples of the theoretical and formal knowledge available to designers.

Assumptions need to be examined in the user models used to inform user interface design.

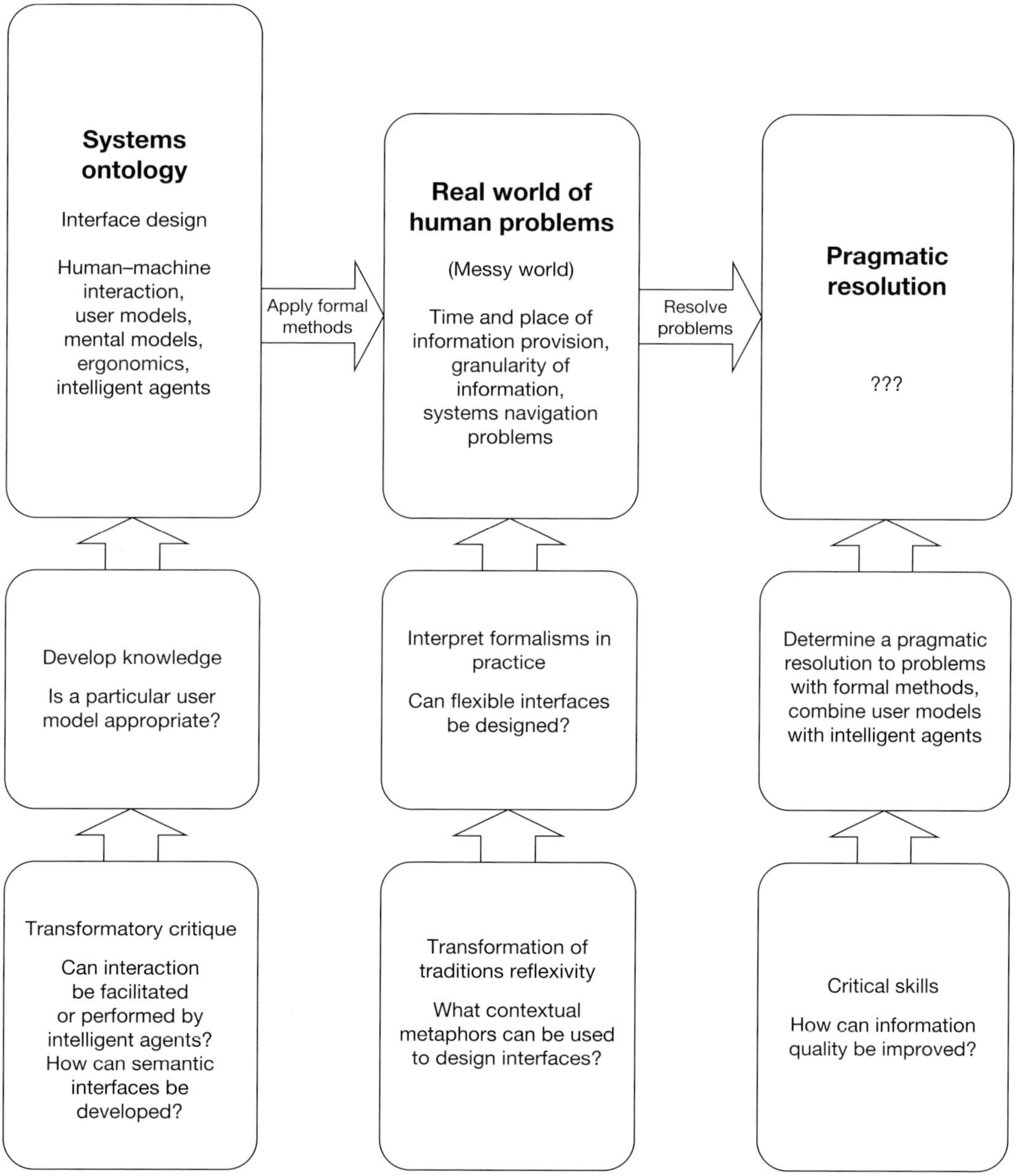

Figure 10.2 Critical framework: human machine interaction systems ontology

Critical awareness is especially needed when using mental models. Mental models are made of users mental activity when interacting with a computer system. Designers need to ensure that the mental models they use are appropriate for the type of user expected to use the system.

Usability engineering is based on specific user actions and not the wider social and organizational context in which IS are used. It is based on interpretations and shared understanding, which assumes meanings. The assumption is that shared understanding will provide agreed goals and metric with which to evaluate usability satisfaction. Such shared understanding is difficult to realize in practice.

10.5.2 Mental models

The application of user or mental models in user interface design affects IS operation and usage. A person's mental model is effectively the system – this is not considered in deploying mental models to design user interfaces and interactivity. Mental models do not have the same level of detail as usability engineering models because a mental model is a users' own interpretations of how a system works. Often people have only a partial understanding of the full capabilities of a system, and the mental models can be unstable or subject to change over time.

Incorrect mental models lead to design error. A mental model can be internally inconsistent because a person does not calculate all the logical consequences of an action. Some designs result in lack of usage because people are unable to navigate through the various user and system interfaces. This is an undesirable result for companies making strategic investments in IS. So usability is a critical design issue. Pragmatic use of user and mental models is necessary to ensure the operational success of IS. When considering the pragmatic resolution element of the critical framework systems designers should be aware of information quality arising form untested mental models.

10.5.3 Semantic information structures and user interfaces

A weakness of both structured user interface design and object-oriented human component interaction design is the lack of semantic and contextual modelling. This is because of assumptions on data and information in their respective systems ontology. Both structured and object-oriented systems ontology lacks semantic information modelling.

Web technology makes use of semantic and contextual mark-up languages to represent the problem domain terminology in a system and its interfaces. XML is an example of a mark-up language. It enables web application analysts to use the terminology of the problem domain and construct data structures to reflect it in the application and user interfaces. Such semantic user interfaces improve the usability features of a system and make it intuitive for people to use it. The semantic web is defined as an extension of the current web in which information is given well-defined meaning, better enabling computers and people to work in cooperation (Berners-Lee *et al.*, 2001).

10.6 Personal Critical Framework development

10.6.1 Personal constructs for human–computer interaction

Table 10.1 is a sample repertory grid for human–computer interaction. Reproduce the grid on a spreadsheet and add further columns and their polar opposites that you consider relevant. To objectify personal constructs, complete the grid by following the details on how to use a repertory grid in section 1.10.1.

Table 10.1 Personal constructs for human–computer interaction

Pole 1	*Interface*	*Data input*	*Data output*	*Display/ presentation*	*User models*	*Mental model*	*Interactional devices*	*Interactivity*	*Pole 2*
Visible									Hidden
Function									Aesthetic
Practical									Impractical
Usable									Unusable
Static									Dynamic
Persistent data									Transient data
Computer-based data/ information structures									Semantic data/ information structures

10.6.2 User models

Questions

1 Define user model and mental model.
2 Determine your criteria for selecting a user model for user interface design.
3 Critically evaluate whether user models add value to user interface design.

10.6.3 Design process

Questions

1 Critically analyse the value of a separate systems design phase in structured systems ontology.
2 Assess the value of the progressive development of a class model from analysis to design in object-oriented systems ontology.

10.6.4 Interactional devices

Label, button, choice, textfield are some examples of user interface devices. For a highly interactive system, like an ambulance dispatch system:

- What kind of interactional devices would you use?
- Discuss critically how the interactivity afforded by your choice of devices relates to the human activity in the problem domain.
- Assess whether structured or object-oriented systems ontology is better for designing interactivity.

10.6.5 Usability

Activity A

Either use the SSADM User Object Action user model in section 10.3.1 (you will need to source it) or using one familiar to you:

- Identify a system interface you use.
- Evaluate the system interface with reference to the identified user model.
- What additional design features would you add? What would you remove? Explain your reasons commenting on interactivity if relevant.

10.6.6 Internet sources

See the notes of the OMG-MCC-DARPA Workshop on Compositional Software Architectures at: http://www-db.stanford.edu/CHAIMS/Doc/Papers/commerce.html.

10.6.7 Further reading

Tim Berners-Lee is the founder of the world wide web. He is now the Head of the World Wide Web Consortium (W3C). For his thoughts on the semantic web see: Berners-Lee, T., Hendler, J. and Lassila, O. (2001) 'The Semantic Web', *Scientific American*, May.

Chapter 11

Systems design

11.1 Learning outcomes

After completing this chapter you should be able to:

- Interpret the distinction between logical and physical systems design.
- Detail the design activities at system level and describe detailed design.
- Apply critical skills to data design.
- Critically evaluate the role of the data management and system interaction components in object-oriented systems ontology.

11.2 Introduction

In structured systems ontology, systems design involves logical and physical design of data covering computer programs, data storage, networks and change management. Computer program specifications are made and diagrams created to define program algorithms. Systems programs require input data and generate output data. Output data may need to be stored, so data storage decisions, whether to use computer files or databases need to be made. The choice depends on the size of data and the purpose for its storage. In object-oriented systems ontology, a class systems model forms the basis of program and other system components design.

Analysts and designers are involved in the implementation of a new IS as part of a change management program. As well as making decisions on hardware, software and computer networks, they consider how to introduce the system to people and organization. Change management is an important issue because large systems have an impact on organization. Analysts especially are involved in managing such organizational change resulting from a new IS.

11.3 From logical to physical data design

In structured systems ontology, implementation-independent design is the logical data design or logical data model built independently of physical computers and software. Implementation-dependent design is the physical design or physical data design of data as it is stored in computers and software. Physical design is concerned with making decisions on the hardware and software required to implement a new IS.

In structured design, physical data design involves converting E-R models into physical formats as part of the Transform Analysis. Physical data design can either be in file format or a database. Where files are used, decisions on the type of files, access methods and their organization need to be made. Each logical entity in the E-R model is translated into a record type and each attribute is translated into a data field. The application program manages the relationships between entities with the keys.

Database storage is common in large companies because organizations are interested in creating corporate information and knowledge. So decisions on the type of database to be used need to be made. Each entity in the E-R model is translated into a table, and each attribute of an entity is translated into a column of the table. Systems designers determine how the entities in the new record types or tables format will be stored. Decisions are made on how to group the entities but relationships are initially kept to reflect the logical model.

The physical data model of the selected database is used to organize the data. The DBMS uses logical pointers to maintain the relationships between entities. If people do not use some of the entity relationships once the system is installed then database administrators can remove them in consultation with them to improve system performance or reduce the cost of storage. The implemented system is then tested to ensure it meets performance and system requirements.

In object-oriented design, the class systems model created during systems analysis is refined and the data management and system interaction components designed. For the data management component, designers consider whether to store persistent data in files or databases, decide between object and relational databases, and model data management objects in UML diagrams. The system interaction component is needed if a new IS is to be networked or if it has to share data with other systems.

11.4 Some fundamentals of hardware and software platform

The term for computers and software combined in an IS is platform. Designers develop options on different platforms that can be used to implement the logical models created during systems analysis. Sometimes there are no options to consider because of investment in existing platforms.

Storing data is an important issue. The quantity of storage required needs to be determined. Analysts and designers consider the existing storage requirements and project what will be required in the future. They would need to consider business growth issues like sales or volume of employees to determine future storage needs. Storage for each entity or persistent object is determined by totalling the bytes required for fields or attributes, multiplying the total by the number of records needed and then dividing by the number of records. The actual implementation may be as data files or a database.

The performance of the system is the measure of its efficiency and effectiveness. Performance is important for safety-critical IS like emergency services. It is also important for critical functions in IS like keeping records of important business transactions. The Taurus stock trading system at the London Stock Exchange and the London Ambulance Service's Computer Aided Despatch System (LASCAD) both failed because they failed to design for the actual processing demand when the systems were made live.

Designers seek flexibility in the choice of DBMS to cater for future needs. Software that is hardware-independent is preferred to ensure maintainability and prevent obsolescence. For eCommerce systems the Java programming language is used because of its platform independence. Designers need to ensure that the functionality of the system is assured given monetary and time constraints. Data storage, processor usage, and access times need to be efficient. Having more data storage capacity than is required will mean additional unnecessary costs, but having insufficient storage space could affect the performance of the business detrimentally.

Where multiple databases are used, designers need to ensure data integrity is preserved across databases. In one case, consultants used two databases for a pharmaceuticals company because one database could not cope with the volume of data. They did not check adequately for data integrity across the two databases. Managers were unaware of this situation and used the data from both databases in operational and strategic decisions. In time, the company had to file for bankruptcy.

11.5 Structured design

People are not involved in drawing up the detailed systems design documents in structured design. The reason is the technical content of the design documents required for the design of data inputs, outputs, processes and data files or databases.

During systems analysis references to physical entities in the problem domain are removed from the systems models. In the design phase decisions about physical details of a new IS need to be made. The data and process models and requirements specification produced during systems analysis is transformed into design documents that detail file and record structures, file access mechanisms, database design and storage media. The E-R model is used to design the database model and the logic models used to inform program algorithms.

During the transition from analysis to design, inefficiencies in the problem domain may be addressed. The logical modelling process may reveal problems in the problem domain, particularly physical aspects that can be addressed. For example, DFD systems models may show duplicate physical file storage, these can be removed in the logical DFD and consolidated in one store.

11.5.1 Detailed systems design documents

Systems design diagrams differ from those used during systems analysis. Structured design based on the SDLC includes both the computer and manual aspect of a new IS. Various design documents are generated including those shown in Figure 11.1.

A structure chart is used to detail the connection between program modules in a system. It is compiled on the basis of the HIPO technique (hierarchical input/process/output). A given process in a system, for instance prepare an insurance policy, will require separate program modules to interact to exchange data. The structure chart depicts the various program modules and the data flowing between them. It provides a hierarchical, clear and easy to understand picture of a new IS in terms of programs and data movements.

Grid chart
System outline
Structure charts
Record specification
File specification
Flowcharts
Hardware specification
Stored database design

Figure 11.1 Structured systems design documents

A grid chart is used to show the use of files for updating or reading data by each program in the system. Each program is listed in a column and each data file is listed in separate columns alongside. Crosses are marked in the rows for each program that makes use of a data file, if files are used for data storage.

11.5.2 Systems design tools

There are commercial software tools available to aid the systems design process. The aim of designers of structured and object-oriented systems design tools is to make the design process efficient and effective. Systems design tools focus on enforcing certain practices, usually associated with a methodology like IE or notation language like E-R. They enforce stipulated design processes, whether structured or object-oriented, program modularity and documentation. Their use produces better quality design because they check for consistency and integrity across different diagram types.

11.5.3 Program design

How the computer will process the data entered into the system is an important part of systems design. The program algorithms convert the data into the required data outputs for other programs or management reports. Flowcharts may be used to design algorithms but they enable any conceivable structure to be modelled – their flexibility is a disadvantage in some cases. Tools available for program design include:

- Editors to create and modify source code and documentation.
- Static analysers to examine source code.
- Dynamic analysers to examine running programmes.
- Language sensitive editors to create syntactically correct source code.

Many systems designers prefer the Nassi-Schneiderman chart because it is precise in describing algorithms produced by a team of software developers normally associated with a system project. The Nassi-Schneiderman chart or box diagrams, illustrated in Figure 11.2 are used to represent procedural constructs consistent with structured concepts. They help to ensure correct program design in terms of:

- Functional design: well-defined and visually presented definitions of scope of repetition or if-then-else functions.
- Prevent arbitrary transfer of control.
- Easily determine the scope of local and global data.
- Easily represent recursion.

Structured program design makes use of modularity. A module is a program that focuses on one function in the system. A computer program consists of lines of code, or algorithm, written in a particular programming language to process data to provide particular outputs. A good program has high internal cohesion and low coupling. Cohesion is a measure of the strength of relations of subroutines within a program and coupling measures the strength of bonds between programs. Structured design rationalizes input and output with processing through structured programming. Architecture efficiency is improved through cohesion and coupling resulting in modules of program code.

11.5.4 File design

A file design specification will define the file medium. The decisions to be made are whether it will be sequential or random access, the file size and back-up procedures. The grid chart shows the programs in the system and which files they access to read data or update records. Flow charts are used to depict procedures and

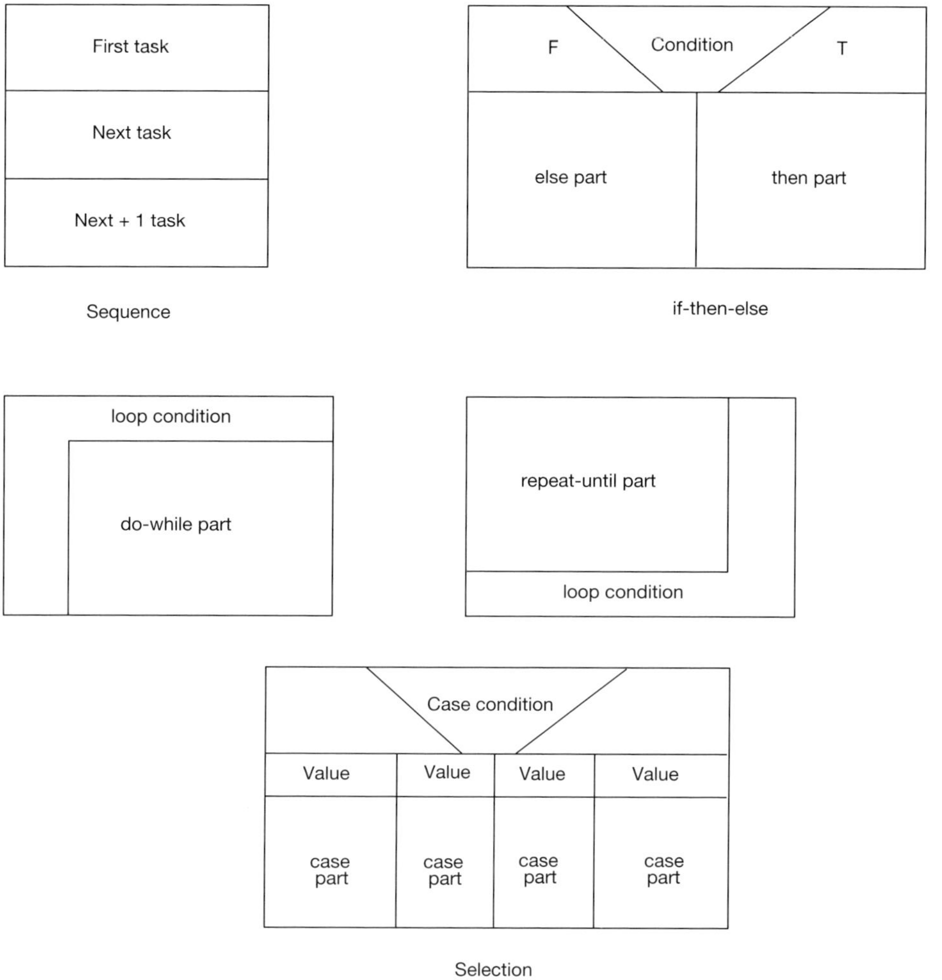

Figure 11.2 Nassi-Schneiderman diagrams

the control of the flow through procedures required to complete a computer run.

11.6 Database design

Database design is separated into logical and physical design to enable data independence. Data independence is desirable because changes in business needs can be enabled by changing the logical database design and not having to redesign the physical computer files. Conversely, this separation enables database administrators to change the physical database design with no impact on the logical design and users. They may need to do so to improve the efficiency of the database.

People are not involved in database design, to them the DBMS is a black box. The DBMS provides physical data independence. As it is not possible to predict the volume of actual queries of the database, estimates are generated and used to anticipate usage patterns.

The structure of the data defined during systems analysis determines the data model of the database and the data models from systems analysis are used to design the logical database. The logical record structure and how the records will be accessed is derived from the E-R model. Each set of E-R diagrams can be converted into a logical record type. Logical record structures are defined at the system level and include the primary keys of records. For example, for a relational database model the conceptual database is determined by converting the E-R model into a relational model. An entity is represented as a relation. A relationship is represented as a relation too with the primary keys of the entities in the relationship.

11.7 Object-oriented systems design

The focus in object-oriented design is on the architecture of the new IS, quality and standards issues. Systems designers create additional classes for files or database storage. Unlike structured systems design, in object-oriented systems design a class model is not transformed but *refined*. In this sense there is no separation between systems analysis and systems design in object-oriented systems ontology. The class systems model produced during systems analysis is enhanced with more detail but remains essentially the same model, though different diagrams are used to produce the logical design or implementation. So it is possible for people to be still involved during design.

Detailed systems design decisions are concerned with designing components that compose the system architecture. During systems design the classes are refined and the classes form the basis of program algorithm designs. Like object-oriented analysis, the focus of design is on classes, reuse and assigning responsibilities to classes – a continuation of class modelling.

Designers take advantage of reusable components. Software development has a tendency to address recurring coding problems. Design patterns are used to reduce the repetitive resolution of the same problem. They provide encapsulated functionality and data in classes that can be reused. Designers do not need to create new design for classes if they already exist or can be bought commercially. Categories of design patterns available are creational, structural and behavioural. Patterns that can be bought commercially are termed components.

11.7.1 Additional design diagrams

During design additional diagrams are generated including collaboration diagrams, state transition diagrams, package diagrams and deployment diagrams. Systems designers continue modelling the system by selecting the appropriate diagrams.

Collaboration diagrams are used to model different objects interacting to exchange data for a specific process, as shown in Figure 11.3. Collaboration modelling is central in systems design, comparable to use case during systems analysis. Use cases are realized through depicting the necessary collaborations. Collaborations diagrams depict associations between objects. Collaboration in UML consists of objects interacting to achieve a specific purpose and can be used to show a particular operation or a whole use case. As they do not depict temporality, sequential numbering is used, and when iteration is required the numbers are nested. Branching can be similarly represented.

A package is a way to organize classes into logically related groups. It can be used for classes identified during systems analysis and systems design. Figure 11.4 shows a package diagram for a database. Class A and Class B can be for two different database connections. For example ODBMS (Open Database Management System) and CORBA (Common Object Request Architecture).

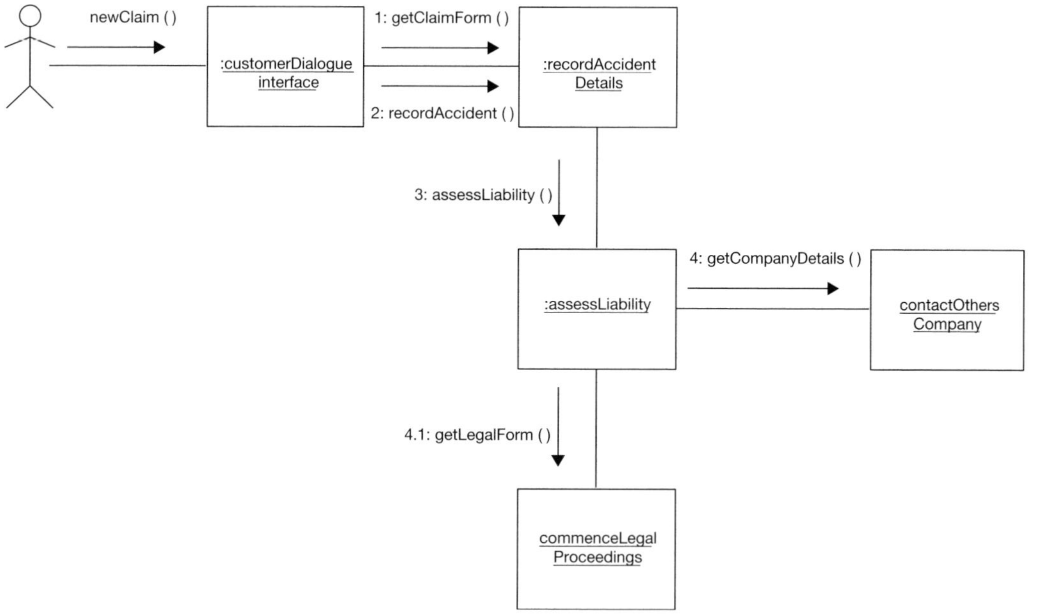

Figure 11.3 Collaboration diagram for insurance claim

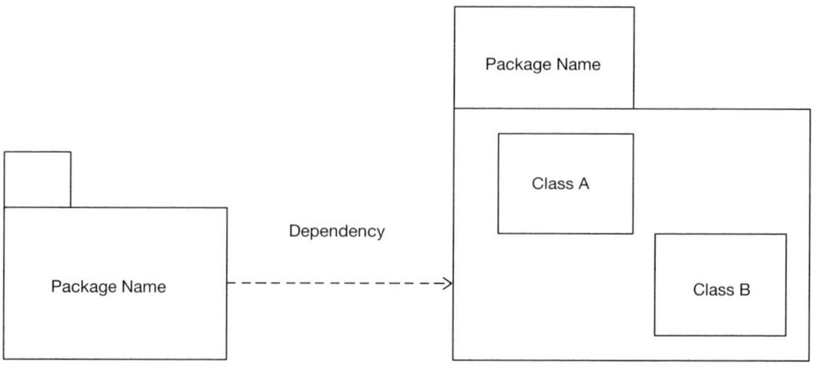

Figure 11.4 Package diagram

Packages are useful for testing systems designs. Unit testing can be done on the basis of packages. This makes it easier to manage a large testing programme.

11.7.2 Data management component

The data management component is concerned with how to store and manage such stable, or persistent data. Database designers identify data storage classes, usually based on an abstract superclass, and determine the collaboration required with business classes (identified during systems analysis) to determine storing and retrieving of objects.

Organizations have both transient data and stable data. An IS may produce both types. An example of transient data for an insurance company may be a quotation for a policy that the inquirer does not draw out. As transient data is not needed beyond its immediate purpose it is not stored. Some data, though, are regarded as valuable and stable. They need to persist in the system. An example is the details of a new policy written for a client.

Designers determine whether to store persistent data in files or databases. The choice is made on the basis of how the stored data will be used. File storage is used if the data is for one application only but it is inappropriate for developing corporate information and knowledge, because file stores would multiply with each application created, and data integrity checks would become problematical. Multiple file stores could not easily facilitate new information requirements arising from changes to business needs or markets and cannot support online data query.

When the data is used for corporate information needs it is normally stored in a database. An object-oriented database contains class objects and their interrelationships based on a class model. A database is used to store data that need to persist and enable sharing of persistent data between objects, applications and systems. Systems designers decide between object and relational databases, and model data management objects in UML diagrams. A design class model is created to depict normalized relations with multiplicity.

A database can be accessed by many applications and deal with multiple and changing queries or interrogation of the data. It facilitates data sharing through the external schema – the view of data in an application, and conceptual schema – logical model of data. The key to achieving such flexibility in managing the data is the DBMS. A DBMS is used to organize and manage the physical data separately from applications that use it. The conceptual schema achieves this separation, with the external schema being the application programs' view of the data and the internal schema being the physical storage of data in files and indexes.

11.7.3 System interaction component

The system interaction component needs to be designed if a new IS is required to interact with other systems. The new IS may take data as input from another system or provide data as output to a different system. The communication systems are modelled with deployment diagrams. A deployment diagram shows two systems communicating with each other as nodes and the protocol and network used for linking as communication associations.

An example is shown in Figure 11.5. It shows the PC client, which is called a node, with an insurance policy system and another component for database connectivity. Components are the various functional units in a system. The PC is networked to a mainframe machine which has the policy database installed on it, which becomes active when it is called from the PC client.

For example, client-server architecture is commonly used in new IS. It consists of PCs, called the client, used by people, and the server that contains the required software. Designers make decisions on what process to allocate to the different server machines available and how the different machines will communicate. This is done using UML physical deployment diagrams. A deployment diagram is used to show the physical relationships between hardware and software components in physical systems design.

11.8 Change management

The success of a new IS depends on people's perception of it and whether they use it. The

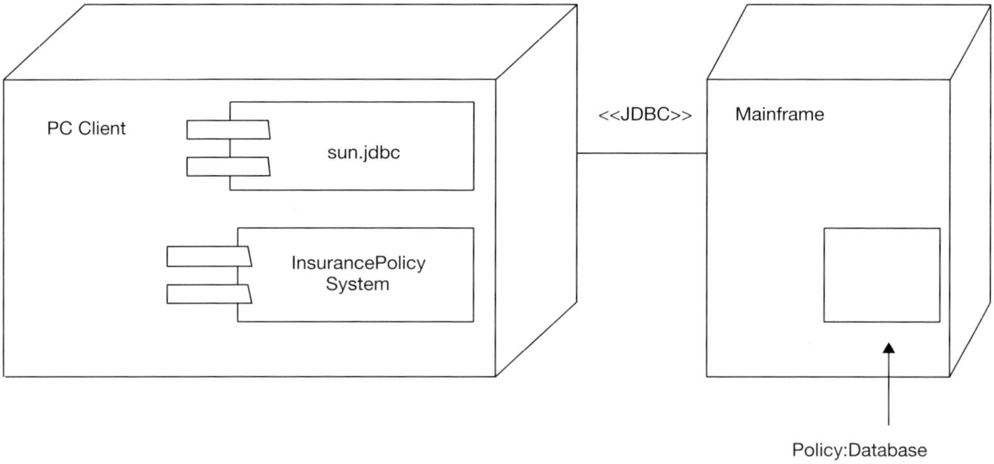

Figure 11.5 Deployment diagram

introduction of a new IS will require reorganization of people's job roles, responsibilities and business processes. Organizational change management is concerned with managing the people issues that arise from such change.

Although not strictly a systems analysis and design issue, change management is now accepted as being critical for the success of a new IS and for realizing its planned benefits. Work packages in system projects may include analysts working in a change management team. People are the essence of change management. The expected benefits from a new IS will only be derived if people accept it.

Analysts and designers are involved in training and educating people about a new IS as part of a change management program. Systems analysts work closely with business analysts to define new job roles and responsibilities or changes to existing ones. A key involvement area for analysts is managing people's expectations of a new IS. Analysts' early contact with people during systems analysis can form the basis of managing such expectations, and analysts are usually in a business team set-up to manage the change.

11.8.1 Change management theory

Change management theory is concerned with how change can be successfully realized in organizations. The issues it addresses include identifying who the relevant people are, how they are motivated, their understanding of the change and identifying and managing resistance.

People perceive any organizational change including a new IS which affects them with some reservation. They do not commit fully because of fear. The change may remove their current status or even their jobs. Change management theorists seek to understand how organizational change can be managed by understanding the role of people in change.

One model of change management is Kurt Lewins' (1958) 'unfreezing, moving and refreezing'. It begins by giving people a reason for change and enabling them to think for themselves why it is necessary, the 'unfreezing'. Senior managers, once they have decided that a new IS is required, need to communicate the reasons to people. Keeping people uninformed of the need will result in anxiety and fear, and even result in losing key people. The 'moving' step is to inform people of the chosen change and prepare them to become party to it. This step is more difficult than the first because it requires people to make the change themselves. Their commitment to the new order needs to be strong because problems may arise that may deflect them. Peoples' involvement and commitment is particularly required during requirements analysis. The final step is 'refreezing'. It requires enabling people to familiarize themselves to the change. People's training and best practices are reinforced, and any problems resolved.

11.9 Systems design and the Critical Framework

Arguably, systems analysis is more important than systems design for understanding and specifying system requirements. Systems design is nevertheless significant because it involves a transformation step in structured systems analysis, which is problematical. In object-oriented systems ontology its significance is in how a class model continues to be progressively developed.

11.9.1 Systems ontology

Figure 11.6 is the Critical Framework populated with critical reflection on systems design. As the bottom layer shows, many questions concerning the four themes of criticality arise from the assumption of a data-centric or object-centric **reality**.

The notion of 'stable data' underpins structured systems ontology. Data are considered sufficiently stable to describe business activities in a problem domain. It is because they are considered stable that they are stored in corporate databases. Decisions regarding file storage or types of databases, relational or object-oriented, depend on the IT/IS strategy of the organization. Where corporate data forms a significant part of strategy then database storage will be required. If unstructured data like image, voice and graphics needs to be stored then an object-oriented database is more appropriate.

Structured systems ontology is a 'functional' characterization of the problem domain, which reinforces the input, process and output view of an IS. It is arguably a 'computerization' perspective, in the sense of a mechanistic characterization of real human problems. Its focus on data analysis gives primacy to functions, data and processes as defining characteristics of a problem domain. This mechanistic characterization of the problem domain leads to a separation of systems analysis and systems design, because the latter is closer to the machine itself and in which the real world human problem is reflected.

Structured systems ontology does not question this separation. Whether systems design can inform systems analysis is both a transformatory critique and refashioning of traditions. Changes in practice, such as prototyping and user interface prototypes during requirements analysis, indicate that the separation is flawed. IS development benefits from intertwined systems analysis and design activities, rather than concrete separation.

The structured systems ontology focus on data, with 'data analysis' being a significant element of database design, has consequences on how an organization is modelled for IS development. A corporate database is regarded sufficient to describe or model *the* organization. The social element of organization is not accounted for in such systems ontology. Also, importance of organizational knowledge and its management means that a corporate database is insufficient as a model of an organization. Organizations are now also building 'knowledge bases', which are significant in their attempts to manage organizational knowledge.

A database stores organizational data important or critical to a business. Modelling an organization as databases is problematical because the

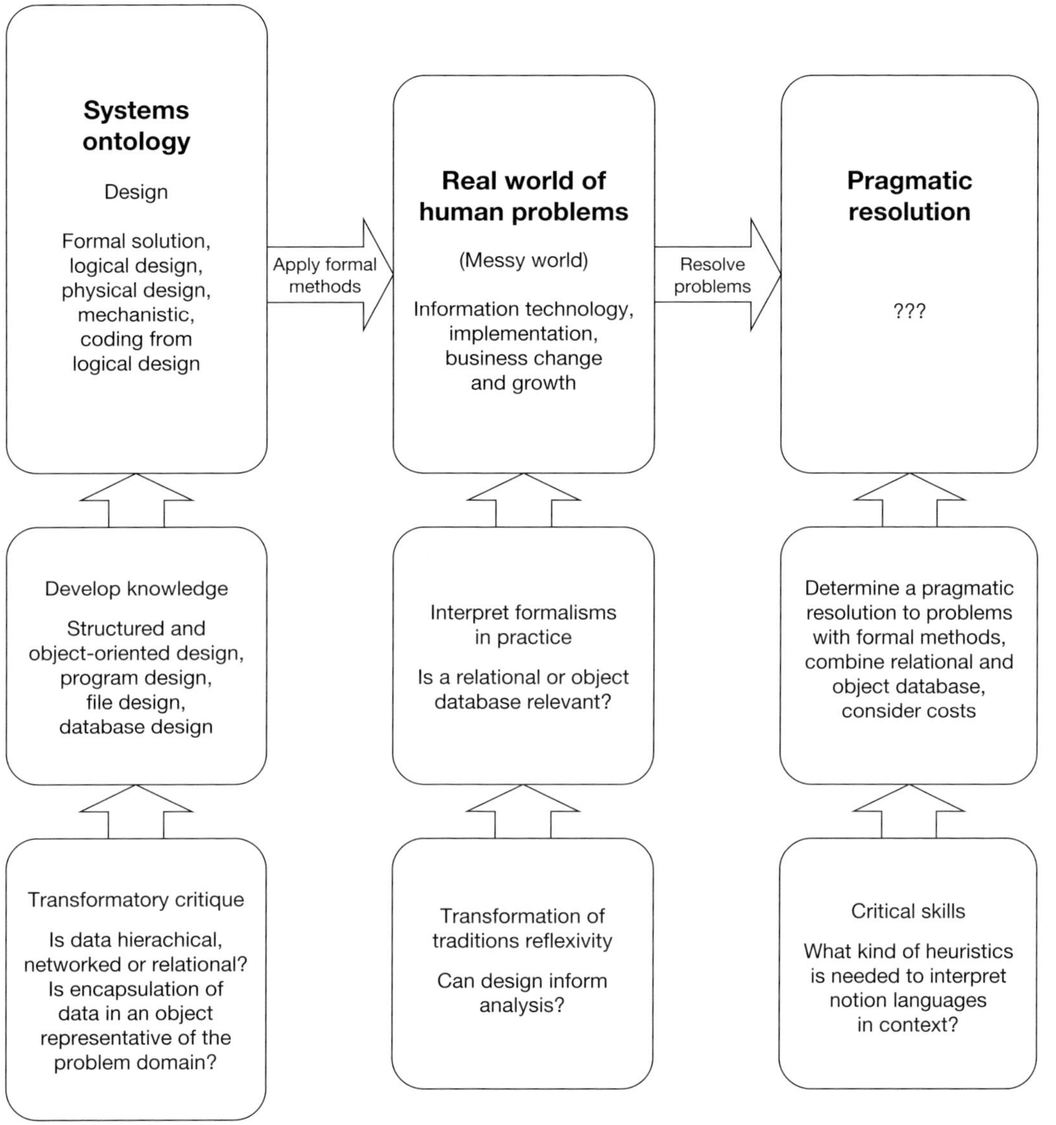

Figure 11.6 Critical framework: ontology and systems design

modelling process needs to reflect human and organizational needs. In pragmatic terms, because of costs a database is only composed of data models of organizational functions like personnel or sales. In a process view of organization it is composed of processual data models.

Systems analysts and systems designers need to decide whether their perspective of the data model from which a database is designed is objective or subjective. From an objective perspective the data model *is* the organization. Such a personal construct will have consequences for corporate database design. From a subjective view the data model is one interpretation of an organization. It does not represent the whole organization for all time or the varied perspectives of its members. Database design from this perspective is likely to be a truer reflection of real human problems.

Problems with Transform Analysis in structured systems ontology are being overcome by emerging systems analysis and design approaches like ASD and XP. They conceptualize systems development as continuous, thus not having to make a transition from systems analysis to systems design.

The critical development of database technology itself reflects the development in ontological knowledge of organizational data, information and knowledge. Various database conceptions are reflected in developments in database types: hierarchical, networked, relational and now object-oriented databases. Database technology is significant in defining systems ontology appropriate for organizations.

Each stage of development in database technology permits richer representation of real human problems. Developments in database conceptions reflect both transformatory critique and reflexivity, depicted in Figure 11.6. Systems design is significant for systems ontology because of the physical capabilities and limits of available IT. Knowledge of program and database design determines how a system is defined and designed, and its design has an impact on the performance of the problem domain or the organization. Analysts and designers need to bring their personal constructs on systems design to the surface and reflect on knowledge and practical validity.

11.9.2 Separation of analysis and design

In structured systems ontology there is a distinct separation of systems analysis and systems design phases. Earliest versions of the SDLC made such a distinct separation, but have now accepted a spiral version of software development. It is arguable whether this separation reflects the problem domain, systems ontology, organization ontology or even the actual work of practitioners. Many developers move between analysis and design, and organizations would rather make IS development an operational issue than a distinct system project.

Though object-oriented analysis and design caters for the non-separation of systems analysis and design, through its progressive development of a class model, on the whole, practitioners persist on completing systems analysis before undertaking systems design. The reason is the timeboxing constraints of system project management and the discipline imposed by project management. In essence, business constraints prevent the fluid development of a class model.

The separation of analysis and design in structured systems ontology raises practical problems once a system is implemented. Often, it is necessary to make amendments to the detailed analysis and design documents because of changes in requirements. Business requirements often change during systems development, and occur because of market or competitor forces.

For both structured and object-oriented systems ontology, another reason for change in requirements is lack of information on what is required. During requirements analysis people in the problem domain are expected to provide a detailed account of what they require. They are unable to do so in the set time for requirements analysis or timebox period allocated in a systems management work breakdown package.

Requests for new requirements or changes to captured requirements arise as new information becomes available to people. Often the changes are only reflected in the design documentation but not the analysis documentation. The requirements specification becomes inconsistent with design documentation. This is a problem of separating analysis and design in structured analysis.

Earlier development in prototyping, RAD and JAD reduced the distinction between analysis and design. The recent move to ASD is a stronger statement for the non-separation of analysis and design activities.

11.9.3 Pragmatism

The pragmatic resolution of which database to choose reveals a mixed approach rather than adhering to a single type of database. Designers need to reflect on how they make their choices. As Figure 11.6 shows, critical thinking leads designers to make use of mixed databases containing relational and object-oriented elements. Such a choice is significant from a systems ontology perspective. It indicates that both models, rather than just the one, better describe the actual problem domain or real human problems.

Systems analysis, design and implementation does not have to be either structured or object-oriented. It can be a mixture in practice. It is possible for practitioners to make use of structured analysis and design but implement an object-oriented system. It is also possible to implement a systems design using the object-oriented approach in a combination of conventional and object-oriented programming languages.

Systems analysis, systems design and programming can each be done in different genres or practices for the same system. So it is possible to do the systems analysis using structured instruments, systems design using object-oriented instruments, and programming in an object-oriented language like Java. Other permutations are also possible, but practicc tends to adhere to one underlying approach in a given system project.

In structured systems design, the technical content of design documents prevents designers from involving people in designing. To ensure that the completed system is valid, it is validated with requirements specification, but normally this is not done thoroughly. In contrast, people remain involved in object-oriented systems design. The problem domain class model is refined during design rather then separately developed, so people can still be involved, though they will not comprehend technical program and database terminology.

IS as a field itself has recognized the importance of change management through reflexivity and refashioning of traditions criticality. Methodologies and in-house expectations now require analysts to be significant members of a change management team. Analysts need to appreciate that a new IS brings organizational change. Though it is important to manage carefully the people aspect of the change, analysts may not be the most appropriately skilled people to do so.

11.10 Personal Critical Framework development

11.10.1 Personal constructs for systems design

Activity A

Table 11.1 is a sample repertory grid for systems design. Reproduce the grid on a spreadsheet and add further columns and their polar opposites that you consider relevant. To objectify personal constructs in systems design, complete the grid by following the details on how to use a repertory grid in section 1.10.1.

Table 11.1 Personal constructs for systems design

Pole 1	*Data model*	*Information*	*Knowledge*	*Databases*	*Logical design*	*Physical design*	*Organization*	*Pole 2*
Objective								Subjective
Stable								Fluid
Mechanical								Organic
Explicit								Tacit
Conceptual								Actual

11.10.2 Logical data design

Questions

1 What is meant by 'data independence' and how does logical data design during systems analysis contribute to achieving data independence?
2 How is the E-R data model used for logical data design?
3 How are logical data models used in defining the conceptual schema in a DBMS?
4 Critically discuss the idea of the progressive development of a class model through systems analysis and systems design.

11.10.3 Physical data design

Questions

1 What is meant by 'physical data design' and what is the role of the DBMS internal schema?
2 What kinds of platform decisions need to be made for physical data design?
3 What value do UML deployment diagrams add to physical systems design?

11.10.4 Database design

Activity A

Databases are used to store corporate data and develop it as a corporate resource or information assets.

1 Identify a database in your organization.
2 Enumerate the kinds of data it stores.
3 Analytically evaluate the corporate information value of the stored data. Is it used to add value to the organization? Is it used to compete with other firms or differentiate your organization from others?

11.10.5 Managing change

Activity A

- Find an example of a system project that has failed because of poor change management. For example, the London Ambulance Service's Computer Aided Despatch System (LASCAD).
- Read any one version of LASCAD. You can use either the official report on LASCAD available at http://www.cs.ucl.ac.uk/staff/

A.Finkelstein/las/lascase0.9.pdf (accessed 26 March 2004) or a personal view available at www.lond.ambulance.freeuk.com/cad.html (accessed 26 March 2004).

- Analyse LASCAD in terms of Kurt Lewins' model of change management.
- Detail the activities you would initiate for LASCAD for each of Lewins' model steps.
- Detail the methods you would use to convince people to accept the change.

11.10.6 Further reading

Lewin, K. (1958) 'Group Decision and Social Change', in Swanson, E., Newcombe, E. and Harley, R. (eds) *Readings in Social Psychology*, New York: Rhinehart and Winston, pp. 459–473.

For an alternative object-oriented systems design see: Shlaer, S. and Mellor, S. (1992) *Object Lifecycles: Modelling the World in States*, Englewood Cliffs, NJ: Prentice Hall.

Part V

Criticality, paradigms and IS development

Part V covers the broader issues in the development of knowledge and practice. The PCF section in each previous chapter focused on reflection and critical thinking on a specific systems analysis and design topic. In this part, the question of how knowledge of systems analysis and design is acquired and accepted is covered and critically considered. Systems ontology is heavily influenced by innovations in software programming. Radical changes in knowledge and practice stem from new ideas, concepts and methods developed by practitioners in software. The structured, object-oriented, and agile concepts all originate in software programming.

Since its introduction in the 1960s, the engineering metaphor in software development has dominated intellectualism and practice. Both structured and object-oriented systems ontology focuses on systems and systemic factors. Chapter 12 is on social action. It examines the non-systemic social, cultural and political factors in IS development. Systems analysis and design is undertaken in a political, organizational and cultural context. This questions the suitability of structured systems ontology because of its assumption about factual design knowledge, and to a lesser extent object-oriented systems ontology. Structured systems ontology has been adapted subsequently to account for some of these criticisms.

Chapter 13 is a general critical reflection on the previous parts of the book. It is a discussion on the assumptions made of users, systems analysis, systems design and organization in structured and object-oriented systems ontology. The chapter is a consolidated critical reflection drawing together various strands of critical argument.

Chapter 14 introduces paradigm as a way of thinking and acting. IS development has progressively resulted in paradigms of thinking and knowledge. IS developers unwittingly accept certain paradigms of knowledge and act upon it. No one paradigm is *the* solution to the problem of IS development. Each paradigm makes different assumptions of systems, instruments, IS, organization, people ('users'), and systems analysis and design, which researchers and practitioners have subsequently challenged. Analysts need to be aware of paradigms and understand that their actions are underpinned by explicit or implicit choices of paradigm.

In terms of the map of criticality presented in Figure 1.1, Chapter 13 covers refashioning of traditions and Chapters 14 and 15 cover transformatory critique. Developing theoretical and conceptual knowledge can enhance a PCF, especially if it is critically evaluated. Such knowledge can benefit the practice of systems analysis and design too if it is based on understanding how knowledge is acquired, and how it is accepted as valid knowledge by a community of practice such as systems analysts and IS developers.

Chapter 12

Social action

12.1 Learning outcomes

After completing this chapter you should be able to:

- Apply refashioning of traditions criticality to systemic problem-solving strategies.
- Analyse and assess the impact of social action on systems analysis and design, generally on IS development.
- Evaluate the effectiveness of using instruments in the context of social action.

12.2 Introduction

Stemming from the SDLC, structured and object-oriented systems analysis and design focus on systemic factors. They characterize IS development, and systems analysis and design, as a systemic problem that can be resolved rationally and objectively. The set of assumptions underpinning objective problem-solving though does not account for organizational and human factors. Research reveals how social action affect IS development. This chapter examines such factors and considers how knowledge of them can be used to improve knowledge and practice.

In addressing the central problem of how to develop IS, practitioners and researchers need to be aware of social theory and social action. The non-systemic factors are organizational, social, cultural and political, and they need to be reflected in systems ontology. They are termed **social action** here. This is essentially human action in a social context, its explanation and knowledge of social factors relevant for IS development. Knowledge of social action is important because it poses potential obstacles to systems analysis, systems design and implementation if it is not understood and accounted for in systems ontology. The relations between technical factors, social action and IS are shown in Figure 12.1.

The figure depicts the relations between technical factors, social factors and IS. It shows that the relation between technology and social action is bi-directional. Technology affects social action and social action affects technology. Information Technology is a keen example of this bi-directional relationship. Both these factors determine the conceptualization and realization of IS.

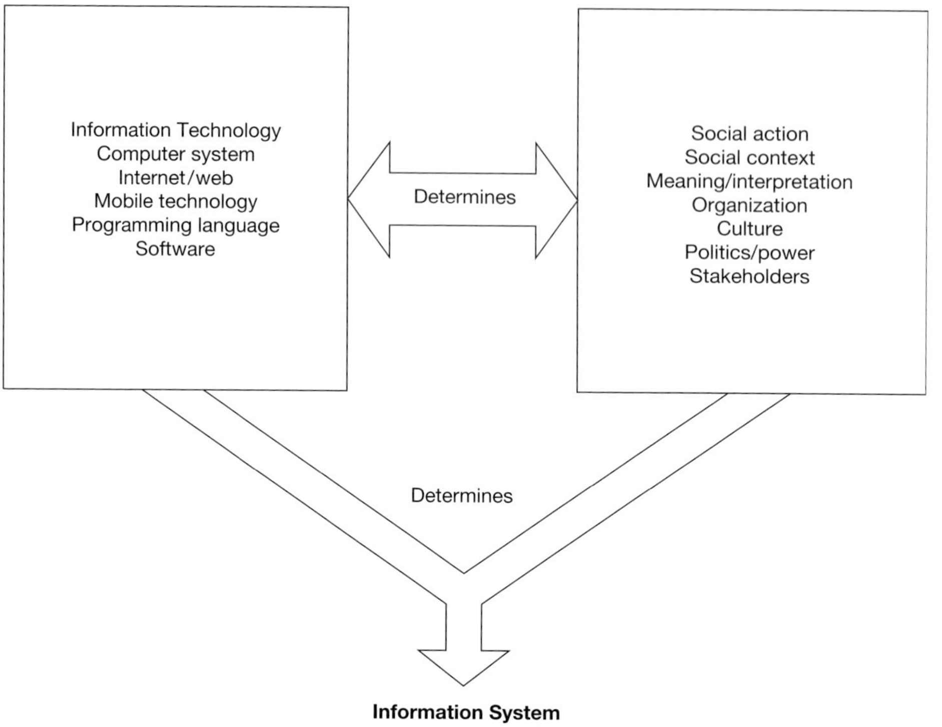

Figure 12.1 Technical factors, social action and IS

12.3 Social theory and IS

Interpretive researchers have framed IS research questions in the context of social theory to understand and explain IS and IS development. For example, the structuration theory of society in sociology has been used to explain IS. Such accounts based on social theory add important critical considerations in conceptualizing and realizing IS. Other academics have combined knowledge of technical and social action in methodologies to improve effectiveness.

The fundamental concept in interpretive IS research is the subjective construction of reality. It asserts that knowledge of reality is socially constructed rather than being an objective entity. Its implication for structured and object-oriented systems ontology is that systems models cannot be construed as being objective. E-R and class systems models become a matter of social construction, rather than being objective or 'correct'. 'Data', 'process', 'class', 'object' and 'relationships', and other systems modelling terminology, is not an objective fact. They do not exist independently of humans – both IS developers, including systems analysts, and people in the problem domain. The subjective construction of reality means that people attach unique meanings to their actions and interpretations of others' actions. IS developers need to account for these interpretations in systems models. Accounting for social action affects what is represented in systems models and who decides what to include or exclude.

12.3.1 Social and cultural factors

The 'social life' of information and knowledge is significant. Each organization is socially and culturally unique. It forms the context for IS development. This context has an impact on methodologies and instruments used and on how effective systems analysts and designers can be.

Interpretive and socio-technical systems ontology recognize social action. It seeks to account for the social, political and cultural factors in IS development. Socio-technical ideas are incorporated in the ETHICS and Multiview methodologies, but have not had an impact on practice. Now the activities of agile software developers are beginning to make social action the focus of systems analysis and design.

Information and knowledge have proved to be great levellers of status in organizations. If a new IS challenges the existing status of individuals or groups it will be heavily contested. They may become 'passive' during the development and withhold information. In some cases groups whose high status is threatened by a new IS will object and create political issue. They will conflict with newly formed status groups privileged by a new IS.

Other organizational issues include culture, values and 'rites and rituals'. Research shows that organizations have 'value systems' that need to be recognized by IS developers, who themselves may have a different set of values. A particular issue is the culture of information, its value, sharing of it in business problems, and capitalizing on it are significant issues in IS development. People may withhold certain information if they deem it politically powerful and significant. Where the culture is one of individualism, competition and non-sharing of information, systems design needs to reflect it in the data and information provided. When the culture is team-based, consensual and cooperative, a different systems design is needed. Another aspect is the behaviour of people in the problem domain and norms shared by groups that may be contrary to those of other stakeholders or even IS developers. Information may be heavily protected.

12.3.2 NIMSAD

There is recognition of normative values in adopting methodologies. The Normative Information Model-Based Systems Analysis and Design (NIMSAD) originates in systems thinking and addresses problem-solving in IS. It is an action research evaluation framework, which enables an understanding of subjective processes.

In the NIMSAD framework problem-solving consists of four elements:

- The problem situation and understanding it as a 'situation of concern'.
- The problem-solver.
- The problem-solving process.
- Evaluation of the above three at three time periods (t_0, t_1, t_2).

It is used to evaluate methodologies-in-action and contains four elements: methodology in context, methodology user, methodology and evaluation. Each of these elements consists of asking a set of questions and interpreting the answers in terms of problem solving.

12.4 Organizational factors

The flow of information is an important defining characteristic of organization. Information flow 'binds' an organization. A new IS results in different information flows that may raise conflicts of interest. Change management is critical in recognizing different interests and finding strategies for reconciliation of conflicting groups of interests.

Stakeholder analysis is used to understand who or what group has an interest in a new IS, what potential conflicts of interest exist, and how they should be managed. Systems project managers use it to identify interested groups, assess their power of influence and develop strategies for managing stakeholder interests and expectations. Analysts need to behave diplomatically with powerful but sensitive stakeholders.

Organizational change resulting from IT is central to the ETHICS methodology. It is particularly oriented towards systems analysis and design that is sensitive to job roles and job satisfaction. Involving people or participation in the systems design decision-making process facilitates this sensitivity in ETHICS.

12.5 Political factors

The politics of information and knowledge is a significant factor in IS development. Practitioners disregard politics because it questions the assumption of objective and rational problem resolution underpinning systems ontology. Structured and object-oriented systems ontology do not explicitly recognize politics in systems analysis and design. They encourage systems analysts to interact with people free of political bias, and interpret the problem domain rationally and logically. Such objective systems ontology does not recognize politics.

Similarly, training courses, and often degree modules on systems analysis and design, do not adequately consider the laden politics implicit in IS development. Consequently, analysts can become pawns in political struggles of organizational stakeholders interested in developing stronger political power through the development of particular IS.

Interpretive systems ontology can allow for political factors. Though not entirely subjective, socio-technical approaches acknowledge social conflict. SSM, Multiview and ETHICS recognize conflict among stakeholders and different world-views of developers and people affected by a new IS. SSM focuses on 'human activity system' and uses systems theory to propose holistic solutions to achieve predetermined purposes. It acknowledges the importance of people and describes their differing perceptions and conflicts in systems models.

Analysts developing inter-organizational IS, or networked organizations, in particular need to be aware of political factors. Such networked IS involve multiple organizations vying for political dominance. In the case of a manufacturing organization and its suppliers, a powerful manufacturer will be interested in dictating terms to suppliers, and the proposed networked systems design will reflect the shift in power to it. Such systems change traditional business practices and have an impact on economic transaction costs and organization structure.

12.6 Social theory in the Critical Framework

Some researchers argue that social action defies rational explanation characteristic of structured and object-oriented systems analysis and design. Systems ontology need not be devoid of social action though. It can account for social action. For example, socio-technical and interpretive approaches recognize varying social perspectives and political conflict among stakeholders. Analysts also need to account for social action in personal constructs because IS development happens in the context of social action. Analysts need to develop appropriate social action personal constructs and consider how they fit into a PCF.

The assumption of rational human behaviour made in structured systems ontology, and to a lesser degree in object-oriented systems ontology, does not account for social action, particularly organizational politics. IS development is heavily influenced by political power. This puts the analysts' task of deploying requirements analysis techniques in a charged political context. Line managers will determine people available for interviews or observation. The information interviewees volunteer will depend on their perceptions of a new IS and its affect on their jobs and status. All these decisions are social and politically motivated in an organizational context.

Figure 12.2 is the Critical Framework populated with critical reflection on social action (non-systemic factors). As the bottom layer shows, questions concerning the four themes of criticality arise. The question of adapting instruments to model social action is raised in the pragmatic component. Developments in ASD, based on XP, can be interpreted as modelling social action. The AgileAlliance Manifesto restates the practice of software development in comparison to structured systems ontology as principles:

- Individuals and processes over processes and tools.
- Working software over comprehensive software.
- Customer collaboration over contract negotiation.
- Responding to change over following a plan.

Significantly, all the things sought in emerging systems ontology are characteristic of the Real World Component of the Critical Framework. ASD is now incorporating them into agile systems ontology. XP uses the stories people can tell of their experiences of the problem domain to develop implementable models. The stories are captured as 'storycard' and stored in structured databases, subsequently are analysed to determine requirements and functionality.

12.6.1 Mutuality and shared understanding

A significant issue in systems ontology not addressed in structured and object-oriented systems ontology is mutuality and shared understanding. Mutual IS development has only recently been accepted through participative IS development or analysis and design that involves 'users'. Even more important is shared understanding of IS development and a new IS itself.

Developers and 'users' do not have a shared understanding of a new IS, it is even lacking among users. They each approach its conception, analysis, design and realization from very different perspectives, and develop it in equally varying contexts. The developer and analyst act from a technically informed viewpoint and the user from an organizational viewpoint, in which IS are a means to an end – organizational goal, task achievement, workflow support or process improvement.

Shared understanding depends on communication and communicative methods that are not sufficiently considered in systems ontology. The instruments used to conduct systems analysis and design originates in a technical context – programmers and analysts technical problems. The instruments are used euphemistically to allay 'users' concerns, and even propounded as enabling 'communication' and participation with them. Practical use, though, reveals how uninformed or unintelligible users remain. Analysts need to explore alternatives that genuinely develop shared understanding of IS development and the IS itself.

12.6.2 Best practices and COTS solutions

The design of software components and Commercial-Off-The-Shelf Software (COTS) is based on best practices. When implemented in a company, business processes in it are altered

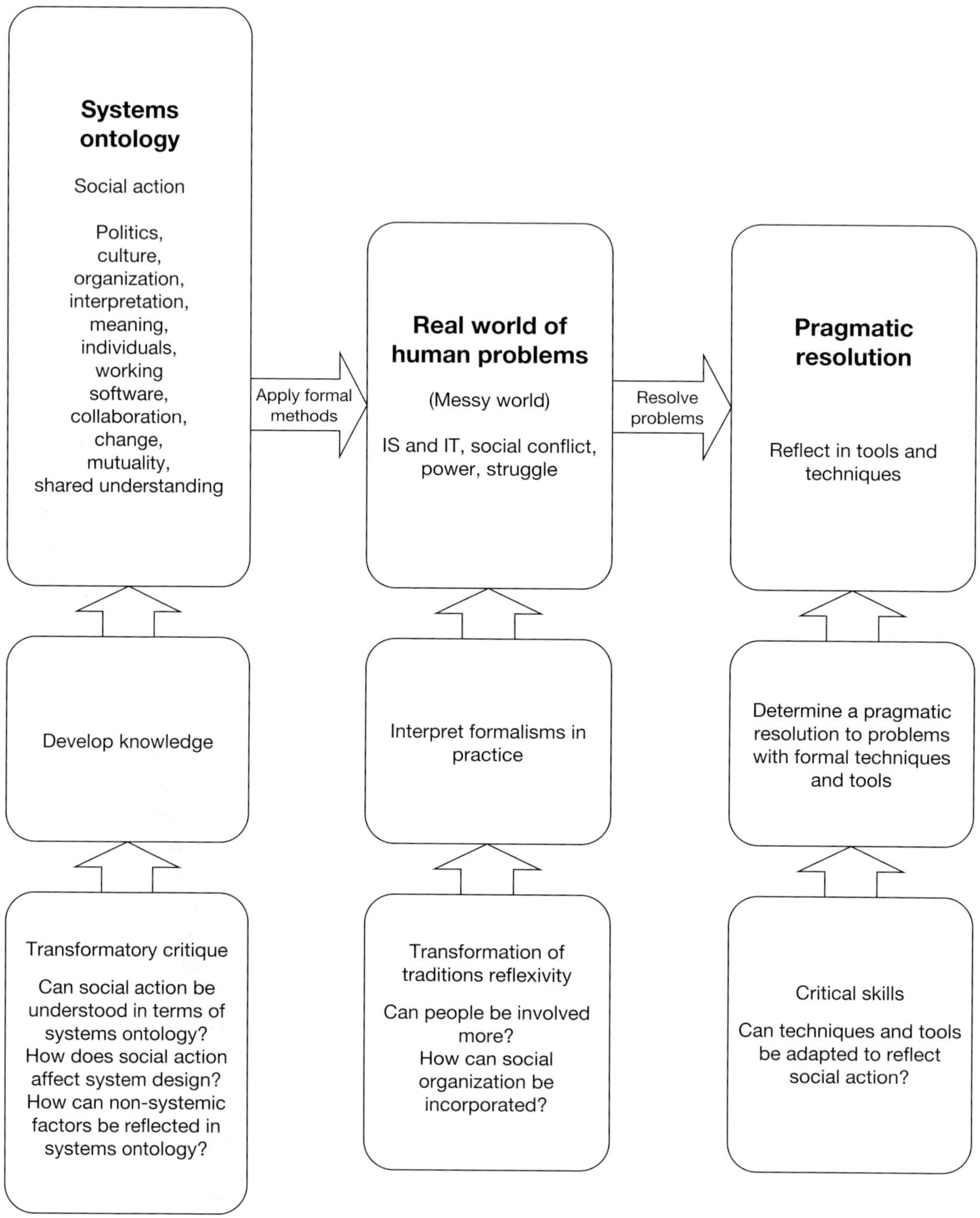

Figure 12.2 Critical framework: social action

to suit the information processing design based on best practice. Solutions based on best practices have resulted in the failure of companies in which they were implemented.

A systems ontology that recognizes the social and cultural uniqueness of organization would form the basis of questioning the use of best practices in systems models and COTS. Analysts are involved in deploying Enterprise Resource Planning (ERP) or Customer Relationship Management (CRM) COTS. Often, the problem domain has to be changed to suit the COTS being implemented. This means that the organization itself is changed. This is a highly questionable practice, which can result in disruptions and even cause poor performance.

The same observation can be applied to IS developed using software components. As solutions components embody certain 'problem domain', for which they were originally developed. The attributes and operations in classes may be too complex to change for the target problem domain. So the target problem domain is often changed to suit the componentized solution.

The same critique applies to methodologies as planned action. Methodologies embody best practices for developing IS in a planned and systematic manner. They are also underpinned with ontological assumptions about the 'problem domain'. Often these remain hidden to practitioners. Research reveals that methodology use in organizations is low. When it is used it is adapted to suit the organization using it or selectively applied alongside in-house practices. When a methodology is deployed, analysts often have to change their practices to apply unfamiliar instruments. They often become blasé as a result.

12.6.3 Deploying instruments

How knowledge of social action can be reflected in systems ontology is a significant issue in IS development. Figure 12.2 provides exemplary critical questions. Transformatory critique is significant because it can result in alternative approaches like ETHICS or XP that account for social action. XP focus on social action with the 'storycard' technique to capture the social aspect in systems design. It may be interpreted as situated action.

Non-systemic factors affect significantly the deployment of instruments. The rational and objective base of structured and object-oriented systems ontology instruments mean they are not capable of acknowledging varying human perspectives and political conflict characteristic of social action found in a problem domain. Instruments based on structured systems ontology particularly need to be adapted.

In terms of the pragmatic component of the critical framework in Figure 12.2, analysts need to consider non-systemic factors when analysing a problem domain. This can be achieved by questioning systems ontology that underplays or negates politics, culture and organization. Alternative systems ontology, for example ETHICS or ASD, can be analysed to evaluate how they seek to reflect such factors in instruments and IS development.

12.7 Personal Critical Framework development

12.7.1 Personal constructs for social action and organization

Activity A

Table 12.1 is a sample repertory grid for organization and non-systemic factors in IS development. Reproduce the grid on a spreadsheet and add further columns and their polar opposites that you consider relevant. To objectify personal constructs in organization, complete the grid by following the details on how to use a repertory grid in section 1.10.1.

Table 12.1 Personal constructs for organization

Pole 1	*Social action*	*Politics*	*Culture*	*Strategy*	*Internal*	*Change*	*Collaboration*	*External*	*Pole 2*
Organized									Unorganized
Planned									Unplanned
Captured									Not captured
Consensus									Conflict
Context									Abstract
Stable									Fluid
Mechanical									Social
Process									Individual
Contract									Customer
Negotiation									Collaboration

12.7.2 Accounting for social action

Questions

1 Discuss what uses systems project managers and analysts can make of stakeholder analysis results to manage political issues in IS projects.
2 Critically discuss whether situated action accounts better for social action than planned action.
3 The E-R, DFD, and class models lack social context. Discuss how you would deploy them in actual situations to take account of social action.
4 Critically discuss structured and object-oriented systems ontology from the ontological perspective of social construction of knowledge.

Activity A

Compare structured systems ontology with ASD. Use the sources for ASD given in section 12.7.4 and your own:

- Is ASD a methodology?
- Is it appropriate to characterize ASD as situated action rather than planned action? Explain your position.
- How does ASD respond to change in system requirements?
- What instruments are used to cater for the agile manifesto claim of 'Individuals and interaction over processes and tools'?
- How does ASD cater for social action?

12.7.3 Political and cultural factors

Activity A

For this activity use the information you have on LASCAD from Activity A in section 11.10.5 and additional information you can collect.

- Identify three political factors that had an impact on LASCAD's development.
- Discuss how each one affected systems analysis and design.

- Analyse the impact of London Ambulance Service organization culture on LASCAD's development.
- Discuss appropriate systems ontology for managing the LASCAD project capable of accounting for social, political and cultural factors.

12.7.4 Internet sources

Surf the Agile Alliance manifesto site at http://agilemanifesto.org/. Analyse how the SDLC systems ontology is replaced with one that reflects social action.

Authoritative information on ASD is available at www.agilealliance.org.

12.7.5 Further reading

Brown, J. S. and Duguid, P. (2002) *The Social Life of Information*, Boston, MA: Harvard Business School Press.

For an understanding of the philosophical basis of methodologies see: Fitzgerald, G. and Avison, D. (2003) *Information System Development: Methodologies, Techniques and Tools* (3rd edn), London: McGrawHill.

Chapter 13

Critical reflection

13.1 Learning outcomes

After completing this chapter you should be able to:

- Evaluate themes in systems ontology: engineering metaphor, organization, instruments, modelling and people.
- Evaluate the value of the concept of computer-independent design in systems ontology.
- Critically evaluate structured and object-oriented systems analysis and design and their respective ontological basis.
- Evaluate the relevance of planned action and situated action in systems ontology.
- Apply transformatory critique to structured and object-oriented systems ontology.

13.2 Introduction

The problem of IS development remains unresolved and is continually being advanced by emerging understanding, knowledge and practice. Structured analysis and design is the earliest attempt to develop consistent knowledge or a shared 'value system', but its ontological basis of data and processes do not account for social action. Object-oriented analysis and design is the most recent body of knowledge. Through certain UML diagrams it can develop class models that reflect 'actors' in the social context, but arguably not social action.

Structured systems ontology characterizes IS development as 'engineering' an artefact. The complex elements of humans, organization and social context are considered subsidiary. These elements form the social context in which an IS is proposed, designed, built and used to achieve organizational goals. Methodologies, techniques and tools, and professional IS developers, including systems analysts, all have to operate in this social context. Researchers and practitioners are increasingly acknowledging its importance in software development and systems analysis and design.

The various strands of criticality in previous chapters on systems thinking, systems analysts, requirements gathering, data and process modelling, logical modelling, class models and systems design all need consolidated critical reflection.

Deeper critical analysis is developed in this chapter of the premises and assumptions of system, humans and organization made in structured and object-oriented systems ontology to engender transformatory critique. The critical coverage will encompass how the problem of IS development is defined, conceptions of IS itself, adequacy of instruments for data, process, and logic modelling, object modelling and systems design.

13.3 Defining the IS development problem

Characterizing IS development is problematic for researchers and practitioners. It needs to account for technological, systemic, human, organizational and social elements. There are no metrics to determine relative weights or importance of these elements but satisfactory characterization of the problem needs to be based on clear understanding of the significance of each one. Research in IS, and recent practice, indicates that the human and social elements need to be central rather than peripheral.

Various methodologies and approaches, including emerging approaches, give varying, and often unequal, weight to one or a combination of these elements. Structured and object-oriented systems ontology emphasize technology. Structured systems ontology is concerned with the efficacy of using computers, so it emphasizes completeness of data, its unambiguity and non-redundancy to enable efficient computer processing. Object-oriented systems ontology emphasizes systemic relevance of objects, the services they provide, and how they collaborate with other objects.

Through research and practice the other elements are becoming recognized as significant. Socio-technical approaches provided the seminal focus on human and social elements, but maintained the supremacy of technical knowledge. ASD gives primacy to individual, social and organizational elements and underplays technical and contractual knowledge and practice.

The ontological basis of systems and systems interoperability is also being recognized, for example in ontological definitions of domain knowledge. Ontology management tools are designed to enable sharing and reuse of ontological knowledge of the problem domain. Critical questions concerning how knowledge is elicited, constructed and formalized inform research into systems ontology.

A significant problem for developing ontological system knowledge is the proposition that social action, or human and social elements, defy rationalization. Nevertheless, since they are constituents of IS, they need to be included in systems ontology and IS development. Systems analysis and design in some form, whether planned or contextual, is a vital element of the problem. Whereas structured and object-oriented systems ontology construe the problem in terms of technological problem-solving, underpinned with rationalism, ASD construes it as situated and contextual activity, underpinned with pragmatism. The emerging ASD alliance recognizes all the elements of the problem but shifts development activity to human and social elements.

Defining the process of IS development is problematical too. The objective and rational basis of the SDLC defines it as a sequential process, consisting of specification, design and implementation. Even with later additions recognizing iteration and spiralling between phases, practitioners maintain a phased approach demarcated into requirements analysis, systems design and implementation. RAD and JAD compress the sequential process activities to speed development. Prototyping iterates among the activities until a satisfactory product is developed. The emerging ASD jettisons the planned action basis of the whole process and its activities, and engages with individuals, seeks working software rather than technical efficiency, regards collaboration as more important, and makes the enabling of change central in the development process.

Modelling is a significant issue in defining the IS problem because it is the basis for developing IS. As an IS consists of all the elements enumerated above, the modelling process needs to cater for them comprehensively and in sufficient detail. The focus on data and processes in structured systems ontology neglects social action. Object-oriented systems ontology considers them peripherally. ASD is based on these factors. A subsidiary problem is the type of models developed. The prevalent form is the passive model type, which is abstracted and detached from reality, rather than the active model type, which can be dynamically linked to social action or human activity. Class models though are capable of dynamic binding.

13.3.1 Abstract forms and reality

Systems analysis and design is a process of abstraction from actual, real human problems with the resultant forms called systems models. A model is composed of pertinent elements, those judged to be relevant by modellers, with other details of the actual situation removed. Such models contain abstract forms. In structured systems ontology the abstract forms are entities, attributes, relations, cardinality or the E-R model. In process models abstract forms are process and logic. In object-oriented systems ontology the abstract forms are classes, objects, attributes, operations, encapsulation, relationships, hierarchy, and polymorphism, or the class model.

E-R and process models and class model are technical notations considered sufficient to represent reality. The problem is not how to duplicate reality, which is not achievable, but its pertinent features need to be reflected in systems models. The very need for an IS is rooted in human and organizational activity, and removing details during modelling raises critical questions. What details should be removed and why remove them are significant issues because they determine how effective a system will be in operation. The corollary issues are what details should be included and why. Systems models that remove details pertinent to business needs result in IS that tend to be underused or not used at all. Conversely, what new details should be added to models to improve business performance is significant too because it affords the opportunity to create a new reality that could provide benefits. eBusiness, eCommerce, and networked organization are examples of such transformations of organizational reality through radical modelling.

Who should decide what details to include or remove is pertinent. Analysts and designers make the decisions in structured and object-oriented systems ontology. In the earliest versions of structured IS development, developers solely decided definitions of the 'problem domain' and the activities emphasizing consequent technical compliance of a new IS. The resulting IS were heavily biased in IT. Subsequent approaches have introduced 'participative development' and 'joint development' to include people, though people still lack skills required to understand systems models and make informed intelligent contribution.

The assumption that IS can be engineered is central in the abstraction process. The SDLC, and methodologies based on it, support the premise that an IS is an engineered product. In methodologies, the engineering analogy is applied to *define* the problem, *specify* what is required, *design* the system and then *implement* the design. This introduces an additional layer of abstraction in structured systems ontology and modelling notations. This is the separation of design from the computer – computer-independent design. It is the distinction between the logical and physical design in structured systems ontology. The logical designs constitute the abstract forms that in turn are converted into physical design – the actual computer system.

Reality is systemless. The notion of system itself is an abstract form, which is used as an 'organizing concept', to organize ideas and thoughts about reality. It is a fundamental element in the definition of the problem of IS development and conceptions and definition of IS itself. Human and organizational need for information is characterized as an 'Information System'. Systems thinking is incorporated into the Multiview methodology. The system abstract form is powerful because it is concise and precise in providing techniques for including and removing details of reality in models. The sophistication of IT, the internet and the web in turn lead to sophisticated interpretations of *systems* in the world.

The process of abstraction is reflected further in systems project management. Project plans define in detail activities of a project team and resources required to produce an IS. Such a plan is an abstract form of organized human activity. Project managers' monitoring and controlling activities are to ensure systematic development within finite resources detailed in the plan. Such plans have often failed in practice because the weight of the detail removed supersedes the elements of reality kept in the plan. The real world mess is stronger than any plan, and planning in the mess remains problematical.

Abstract forms fail to recognize social action – critical social and organizational factors. A major assumption underpinning E-R, process and class models, and plans, is that human and organizational behaviour is objective and rational. Consequently, the abstract forms neglect social and political organizational behaviour and culture. In organizations that serve humans, the police or healthcare for example, there is an additional human factor that researchers in labour have identified – emotion. Object-oriented systems models have the potential to depict 'aspects' and 'roles' in organization.

13.4 Further development of the Critical Framework

The original Critical Framework presented in Chapter 3 can itself be further developed with critical questioning of systems ontology. Through transformatory critique systems ontology can be improved to reflect human intention, purpose, mutuality, shared understanding and organization. Such critique can lead to the development of ontological knowledge of systems that integrates technology and social action.

The questions concern the value and validity of existing ontological knowledge and knowledge yet to be acquired. For example, organization theorists describe organizations as having a boundary, goals and activities. Systems ontology shares two of these descriptors, boundary and purpose (goals). It does not have activities – which is present in SSM as 'human activity system'. Other ontological issues in IS development include: the value of the SDLC and the structured and object-oriented systems ontology based on it, and the effectiveness of concepts, instruments, and models.

Figure 13.1 is an improved critical framework reflecting technology and social action perspectives. It is populated with paradigmatic critical reflection. As the bottom layer shows, questions concerning systems ontology, real situations and pragmatic resolutions arise. Systems designers have emphasized system integration as a critical issue, but the focus remains on systemic issues. More important is integrating IS and organization. This is one research direction taken in Deferred System's Design (see section 15.6).

Paradigms and paradigmatic knowledge revolutions are significant for improving the Critical Framework. Paradigms of IS development reflect what knowledge is accepted as valuable and valid by practitioners. The SDLC,

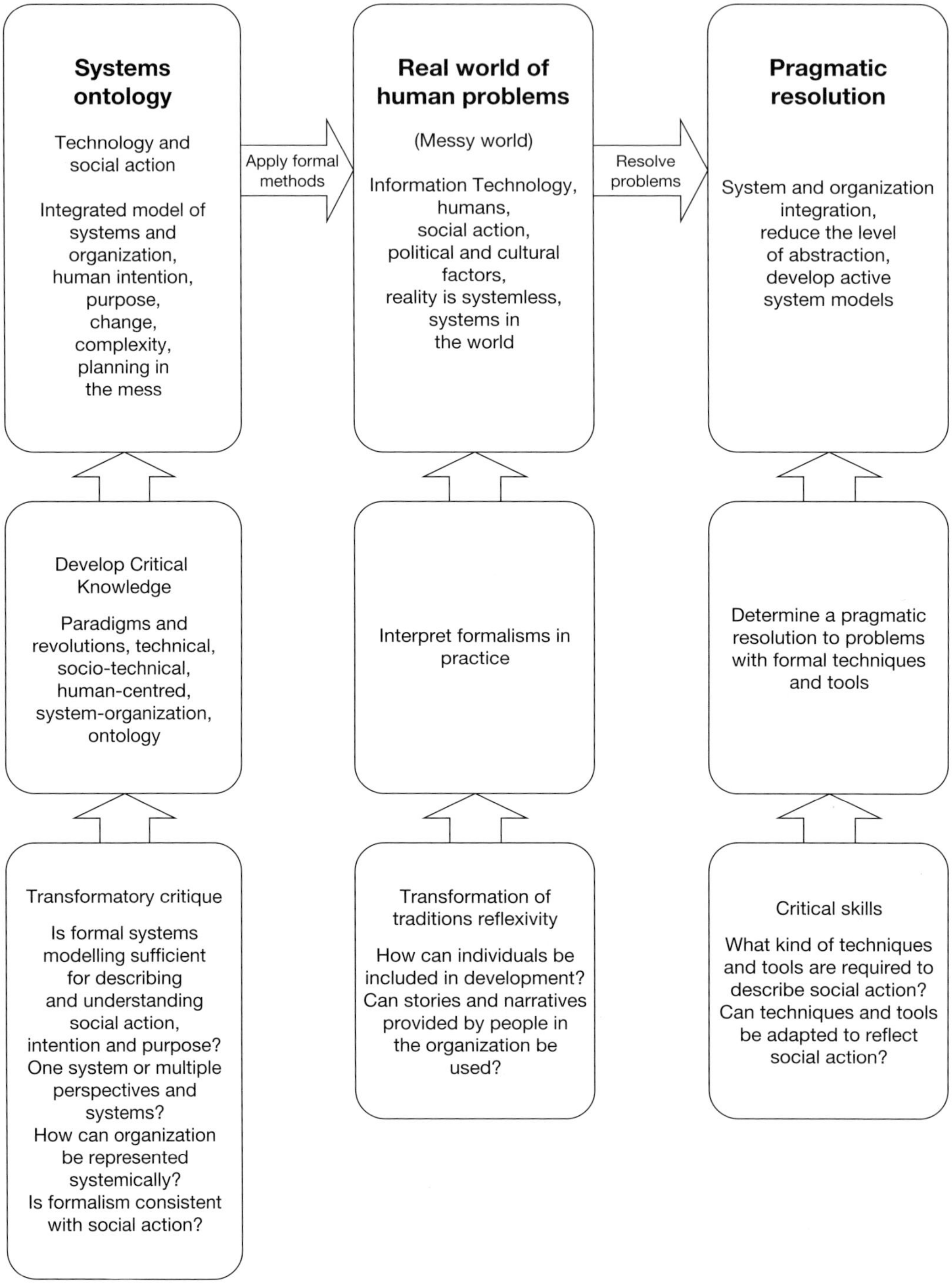

Figure 13.1 Improving the critical framework

and IS methodologies based on it, form a paradigm of objectivity and rationalism in IS development. The objective systems ontology paradigm can be described as a 'clean room', where non-systemic factors are cleaned or removed. The clean room makes it possible to speak of a 'problem domain', 'optimum solution', 'an agreed system', 'logical' and 'physical design', and other such systemic and technological terminology. This paradigm is critically questioned by socio-technical and interpretive IS development, and recently by ASD.

A fundamental assumption of the SDLC is that *a system*, rather than different perspectives of the system or even different IS, is to be produced. The DBMS conceptual schema enables various perspectives, but the underlying physical system and conceptual schema are not rooted in social action. In the SDLC a system agreed by all the parties concerned is to be produced. Consequently, systems analysis and systems design is for one physical system. Such conceptualization leads to systems ontologies that assume a single system. This assumption is misplaced for certain kinds of IS and fails to account for differences among stakeholders in organizations. An exception is the Multiview methodology, which develops multiple perspectives systems ontology.

The single system view prevailed because of contemporary technology during the development of structured systems ontology. Recent object-oriented technology is capable of the alternative concept of multiple perspectives on a system. Aspect-oriented systems ontology recognizes multiple perspectives. Objects are versatile and can be modelled to reflect 'aspects' and 'roles', as well as behave polymorphically.

13.4.1 Organization, organizing and systems in the world

The value and validity of ontological knowledge of systems can be improved by developing knowledge of how to develop IS that reflect organizing and organization simultaneously. The terms 'system' and 'problem domain' are abstract forms that do not reflect the present continuous tense that is organizing. Organizational theorists think of organizations as social units with goals, boundaries and activity. Some IS researchers argue that organizations evolve and emerge. Such various explanations of organization undermine the premise of planned action in IS development.

Neither structured nor object-oriented system ontology recognize emergent organization. Emergent organization is the recognition that things can happen in organization which cannot be accounted for in predetermined strategies or plans. For example, a company takeover is an emergent factor if it is unexpected or at an operational level realization of a defective product. Just as organizations have to respond to such emergence, systems too need to reflect them. Passive systems models are unable to cater for such emergent change.

An 'Information System' needs to fuse organization, organizing and system. IS developers emphasize the technological system, the human and organizational element is generally peripheral. To improve systems ontology, the distinctions between technological systems, IS development and business operations need to be critically questioned. Thinking of IS development in terms of systems project management reinforces such distinctions. Some researchers and practitioners argue that systems ontology will better reflect organization and organizing when the activity of developing IS is conceptualized as being an everyday operational issue rather than 'projectizing' it. The development of active models reduces the distinctions and tends towards this view.

Adaptive and evolutionary IS aim to reflect organizing, rather than organization. Systems models for adaptive and evolutionary IS are dynamic, they are designed to reflect actual situations dynamically. Object-oriented systems ontology can cater for such systems through its dynamic binding of objects and deferred objects.

13.4.2 Human intention, purpose, change and complexity

Specifically, the critical framework can be enhanced to reflect social action, as shown in Figure 13.1. Systems ontology needs to account for human intention, purpose, meaning, change and complexity. Systems analysis and design, as practised in structured and object-oriented approaches, does not sufficiently acknowledge social action, of which change and complexity have an inordinate effect on IS development.

The assumption of a static problem domain is erroneous in systems ontology. In business organizations uncertainty can arise from the actions of markets and competitors or internal strategic decisions, which can make the development of systems models, especially for strategic IS, problematical. The effectiveness of instruments is compromised by organizational change and complexity. Such instruments assume stability in real situations.

Similarly, the assumption of a static IS environment is erroneous. Uncertainty and complexity also arises when intended organization design is exposed to its environment. The intended organization design, containing processes and structures designed to achieve goals, needs to be responsive to its environment. The system concept acknowledges the environment, demarcated by the system boundary and feedback. Systems ontology though, whether structured or object-oriented, is weak in suggesting concepts and actual mechanisms for managing environmental effects, because it assumes a static systems environment.

Complexity also arises when formalism like organization design, systems designs or project plans designed to achieve human purpose are applied to actual situations. Applying formalism in practice is difficult because of human and organizational factors, and the variance between what is depicted in the formalism and actual reality. System project plans tend to be revised because the actual situation differs from the projected required situation and systems designs fail because they do not reflect business requirements.

A prime weakness of structured and object-oriented systems ontology is the lack of understanding of human intention. They do not seek to develop knowledge of human intention. The human element in IS though, presupposes intention. In strategic IS the intention is to develop IS that are difficult to imitate by competitors, otherwise any competitive advantage gained would be lost. In decision support systems the human decision-maker is the source of intention, which is usually concerned with decisions on allocating scarce resources or investments. Human intention is critical to systems analysis and design when humans interpret information. Human intention is the most pronounced and the most difficult to reflect in systems ontology.

13.5 Personal Critical Framework development

13.5.1 Personal constructs for criticality

Activity A

Table 13.1 is a sample repertory grid for criticality. Reproduce the grid on a spreadsheet and add further columns and their polar opposites that you consider relevant. To objectify personal constructs in learning, complete the grid by following the details on how to use a repertory grid in section 1.10.1.

Table 13.1 Personal constructs for criticality

Pole 1	*Transformatory critique*	*Refashioning of tradition*	*Reflexive criticality*	*Critical skills*	*Socially constructed knowledge*	*Negative*	*Pole 2*
Question							Fact
Reflect							Internalize
Existing thing							Alternative thing
Interpret							Accept
Analyse							Use
Evaluate							Do not judge
Explain							Ignore
Self-regulate							Do not reflect

Activity B

A PCF enables a person to develop a notion of self and identity. Its constituents provide a basis for self-expression and action in real situations. For example, an analyst convinced of the value and validity of structured systems ontology will deploy structured instruments believing that they will produce the required results. Such an analyst would probably propound structured analysis and design and propose it to develop IS. Drawing on the PCF you developed through the previous chapters:

- Improve your PCF by questioning the correctness, value, and validity of your personal constructs. Would you defend your personal systems ontology? Would you put into practice the methods, techniques and tools you accept as valid and how would you do so professionally?
- What kind of emotion, commitment and professionalism does your PCF result in for you?
- Does your PCF provide you with a 'social identity' as an analyst?
- Would you call yourself a 'structured' or 'object-oriented' analyst?

Questions

1 Assess the practical value of the system concept for IS development.
2 Discuss whether the activities involved in 'problem-solving' are capable of accounting for the human element in IS.

13.5.2 Developing systems ontology

Question

Discuss what you would do to improve structured and object-oriented systems ontology in terms of:

- How knowledge of systems is acquired.
- How such system knowledge is formalized.

Consider the critical thinking skills enumerated in section 1.8.1.

13.5.3 Structured systems ontology and social action

Activity A

Table 13.2 contains descriptive statements of structured systems ontology. Reproduce it in a word processor. Enter your comments in the second column on how it can better reflect the social action.

From your readings, list other claims made by the advocates of structured systems ontology. Reflect on these claims. Are they valid claims? What does practising structured systems ontology reveal?

13.5.4 Object-oriented systems ontology and social action

Activity A

Table 13.3 contains statements descriptive of object-oriented systems ontology. Reproduce it in a word processor and enter your comments in the second column on how it can better reflect social action.

From your readings list other claims made by the advocates of object-orientation. Reflect on these claims. Are they valid claims? What does practising object-oriented systems ontology reveal?

13.5.5 Analysis of systems ontology and real human problems

Activity A

Table 13.4 lists many of the critical issues affecting structured and object-oriented systems ontology in the first column. The issues are divided into practical, philosophical and social action. The issues concern how the problem of developing IS is defined. When completed the table will be a comparison of the SDLC-based structured approach and object-orientation with the real or messy world. The category of criticality is also added to enable you to respond in terms of how you would be critical of systems ontology and the messy world.

- Reproduce the table in a word processor and fill column two with appropriate descriptions

Table 13.2 Questioning structured systems

Statement	*Reflecting social action*
A computer system (IS) can be engineered.	
A computer-independent design is possible.	
Objective knowledge about the problem domain (business area) is possible because things exist independently of humans.	
Structured techniques make communication between analysts and users easier.	
Data, process and logic modelling techniques can reflect the problem domain.	
Entities describe things of interest in the problem domain.	
The DFD notation language is sufficient for developing systems models of business processes.	

Table 13.3 Questioning object-oriented systems

Statement	*Reflecting social action*
A class is a set of objects with common characteristics in the problem domain.	
An object is an abstraction from the problem domain, capable of encapsulating data and operations, processing information and collaborating with other objects.	
Generalization is useful for building logical structures to represent similarity or differences between classes.	
Objects communicate with each other by sending messages.	
Message passing is enhanced by polymorphism.	

Table 13.4 Comparative analysis of systems ontology with the messy world

	Systems ontology	
	Structured and object-oriented systems analyses and design	*Messy world of social action (provide evidence)*
Practical		
Praxis		
Methods to achieve aims		
The role of projects in IS development		
Knowledge of the application domain, characterization of the application domain		
Capable of techniques and tools		
Communication between professional developers and users		
Understanding of data and information		
Formalism, knowledge and theory		
Philosophical		
Premises/assumption		
Ontology		
Epistemology		
Social action		
Resultant system		
Organizational politics		
Organizational culture		
Information		
Criticality		
PAC cycle		
Personal Critical Frameworks		
Critical knowledge and practice framework		
Critical thinking		
Critical skills		
Reflexivity		

of structured and object-oriented analyses in response to each item in the first column.

- Use lessons learnt from this textbook, case studies or experiential knowledge to describe the real or messy world in column three. You may also enter a description of your personal construct.
- Compare the two columns you filled in terms of knowledge gaps and practice implications.

Question

Comparatively analyse the effectiveness of structured and object-oriented systems ontology for representing social action in IS.

13.5.6 Internet sources

See Cambridge University's Department of Engineering page for succinct overview of SSM at: http://www-mmd.eng.cam.ac.uk/people/ahr/dstools/control/softsm.htm.

13.5.7 Further reading

For a special issue dedicated to social computing see: 'Communication of the ACM', *Social Computing*, 37(1) January 1994.

For systems thinking applied to human activity systems, which accounts for social action see: Checkland, P. B. (1981) *Systems Thinking, System Practice*, Chichester: Wiley.

Chapter 14

Ways of thinking and acting

14.1 Learning outcomes

After completing this chapter you should be able to:

- Revisit previous critical perspectives in the context of paradigmatic understanding.
- Assess the impact of IS paradigms on ontological knowledge and practice.
- Analytically evaluate paradigms of IS development to inform practice.
- Critically evaluate how IS paradigms affect praxis or thinking.
- Apply transformatory critique to your knowledge and practice based on paradigmatic knowledge and understanding.

14.2 Introduction

Knowledge of how to develop IS or what constitutes IS remains incomplete. An examination of the intellectual traditions in IS research and practice reveal systems and ontological assumptions underpinning them that have led to certain approaches, some more successful than others. Structured and object-oriented systems ontology contains assumptions and assertions that do not account adequately for actual IS development. This chapter is a deeper theoretical underpinning, taking a discursive, critical perspective of systems ontology, the SDLC, and systems analysis and design. It considers paradigms of systems analysis and design, and examines their assumptions about technology, humans, social action, organizations, data, information and IS.

A critical analysis of the ontological basis of IS is important for evaluating knowledge and practice. IS paradigmatic knowledge determines praxis. Analysts' ways of thinking and acting depend on theoretical and practical knowledge reflected in an implicit or explicit paradigm. The relationship between thinking and acting is understood through praxis, which is informed by knowledge that is internalized, accepted as valuable and valid. Praxis assumes relations that are hidden and assumed, and enacted automatically in actual situations. Through reflexive praxis, based on the critical analysis of systems ontology, analysts can uncover implicit paradigmatic assumptions underpinning their actions and evaluate their relevance to knowledge and practice.

14.3 Criticality through paradigmatic analysis

An understanding of paradigms in IS is necessary for criticality and informing reflexive practice. It is an aid to understanding what constitutes progress. Paradigms in particular characterize progress as 'revolutions' in knowledge. Progress though can be towards certainty, or even uncertainty. Paradigmatic analysis can be used to understand the obstacles to progress, some of which are internal to a paradigm and others beyond it. Notions of progress have complexity as a central theme, where increasing knowledge eradicates the idea that something is simple.

A **paradigm** is a set of beliefs, assumptions, values and shared knowledge accepted by a community of practice. IS developers, including systems analysts, form such a community of practice. The community accept standards and measures of validity for knowledge, how it is acquired, evaluated, and applied. The paradigm focus is on what constitutes knowledge, what is legitimate to investigate, and what methods of investigation are valid.

Practitioners do not think about paradigms or act contrary to an accepted explicit or implicit paradigm. For them a paradigm is 'reason in practice'. There are exemplary forms of reasoning present in different systems analysis and design approaches and practice. Knowledge and understanding gained from a particular paradigm is accepted as valuable and valid, and held by an analyst to be correct. It is this belief in the correctness of knowledge, or certain ways of reasoning and acting, that determines how a particular IS development problem is thought through and what action is taken to resolve it. Alternative conceptualization of the problem and its resolution is resisted because it is not part of the accepted paradigm.

Uncritical acceptance of knowledge and rote practice stems from a lack of awareness of a paradigm of knowledge and practice and a lack of awareness of alternative paradigms. Knowledge and practice can be improved when an analyst becomes conscious of an accepted paradigm and begins to question it in the context of available alternatives. Paradigms can be utilized to improve reflexivity and, through personal constructs, can be validated empirically.

The value of paradigmatic analysis is in understanding how knowledge is acquired and accepted as valid. It enables questioning of reason in practice or criticality and uncovers assumptions of knowledge. Paradigmatic analysis uncovers how knowledge is acquired. Such knowledge is important for understanding epistemological knowledge – methods used to acquire and validate knowledge. It uncovers assumptions made of the physical and social world, which is relevant in developing ontological knowledge of systems. Significantly, paradigmatic analysis is valuable because it reveals the knowledge base that informs and determines how analysts act.

14.4 Paradigms of IS development

The discussion on paradigms of IS development is based on Hirschheim and Newman's (1989) work, which draws on social theory to account for knowledge and practice in IS development. They describe four paradigms of IS development and cite exemplars.

Functionalism The systems analyst is designated an 'expert' developer. IS development is conducted from without with accepted formalism and 'planned intervention' using 'rationalistic' tools and methods. People, hardware, software, organizational rules and objective entities constitute elements in defining IS. Structured systems analysis and design and Information Engineering are classified in the functionalist paradigm. They assume knowledge is objective and that the social world is ordered. This is also true of object-orientation, though it is underpinned with classification theory, which assumes people order their experiences of the world.

Social relativism The systems analyst is designated a 'catalyst or facilitator'. IS development is conducted to improve subjective and cultural understanding within the application area. Evolutionary social change is recognized in the IS development process. Subjective meanings, symbolic structures and metaphors constitute elements in defining IS. Exemplars are enthnographic approaches and the FLORENCE project. Social relativist IS development assumes that knowledge is subjective and can be ordered.

Radical structuralism The systems analyst is designated a 'warrior' for social progress and is partisan. IS development is conducted politically from without using ideology to affect conscience and develop consciousness. The tools and methods are adapted to suit different class interests. The same elements used to define IS in the functionalist paradigm are used but objective entities are interpreted politically to serve economic class interests. Trade Union led approaches, UTOPIA and DEMOS projects are exemplars. Radical structuralist IS development assumes that knowledge is objective but is embedded in social conflict.

Neohumanism The systems analyst is designated 'emancipator' or 'social therapist'. IS development is conducted from within to improve human understanding and bring to the surface rationality underpinning human action. Its purpose is to emancipate suppressed interests and to liberate people from natural and social constraints. The same elements used to define IS in the functionalist paradigm are used, but objective entities are interpreted to develop 'technical knowledge interests' to enable control over nature and social situations. Language and its intersubjectivity is pertinent in defining IS. Exemplars are critical social theory based approaches, SAMPO project. Neohumanistic IS development assumes that knowledge is subjective and stems from social conflict.

The focus of structured and object-oriented systems ontology is on objective modelling. They operate with the premises of rational or economic human behaviour and do not account for social action. They focus not on individuals and groups but technological systems. So they can be classified as the functionalist paradigm.

Explanations of IS range from nomological to behavioural, to pattern-based. Object-oriented systems analysis and design is an example of the latter. Another category of probabilistic-statistical is evident, where IS researchers try to explain IS in terms of statistical probabilities. Probabilistic-statistical explanations, contrary to their proponents' views, do not establish causal relationships characteristic of nomological explanations.

14.5 Paradigms, thinking and acting

Action is concerned with making a difference to a situation for the benefit of the individual or others. Humans act in order to achieve desired aims. Analysts' action is concerned with improving organization performance through systems, and ultimately with applying their expertise to help an organization achieve its aims. Their work makes a difference, often significant, to the working lives of people. Crucially, it has an impact on the effectiveness and success of organizations.

It is possible to show how the assumptions underpinning a particular paradigm affect the actions of its adherents. Analysts educated and trained in structured systems analysis and design unwittingly subscribe to the functionalist paradigm. Structured systems analysis and design leads to prescribed action. A tenant of the functionalist paradigm is that an expert is best placed to make systems design decisions. In structured systems ontology the IS developer – including the systems analyst – is the expert who makes design decisions. This assumption dominates requirements analysis, systems modelling, and systems design, and certainly implementation. The analyst usually makes policy decisions that affect processes and structures in organizations. The role of the expert analyst causes misunderstandings and difficulties between developers and users. Analysts who identify with the functionalist paradigm need to analyse what can be done to reduce friction.

Another tenant of the functionalist paradigm is objectivity of knowledge. This requires systems analysts to be apolitical. Apolitical behaviour leads to neglecting the real political issues pertinent in IS projects, requirements analysis, modelling and systems design. The focus on data in structured systems analysis results in systems designs that lack stakeholder champions. In data-centric approaches an IS is regarded as a product of the data captured, stored and processed with computer algorithms. Its proponents claim it has better design stability and low data redundancy because rigorous mathematically based formalism is applied objectively.

In process-centric approaches an IS is regarded as a product of the processes it needs to support with the danger of limited design stability and propagation of data redundancy.

Action in object-oriented systems ontology is open to interpretation, rather than prescribed as in structured systems ontology. Object-oriented systems analysis and design leads to interpreted action. The UML is not a method or methodology, as such it does not prescribe analysts' action. UML diagrams are deliberately designed to be interpretable in context. For example, the basic features of an object used in most systems models are name, attribute and operations. An object can have any other features deemed appropriate for a problem domain. This is possible in object-oriented systems because of meta-models, which define notation.

14.6 Primacy of method

Conceptions of IS development and methodologies in the functionalist paradigm logically result in the primacy of method and instruments, usually underpinned with formalism informed or derived from predicate calculus. Methods for acquiring knowledge of systems, methods for determining system requirements, methods for analysis and design, and implementation are all developed because of the belief in the value and validity of objective knowledge and a reality separate and independent of IS developers. The methodical perspective seeks to define precisely activities involved in developing IS and to ensure compliance to those activities – standards.

The SDLC is an exemplar of the methodical approach. The development process is demarcated into distinct phases, with each phase containing multiple activities. The outcomes of each phase and activity are the 'deliverable' that forms both an input into another activity and the simultaneous definition of the resultant IS. Structured and object-oriented systems ontology emphasize this kind of methodical IS development.

The primacy of method is further enforced as projects that control costs. The cost of developing a system can be detailed in a systems project plan with distinct activities or work packages stipulating human and financial resource requirements. An exemplar in software development is CMM, with its emphasis on planned and repeatable systems development activities. Practitioners, and some researchers, have lately questioned the primacy of method both in IS development and projects.

14.7 Stories and narratives

Conceptions of IS development and methodologies in social relative, radical structuralism and neohumanism paradigms lead to 'sense-making', meaning, stories and narratives, where method and process is secondary. Social relativism and neohumanism assert that reality does not exist separately from people, so IS development proceeds by 'sense-making' and sharing meanings among different groups. In radical structuralism class conflict and partisan behaviour are characteristic of the analyst.

ASD and XP make use of techniques to generate stories from people about their experiences. The stories are written by 'customers' and used to define the scope of a system project. The important difference from structured or object-oriented systems ontology is that system requirements are sourced directly from the experiences of people in the problem domain through their stories, written on a StoryCard. These StoryCards are kept throughout the development to keep the team focused on the customer's requirements.

Interpretive research in IS also draws on social theory to develop understanding and knowledge of the application of IT and development of IS. Interpretive researchers build on the 'subjective construction' of reality, sense-making, and meaning. Though the concept of interpretation is accepted in the IS research community, as yet there are few practical instruments available. An earlier example though is NIMSAD (Normative Information Model-Based Systems Analysis and Design). It is an evaluation framework for methodologies based on the social relative paradigm. It is used to evaluate problem-solving in IS.

14.8 Paradigmatic critical frameworks

The role of a Critical Framework is central in further developing knowledge and practice. Criticality among researchers and practitioners leads to progress in knowledge. Similarly, a PCF that is static will become obsolete. It should be revised, amended and improved over time as new knowledge and advances in systems analysis and design are made, and as personal understanding of systems ontology improves.

Figure 14.1 is the Critical Framework populated with paradigmatic critical reflection. As the bottom layer shows, many questions concerning how knowledge is acquired, accepted and practised arise from entities presupposed to exist in reality.

Questions on when an assumption or element of a PCF should be introduced, amended or shed can be addressed through paradigmatic analysis. **Paradigmatic critical frameworks** can be of two types. A critical focus on what constitutes knowledge and how it is acquired within a paradigm is one type. Some researchers and practitioners work within a paradigm to improve knowledge and practice. For example, within the functionalist paradigm when an existing method or methodology is found to be ineffective an alternative is devised. For instance, Information Engineering (IE) was designed to fill gaps in structured systems analysis and design.

A critical focus on what constitutes knowledge and how to acquire it across paradigms is the other type. Some researchers and practitioners work across paradigms. When the limit of a particular paradigm is reached an alternative paradigm is explored, either to cross-fertilize or to step over into the alternative. For instance, Multiview is an example of cross-fertilization, as it is rooted in the functionalist paradigm but makes use of social relativism. It

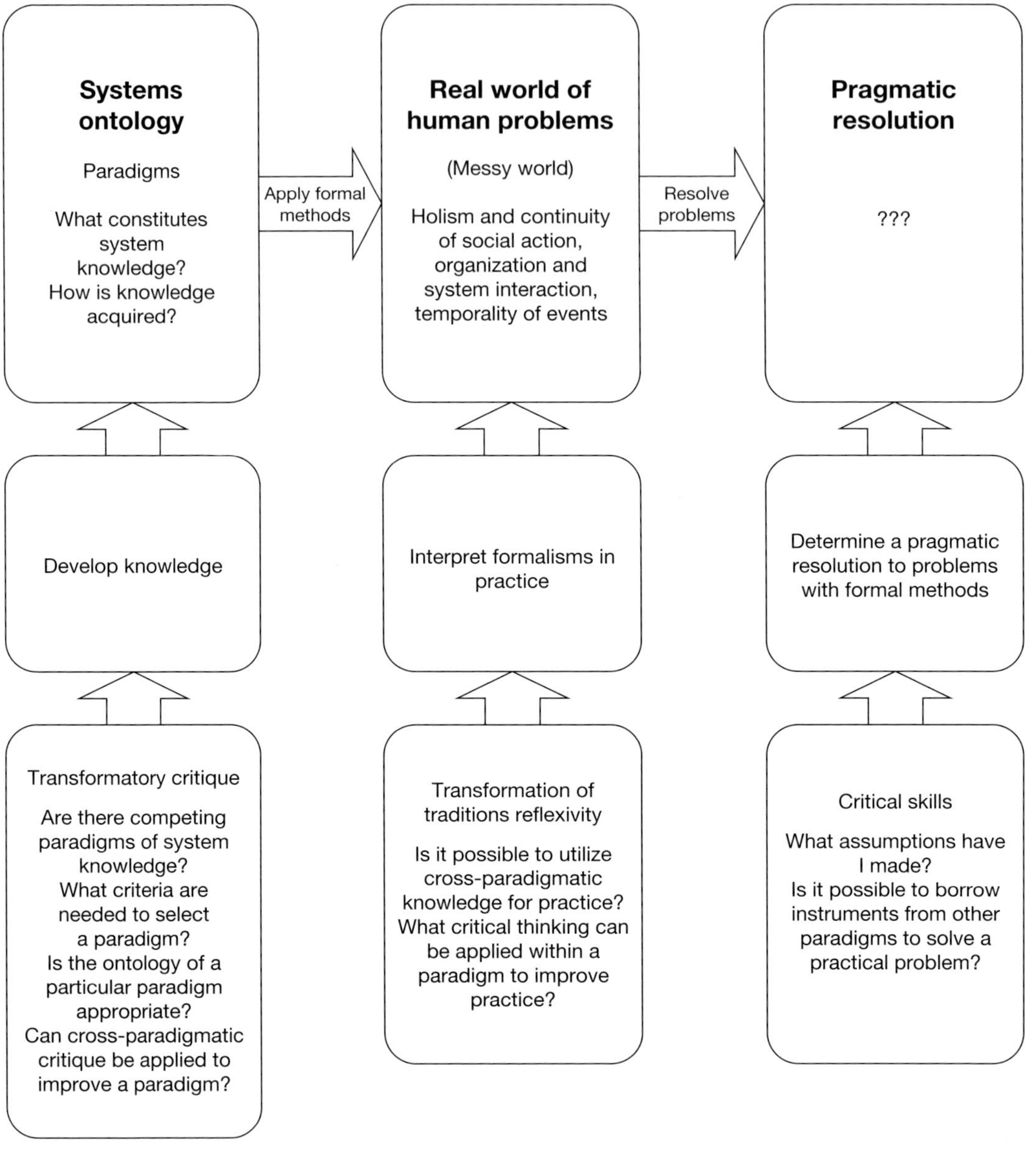

Figure 14.1 Paradigmatic critical framework

seeks to cross-fertilize functionalist thinking in STRADIS and IE with social relativist ideas found in SSM and ETHICS.

Criticality within and across paradigms was covered in Parts II, III and IV of this book. The purpose of the critical framework presented in Chapter 3, and revised in Chapter 12, was to afford within and across paradigmatic criticality. It is possible to map Barnett's (1997) types of criticality to reflect within and across

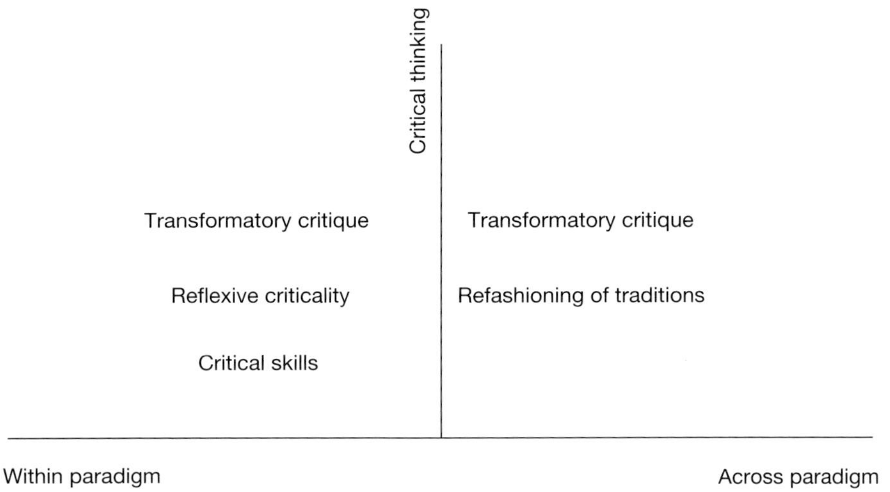

Figure 14.2 Critical thinking and paradigms

paradigm criticality, as shown in Figure 14.2. The figure shows what types of criticality can be used within a paradigm and what types across paradigms.

Elements of critical thinking discussed in Section 1.8.1 apply to both within and across paradigms. Interpretation, analysis, evaluation, inference, explanation and self-regulation are needed to be critical of a paradigm and for thinking beyond a particular paradigm. Self-regulation is particularly pertinent for cross paradigmatic analysis because it enables 'questioning, confirming, validating or correcting either one's reasoning or one's results'.

14.9 Personal Critical Framework development

Knowledge that is not reflected upon and practice that is not questioned becomes rote. The formation of personal constructs should be based on paradigmatic critical analysis. Personal constructs contain assumption of reality made by the owner. They need to be uncovered by understanding paradigms of practice.

14.9.1 Personal constructs for praxis

Table 14.1 is a sample repertory grid for praxis. Reproduce the grid on a spreadsheet and add further columns and their polar opposites that you consider relevant. To objectify personal constructs in praxis, complete the grid by following the details on how to use a repertory grid in section 1.10.1.

14.9.2 Defining the problem of IS development

Activity A

Consider the four paradigms in section 14.4. Form into two groups, based on your readings, each group is to:

- Explore the functionalist basis of structured and object-oriented systems ontology for defining the problem of developing IS. What functionalist paradigm ideas do they use to define IS and determine the IS development process?

Table 14.1 Personal constructs for praxis

Pole 1	*Act*	*Think*	*Decide*	*Concept*	*Framework*	*Valid*	*Acquire knowledge*	*Accept knowledge*	*Question knowledge*	*Paradigm*	*Pole 2*
Fact											Belief
Value-free											Value
Experience											Supposition
Objective											Constructive
Assert											Persuade
Expert											People

- Evaluate the appropriateness of the solutions proffered by each of the ontology.
- Appoint a raconteur in each group to relate the discussion to the plenary.
- Groups to come together in plenary and explain and discuss their respective views.
- Appoint a scribe to note groups' comments on flip charts or white boards.

This activity should result in a rich discussion of individuals' knowledge and the variance in it between individuals. This variance is important because it allows individuals then to question their personal knowledge relative to others.

Activity B

Making Your Paradigm Explicit practitioners often act on the basis of routine behaviour and implicit assumptions. When asked to explain their behaviour they find it difficult to objectify the knowledge they use to underpin their actions. The PCF is a device to objectify the paradigm or framework you use to develop practice knowledge. Reflect on your own behaviour in relation to an interest or work experience. Identify and list the assumptions you made on:

- people;
- your interest or work area;
- the instruments you used;
- your expected results.

14.9.3 Thinking and acting the structured way

Activity A

Hirschheim and Newman's (1991) paper critically appraises structured IS development.

- Source the paper and read it. (Full reference is given in section 14.9.6.)
- The chapters on requirements analysis and process and logic modelling should be revisited after reading the paper.
- Take the authors' arguments and apply them to the structured techniques stipulated for requirements determination, data, process and logic modelling.
- Analytically evaluate the flaws you detect in the structured methods, techniques and tools.
- Discuss how representative structured systems models are of real situations. (You may want to re-read Chapter 12 on non-systemic factors.)

Questions

1 The role of the analyst as 'expert' causes misunderstandings and difficulties between developers and users. What can be done pragmatically to improve this situation?
2 Discuss whether an organization is an independent, objective fact and whether it can be modelled as such.
3 Critically evaluate the relative contributions of structured and object-oriented systems ontology to resolving the problem of IS development.

14.9.4 Thinking and acting in object-orientation

Activity A

Object-oriented systems ontology is based on classification theory (see section 10.2). It asserts that people use cognitive processes to organize thinking of their experiences. The three methods used to organize thinking are:

- Differentiation of experience in terms of objects and their characteristics.
- Making distinctions between whole objects and components parts.
- Formation of classes or groups of objects and distinction between classes.

Recount some of your experiences and apply the above to them to determine objects, attributes, wholes, parts and classes. What difference does it make to your knowledge to become aware of your cognitive processes?

Question

1 Critically evaluate the contribution of object-oriented analysis to resolving the problem of IS development.
2 Critically assess whether the class model can be a sufficient representation of real human problems.

14.9.5 Internet sources

For a philosophical basis of paradigmatic knowledge and the original ideas of Thomas Kuhn see the essay by Pajares, F. at: http://www.emory.edu/EDUCATION/mfp/kuhnsyn.html.

14.9.6 Further reading

For a complete discussion of the four paradigms of IS development see: Hirschheim, R. and Klein, H. (1989) 'Four Paradigms of Information Systems Development', *Communications of the ACM*, 32(10): 1199–1215.

For an elaborate critical discussion of assumptions underpinning IS development based on the functionalist paradigm see: Hirschheim, R. and Newman, M. (1991) 'Symbolism and Information Systems Development: Myth, Metaphor and Magic', *Information Systems Research*, 2(1): 29–62.

Hirschheim, R. and Newman, K. (1989) 'Four Paradigms of Information System Development', *Communications of the ACM*, 32(10): 1199–1215.

Estes, W. K. (1997) *Classification and Cognition*, Oxford Psychology Series, Oxford: Oxford University Press.

Part VI

The future of IS development

Part VI examines emerging systems ontology knowledge and practice. It takes a future-oriented stance on systems ontology and systems analysis and design. The future perspective is important for two reasons. First, the act of developing an IS is itself future-oriented. It is a creative act. Most new IS are creations of new business processes or order in organizations. This is significant because new processes and order contribute to business performance and achievement of goals. The global classic example of new order is the web.

Second, the future of IS is important because of its impact on organizations and society during the 1990s. The acceptance of IS in organizations necessitates appropriate conceptualization of further developments in IT and IS. Unlike other technologies, for example analogue, the progress of digital technology and its application to develop IS shows no signs of slowing.

Chapter 15 is on new and emerging systems ontology and different perspectives on systems analysis and design arising from new software programming strategies and system concepts. These developments will determine future ontological knowledge of systems. Some perspectives are emerging and others are established but are not the dominant form of practice. In terms of criticality, Chapter 15 covers transformatory critique. It focuses on making systems analysis and design inclusive. A more complete systems ontology needs to take account of system ontological knowledge that is marginal but which may be valuable in certain situations, or new approaches that provide paradigmatic shifts, and reveal new features of systems ontology.

Chapter 15

Making systems analysis and design inclusive

15.1 Learning outcomes

After completing this chapter you should be able to:

- Make the Critical Framework and PCF inclusive of emerging and marginal systems ontology knowledge and practice.
- Evaluate the usefulness of emerging and marginal systems analysis and design.
- Apply reflexive criticality and transformatory critique to systems ontology.

15.2 Introduction

This chapter will consider emerging and marginal approaches to systems analysis and design. It covers methods that are marginal but which may be valuable in certain situations or new approaches that provide paradigmatic shifts. These approaches provide an alternative conception that could benefit the IS development process and improve conceptions of IS. For example, emerging approaches like ASD and adaptive and evolutionary systems introduce new elements into systems ontology that require different systems analysis and design.

The future of systems analysis and design is likely to be non-SDLC, as emerging conceptions of IS development reveal. The web itself, rather than applications for it, is an example. The internet is another example. These types of systems and hardware enable radically different conceptions of IS, as evident in the eCommerce and web revolutions.

The fundamental challenging question for analysts is: Who is an appropriate analyst? For examples, individuals determine the web IS as a whole rather than a team of designers. Its analysis and design is not undertaken, as we currently understand the practice, in structured and object-oriented systems ontology.

15.3 Alternative and emerging thinking

The alternatives and emerging thinking can be conceptualized and analysed in terms of amethodological IS development, situated action and evolutionary development. An appropriate conception is amethodogical IS development. An amethodological orientation consists of doing development activities that are pertinent in context. So, analysing system requirements can be done at any time or design decisions can be taken before requirements analysis is complete. This is in contrast to the planned action of the SDLC and systems project management. A prime characteristic of the alternatives is the re-conceptualization of thinking and acting on IS development. Much of the re-conceptualization has resemblances to Suchman's (1994) work on interface design. She uses a navigation analogy contrasting the western and 'Truckese' navigator to distinguish between planning and then acting – planned action – and setting an objective and acting to achieve it – situated action. She states:

> [T]he European navigator begins with a plan – a course – which he has charted according to certain universal principles, and he carries out his voyage by relating his every move to that plan. His effort throughout his voyage is directed to remaining 'on course'. If unexpected events occur, he must first alter the plan, then respond accordingly. The Truckese navigator begins with an objective rather than a plan. He sets off toward the objective and responds to conditions as they arise in an ad hoc fashion. He utilises information provided by the wind, the waves, the tide and current, the fauna, the stars, the clouds, the sound of water on the side of the boat, and he steers accordingly. His effort is directed to doing whatever is necessary to reach the objective. If asked, he can point to his objective at any moment, but he cannot describe his course.
>
> (Preface)

Thinking and action in terms of situated action is a distinguishing feature of amethodological approaches that have jettisoned the SDLC, and in some cases, the functionalist paradigm. The 'plan' of distinct systems analysis, design and implementation phases of the SDLC are merged and done continuously rather than at discrete times. Examples of amethodological approaches are Prototyping, RAD, JAD and ASD. Those covered here in terms of future developments are component-based development, ASD and deferred system's development.

15.4 Component-based development

Rather than compress the phases of the SDLC, Component-Based Development (CBD) radically eliminates it and the need for traditional systems project management. CBD is the idea that software developers will write components for functions, which will develop into a library of component software traded in a market. IS developers can then buy the appropriate components and put them together to develop a required system. CBD can be classified in the functionalist paradigm.

The SDLC is eliminated. CBD also reduces the boundary between IS development as system projects and the actual operations of the organization. Since components can be purchased and 'plugged' together it is not necessary to temporally separate IS development from the actual operations of an organization. In this view, IS development can become a business operational issue rather than a special system project. The required component market though for CBD to be industrially viable has yet to materialize. CBD is practiced only within organizations that have developed their own library of components.

Systems analysis and design in the context of CBD would still involve requirements analysis to determine what the proposed system is required to do. The systems analysis and design activities would be at high level to integrate system components. There would be a need for a class model, consisting of packages diagrams, and other UML diagrams to determine overall system behaviour. Additional analysis and design steps would be needed to configure the components into the required functionality.

15.5 Agile software development

The agile systems ontology is radically different. Activities for systems analysis, design, implementation and systems project management are still required in it, but they are not sequential or prescribed stringently. ASD is concerned with giving primacy to individuals, collaboration and working software. It can probably be thought of as being social relativist paradigm.

The significant change compared to structured and object-oriented systems ontology is to view people as collaborators in IS development. This is in contrast to acting on a contractual basis with clients. Important stakeholders are identified and, with business analysts, included in the IS development team. The stakeholders set the priority list of what functions to develop.

Analysis, design and implementation are concurrent in ASD. Systems analysis is not an initial discrete step. So analysis is done iteratively as required and in small packages of work. Systems analysis and systems design are intertwined and depend on each other, each being done when required. The outcomes of systems analysis do not have a direct relation with systems design or implementation. The analysis does not have to map directly onto software. Implementation decisions are used to inform systems analysis.

A radical change in ASD is that process and methods are secondary to individuals' needs. Systems analysis and systems design activities are not done through stipulated and prescribed processes and methods – instruments are secondary and devised as required. Importance is given to individuals rather than the efficient deployment of methods.

System project management is radically different too. The software development tasks are broken down into small tasks. The tasks are defined so that each one can be completed in a month or less. The priority for determining which task is developed is set by stakeholders and business analysts, rather than the project manager. This aspect of ASD can be regarded as making IS development into an operational matter. Analysis, design and implementation activities are repeated monthly for each task undertaken.

15.6 Deferred system's design

Deferred system's design (DSD) is still in the research and theoretical stages, with some actual application of the concept in commerce. It is a theme of research in a UK university and acknowledged in practice by systems analysts and project managers in UK local government.

Figure 15.1 is the Deferred-Specified IT/IS matrix (Patel, 2003) adapted here in terms of systems analysis and design. The matrix is a synthesis of technological factors and social action. It depicts four systems ontologies in each of the four quadrants, based on a four-dimensional analysis of technology and social action. The four dimensions are planning or rationalism, emergence, deferred design decisions and diffused management. When correlated these four dimensions result in four types of systems ontology: deferred systems, real systems, autonomous systems and specified systems. The correlations can be interpreted for organizations too, as shown in the figure. The matrix thus matches semantically theoretical systems ontology and organization ontology for four types of systems and organization.

The basic theoretical proposition of DSD is that systems ontology needs to be a synthesis of planned action, emergent technical and social factors, and deferred action. The top left quadrant of the matrix depicts real situations where planning potential and capability is low,

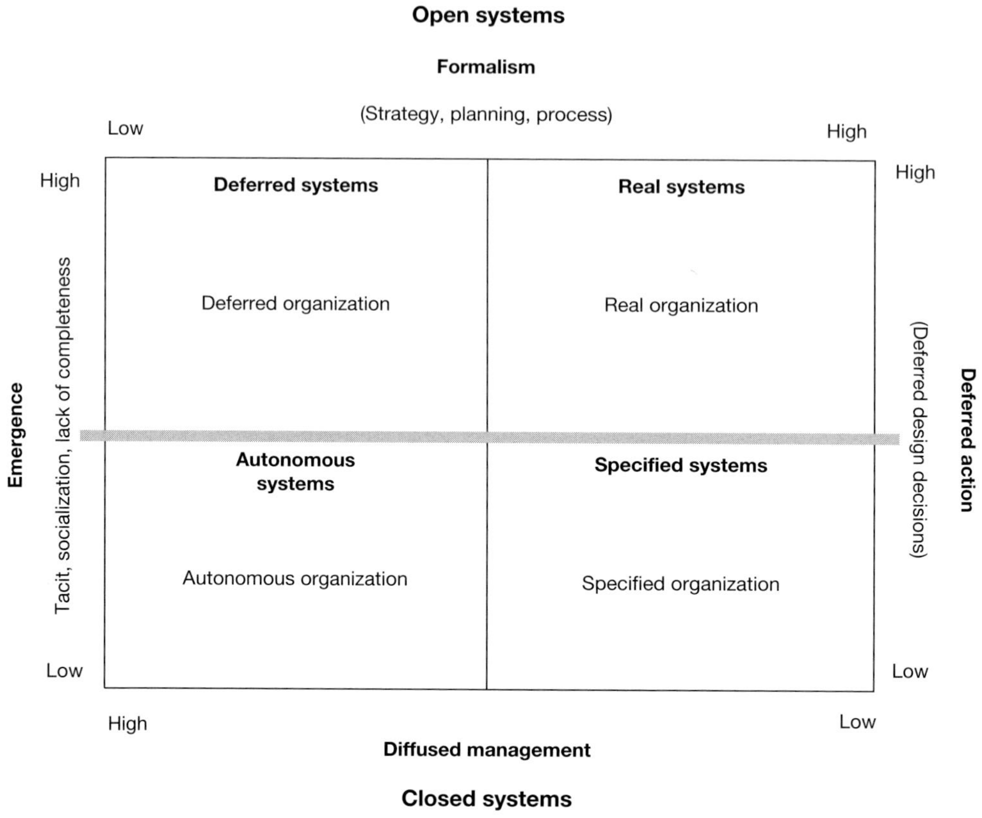

Figure 15.1 Deferred systems and organization

Source: Adapted from Patel, N. V. 'The Logic of Deferring the Design Process', in Patel, N. V. (ed.) *Adaptive Evolutionary Information Systems,* Hershey, PA: Idea Group Publishing.

emergent factors high, and therefore the need for deferred action is high. In such situations systems design needs to be deferred to people who make use of the system in context. The web is an example of a deferred system. This is to be contrasted with Specified System's (SSD) shown in the bottom right quadrant. It depicts real situations where planning potential and capability is high, emergent factors low, and therefore little need for deferred action. Problem domains in structured and object-oriented ontology are assumed to be of this type. (For details of the other two types of systems ontology and detailed explanation of the dimensions see Patel, 2003).

The deferred systems ontology in the matrix is based on an interpretation of social action as being simultaneously planned and emergent. Even the act of planning contains emergence (see section 15.8.5; Harris and Patel (2001)). This is in contrast to systems ontology that assume plans only or emergency only. Recognizing the possibility of planned or rational action and emergent or deferred action in human activity affords the four types of systems ontology and organization ontology to be correlated. Systems ontology above the grey line recognize a fluid boundary, so they are termed open systems. Those below the line have a rigid boundary, hence they are termed closed systems.

The deferred systems ontology has been applied to develop banking systems and interpret systems analysis and design for web applications in British local government. In banking, a system was developed that enabled clerks to change system functionality using the deferred design decision principle (right dimension). Clerks were given software tools to change system functions when required. This is significant for systems analysis because it removes systems design decisions from analysts and places them in the hands of people in the actual human context. People in the actual context who make systems design decisions are called action developers in the deferred systems ontology. Practising systems analysts interpret their brand of systems analysis deployed in the UK local government LEAP project as deferred systems ontology. They have developed systems analysis techniques that place systems design decisions in the hands of action developers.

15.7 An inclusive Critical Framework

The functionalist paradigm assumes organization to be an objective fact, but this ignores the social action that is an organization. The functionalist formulation of the IS development problem surgically removes sociological and political aspects of organization and organizing human behaviour. It claims that objective, optimal and efficient instruments can be devised, and economically used, to solve optimally the objective formulation of the IS development problem.

An inclusive Critical Framework means understanding IT, social action and the application of IT to organization to develop IS. It requires an interdisciplinary approach that includes social theory and technology, and IS – in terms of the knowledge of how to apply IT. An organization is a social unit with goals, a boundary and activity. As such, people within it attach meanings to their actions, struggle for power and resources, and develop cultures, norms and values.

Social theory can be used to enhance the Critical Framework and to re-conceptualize systems analysis and design. The functionalist IS paradigm provides an initial but incomplete understanding of IS. Its premise of acting only on perfect knowledge is questionable. Even if it were possible, it would still not account for phenomenological aspects of organization and information.

In functionalist systems ontology analysts are required to capture facts objectively. The elusive trinity of unambiguous, complete and non-redundant data in structured systems ontology is achievable for some practitioners. They argue that the rational basis of structured systems ontology provides an initial understanding that could be used to move onto higher levels of understanding. The revised inclusive Critical Framework in Figure 15.2 depicts that the real world of human problems contains non-systemic factors that lead to the opposite premise. There are no objective facts that can be captured, only phenomenological interpretations by humans of social action, including data, information and knowledge.

System project management itself is similarly interpretive. It does not have a unified value system in practice. Project managers' priorities for effective, efficient and successful project management are different from stakeholders need to retain a stake in the development. Analysts' concern with users is not shared by software programmers' need to write technically efficient and elegant software algorithms. Not only is the value system not unified it is highly uncertain too. IS projects in practice have shifting objectives, resources and timescales.

Analysts need to develop appropriate systemic and social action personal constructs. An

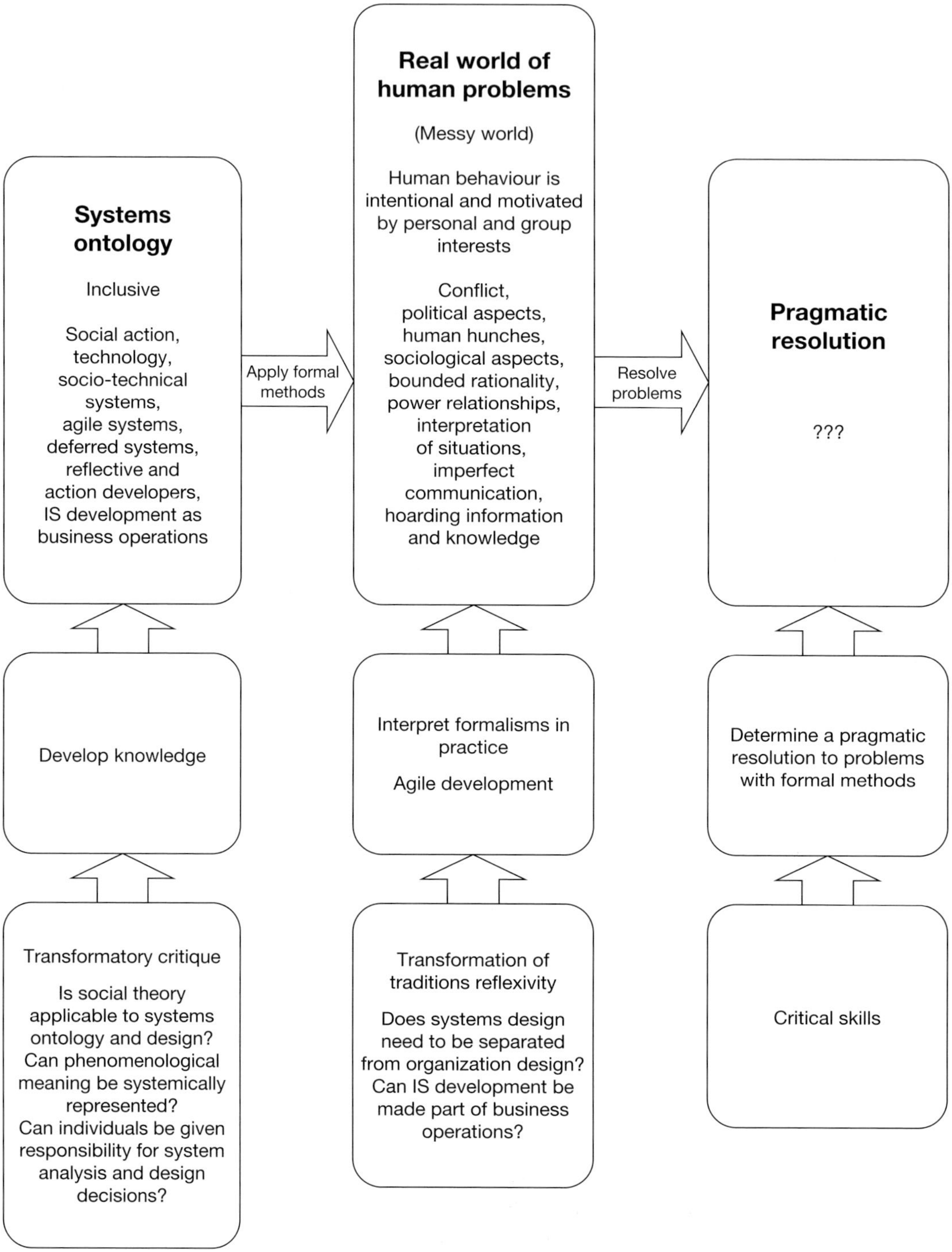

Figure 15.2 Critical Framework: inclusive systems analysis and design

effective PCF is one that is inclusive and progressive. Its continual development should be based on critical systems ontology.

15.8 Personal Critical Framework development

15.8.1 Personal constructs for future-orientation

Table 15.1 is a sample repertory grid for a future-oriented view. Reproduce the grid on a spreadsheet and add further columns and their polar opposites that you consider relevant. To objectify personal constructs, complete the grid by following the details on how to use a repertory grid in section 1.10.1.

15.8.2 Interpretive IS

Questions

1 Evaluate the potential contribution of social theory in the conceptualization, definition and development of IS.
2 Critically discuss how stories and narratives provided by people can be used to determine system requirements.

15.8.3 Component-based IS development

Activity A

The promise of CBD is not realized. Examine the role of systems analysis in CBD:

- Search the internet to identify CBD commercial and interest sites.
- Deduce the proposed role of systems analysis.
- Is there a viable market for components?

15.8.4 Deferred systems ontology

Activity A

Identify an existing IS and examine:

- Its planned basis – what is it expected to do?
- Emergence in its context that was not in the original plan and which needs to be part of its functionality.
- With reference to the second bullet point, what kinds of design decisions can be deferred to people in the context?

Table 15.1 Personal constructs for future-orientation

Pole 1	*Anticipate*	*Experience*	*IT*	*IS*	*Organic*	*Plan*	*Pole 2*
Unpredictable							Predict
Social							Technological
Interpretive							Objective
Revise							Freeze
Pre-design							Defer design
Story							Fact
Critical							Accept

15.8.5 Internet sources

For an example implementation of deferred classes see http://docs.eiffel.com/general/guided_tour/language/invitation-13.html.

See www.agilealliance.org and www.agilemanifesto.org/history.html for ASD perspective and philosophy.

'Only that programming is vital which finds its own elements in the people who use it.' For computer programmers working on the boundary of the future of computer programming see http://www.sgi.com/grafica/future/.

15.8.6 Further reading

For chapters on deferred system's design see: Patel, N. V. (ed.) (2003) *Adaptive Evolutionary Information Systems*, Hershey, PA: Idea Publishing.

Harris, H. J. and Patel, N. V. (2001) 'A Narrative Analysis of Information System Development in a Local Government Organization: Conversations Reflecting Deferred System's Design', ACM Conference on Object-Oriented Programming, Systems, Languages, and Application, Tampa Bay, Florida.

Patel, N. V. (2004) 'Deferred Systems: Deferring the Design Process and Systems', *Journal of Applied System Studies*, 5(1).

For a comprehensive view of CBS see: Szyperski, C. (1999) *Component Software: Beyond Object-Oriented Programming* (1st edn), Perth, Scotland: Pearson Education.

For an application of DSD see: Stamoulis, D., Theotokis, D., Martakos, D. and Gyftodimos, G. (2003) 'Ateleological Development of Design-Decisions-Independent Information Systems', in Patel, N. V. (ed.) *Adaptive Evolutionary Information Systems*, Hershey, PA: Idea Group Publishing.

Glossary

Algorithm A computer program algorithm is the sequence in which a computer is instructed to execute program code. Process logic models in structured systems ontology and operations in objects in object-oriented systems ontology are used by systems designers to develop program algorithms.

Application domain A term used in systems analysis to describe an organization or its parts for which a new IS is to be developed.

ASD Agile Software Development is a radically different perspective on software development. It is concerned with individuals and processes over processes and tools, working software over comprehensive software, customer collaboration over contract negotiation, responding to change over following a plan. Its systems analysis and systems design activities are non-prescriptive, non-linear, and agile

BPR Business Process Re-engineering is the recognition of business activities in terms of critical processes and their transformation to deliver improved operational efficiency and effectiveness. BPR is normally associated with the application of IT to achieve such performance improvements.

Business case A rationale for deciding to proceed with the development of a particular IS.

Business model A set of business ideas related to achieve predetermined objectives, usually in terms of profits, customer satisfaction, reduced costs, product or service differentiation, or other business measure.

CASE Computer-Aided Software Engineering are computerized tools to support the development of systems models during systems analysis and systems design. The advantages of using CASE tools are speed, accuracy and consistency in the development of systems models.

Client A person, group, department or organization for which an IS is to be developed.

CMM Capability Maturity Model is a standard for developing high-quality software. The CMM standard is set by Software Engineering Institute at the Carnegie Mellon University, who developed it for the US Department of Defense to enable them to evaluate an organization's ability to produce software under contract.

Critical Framework The Critical Framework for Knowledge and Practice development is an intellectual tool for gathering, interpreting, analysing, evaluating and critiquing ontological knowledge of systems and its practical application. It serves to enable the development of personal constructs for a PCF.

Critical systems ontology Critical systems ontology is the interpretation, analysis, evaluation, inference, explanation, questioning and critical study of system. It is the notion that current ontological knowledge of systems can be improved, and that it can be 'other than it is' to be practically effective.

Critical thinking Critical thinking is a set of cognitive process required to develop critical thinkers capable of self-regulation.

Criticality Barnett (1997) defines criticality as: 'a human disposition of engagement where it is recognised that the object of attention could be other than it is'.

Data Data are elements of transactions or transactions arising from organized activity. In computing terms data are items inputted into a computer system and processed by it to produce other data or information.

Deliverable A diagram or document resulting from a method or phase of a methodology or the SDLC. It is a concrete thing that can be used to define or develop an IS.

Engineering metaphor The engineering metaphor is used in software development. It is used to define efficient and effective processes for developing software.

Epistemology Epistemology is theory of knowledge and how it is acquired. Usually it refers to the method by which knowledge is acquired. The epistemological methods can be objective or interpretive.

Estimation models An estimation model is used to plan the resources required in a system project. An estimation model provides rigour and exactness based on statistical analysis. An example is Function Point Analysis.

e-Tailing Online retailing is called e-Tailing.

ETHICS The Ethical and Technical Implementation of Computer Systems. ETHICS is a socio-technical methodology. It is the seminal methodology to consider the social factor in IS development.

Ethics Ethics is concerned with making judgements concerning right and wrong behaviour.

Evidence-based reasoning Knowledge and practice that is supported with evidence is evidence-based reasoning.

Formalism Formalism in systems analysis and design is the notation and its structure used to represent things of interest in a problem domain. It is used to develop formal systems models.

Information Information is used by decision-makers to allocate, monitor and control resources in an organization. Information in terms of computing is processed data.

Interpretivism Interpretivism is an epistemology used in IS research. It is applied to understand and develop knowledge of subjective and phenomenological aspects of IS. A person using this method is called an interpretivist or subjectivist. Interpretivism is based on phenomenology, the idea that detailed descriptions of human experience alone constitute knowledge. Phenomenology does not focus on explanations.

JAD Joint Application Development is the notion, methods, techniques and tools that enable people in the problem domain and professional IS developers to work together to develop an IS.

JSD Jackson System Development is the set of structured techniques proposed to develop IS.

JSP Jackson Structured Programming is the set of structured techniques proposed to develop software programs.

Method A method is a set of activities designed to achieve a specific systems analysis or systems design task.

Methodology A methodology consists of a prescriptive staged process for IS problem definition and development with associated techniques and tools. The stages may be variously divided in different methodologies, but they all contain systems analysis and systems design.

Notation language A notation language is a formal set of symbols with precise meanings to represent things of interest in the problem domain. It is used in structured systems ontology to develop models of the current and new IS. Modelling notations help to objectify the problem domain and facilitate communication between analysts themselves, with clients and software programmers.

Object-oriented analysis Object-oriented analysis is the specific systems analysis and design technique proposed by Coad and Yourdon.

Objectification Something is objectified when it is recorded on physical media. Mental constructs recorded on physical media are objectified. Objectification is used to develop a PCF.

Objectivism Objectivism is the doctrine that reality is objective or independent of the observer and that sense data directly correspond to it. A person who practices objectivism is called an objectivist. Structured systems ontology, and to some extent, object-oriented systems ontology, is based on objectivism.

Ontology Ontology is concerned with the nature of being. It is the assumptions or presuppositions in a theory to explain a phenomenon. The notion of 'structure' or 'object' in systems has a set of ideas that describe systems based on certain assumptions.

Organization A sociological definition of organization is that it has a goal, boundary and human activity. It is composed of people and the work they do to achieve predetermined business aims.

Organizational knowledge Organizational knowledge are the experiences and actions of individuals and groups, or communities of practice, involved in achieving specific objectives.

OSS Open Source Software is a movement in software development that makes initial software code open for the public, usually interested software programmers and organizations, to help in its further development. Open source software is licensed by the Open Source Initiative (OSI) for commercial exploitation.

Paradigm A paradigm is a set of beliefs, assumptions, values, and shared knowledge accepted by a community of practice. It forms a body of knowledge and informs practice.

Paradigmatic critical framework Paradigmatic critical frameworks can be of two types. A critical focus on what constitutes knowledge and how it is acquired within a paradigm or a critical focus on what constitutes knowledge and how to acquire it across paradigms.

Participative Design See JAD.

PCF Personal Critical Framework is an intellectual device for acquiring, interpreting, evaluating and inferring personal knowledge appropriate for action.

Persistent data Data that continues to exist when the system is not live and needs to be stored for further use.

Personal construct Individuals experience reality and then develop constructions (personal constructs) to help them *anticipate* it. Personal constructs are used to interpret and explain events. They are part of a personal construct system, which is a set of personal constructs and the relations between them that provides the *unity* in the experience of individuals.

Planned action Action that is predetermined with expected outcomes to achieve known objectives and enacted in actual situations to realize the expected outcomes.

Praxis Praxis is the art of acting in given conditions in order to change them.

PRINCE **Pr**oject **M**anagement **I**n **C**hanging **E**nvironments is a systems project management method. It is devised to determine the scope, plan, monitor and control and IS development.

Problem domain A systemic term to describe an actual situation for which an IT solution is required.

Problem-solving Problem-solving in systems analysis and design requires the application of rationality, logic and formalism to a human area of concern to seek a resolution.

Prototype An initial IS product used to determine system requirements by testing it in the problem domain with users. Prototypes are developed using RAD.

RAD Rapid Application Development is a process for rapidly developing software with users participating.

Reality The actual condition in which human action happens.

Repertory grid A repertory grid is used to elicit, assign, and analyse knowledge in terms of personal construct and can be used for self-help. It can be used to assess and evaluate personal experiences and knowledge.

Scientific method An empirical epistemological method for acquiring, testing and verifying empirical knowledge.

SDLC Systems Development Life Cycle is a method demarcating distinct phases of IS development activities.

Situated action Situated action depends on its material and social circumstances.

Social action The activities of humans in a group oriented towards the achievement of objectives.

Socio-technical The combination of social and technical factors in an IS methodology.

Software engineering The idea of applying engineering precision and discipline to software development.

SSM Soft Systems Methodology is a continuous learning cycle for people in situations of social concern.

Stakeholders Stakeholders are groups of people or a person who has a specific interest in a system project because they are the sponsor of it, or will be affected by it, or will have to work with a new IS.

Structured systems analysis Structured systems analysis is the practice of systems analysis based on the discipline of formal structure and problem-solving in the process of conducting systems analysis.

Structured systems design Structured systems design is the practice of systems design based on the discipline of formal structure and problem-solving in the process of conducting systems analysis.

System A system is an abstract concept used to structure a problem and seek its resolution in terms of a boundary, sub-elements, control and feedback.

Systems analyst A systems analyst is someone who investigates and makes systems models of a problem domain.

Systems ontology Systems ontology is knowledge of what constitutes the nature of systems. The term systems ontology is used to describe knowledge and practice relating to computer-based IS.

Technique A specific intellectual device to achieve a concrete result used to develop an IS.

Temporary data Data that exists only during the time the system is live or during run-time.

Time boxing The demarcation of systems development and project management activities into discrete time periods for completion.

Tool An implement used as a medium for achieving a specific objective.

UML Unified Modelling Language is a graphical notation for systems modelling used on its own or as part of a method or methodology to express concepts for IS in object-oriented terms.

Usability Usability is the consideration of efficient and effective means in user interfaces to enable specific users to achieve specific tasks.

User A label given by technically qualified IT professionals for people who use computer-based IS. The term 'user' is also used by IS researchers and others.

Index